Global Sustainability

Benno Werlen
Editor

Global Sustainability

Cultural Perspectives and Challenges for Transdisciplinary Integrated Research

Springer

Editor
Benno Werlen
Department of Geography
University of Jena
Jena, Germany

ISBN 978-3-319-16476-2 ISBN 978-3-319-16477-9 (eBook)
DOI 10.1007/978-3-319-16477-9

Library of Congress Control Number: 2015941901

Springer Cham Heidelberg New York Dordrecht London

Printed on acid-free paper

Springer International Publishing AG Switzerland is part of Springer Science+Business Media (www.springer.com)

Preface

We live in the age of the anthropocene. Nature is no longer nature, but everywhere bears the deep imprint of human activity. Of course, human beings have long influenced the habitats in which they live. Some of the early civilizations, for example, by felling forests, or causing watercourses to dry up, created deserts where none existed before. Yet none intervened into the natural world by more than a small fraction of the degree to which our civilization does, and on an everyday basis. Climate change is perhaps the fundamental example of this transformation, caused by the burning of fossil fuels on a massive scale. The speed with which we are using up the world's oil reserves, for instance, contrasts in an extraordinary way with the length of time it took for them to be laid down – which happened over many millennia. At current rates of consumption, we will burn our way through most of them in less than 200 years from the first time at which their commercial exploitation began.

It is an awesome thought that we are busy changing the world's climate – and on a permanent basis, since we know of no way of getting the greenhouse gases that are causing the world to warm up out of the atmosphere once they are there. Some will persist for centuries. If left unchecked, current trends will produce more and more volatile and extreme weather patterns the world over. As I write, a typhoon with gusts of wind reaching more than 100 miles per hour has just hurtled through Japan, leaving a trail of destruction in its wake. In the future, without practical policy intervention, such storms will increase in average intensity. To be sure, there are uncertainties around just how extreme changing weather patterns will be and over what time period. The climate change skeptics fasten on this and declare that the risks are exaggerated. Yet uncertainties cut both ways. The risks could just as well be greater than the majority of climatologists believe, and develop at a faster pace. James Hansen, from NASA, one of the world's pre-eminent climate scientists, believes such to be the case. He sees some truly terrifying possibilities waiting down the line if greenhouse gas emissions are not curbed, and in relatively short order.

There are three models of the impact of human activities upon the natural world – or what used to be the natural world. The skeptics argue that the earth is robust. Nothing human beings can do will influence it very much. Nature will go its own way regardless of what we might do. Those in the green movement, to the contrary, see the planet as fragile and vulnerable in the face of the human onslaught. Such a perspective is worrying enough. Yet there is a third view, associated with authors such as Hansen. This is that the earth is like a wild beast, and we are busy prodding it with sticks. It will react with increasing violence to our interventions. Every typhoon, hurricane or large earthquake is a reminder of the awesome forces in play and we toy with them at our collective peril.

So far, in spite of endless conferences organized by the UN, policy interventions by some governments and a proliferation of bottom-up approaches, we have had very little success in bringing down the level of carbon emissions globally. As measured by the Mauna Loa Observatory in Hawaii, the level of Co_2 in the atmosphere is increasing relentlessly year upon year. In 2014, it passed 400 ppm for the first time – the highest level observed for at least 800,000 years and probably far longer. The Arctic ice is melting at a rate not thought possible by most observers until recently. Some scientists believe that the Arctic could be ice free in the summer within some 15 years. On the face of things, the visible shrinking of the Arctic ice should be a wake-up call to the world to take more radical action to cut emissions. Not a bit of it. There is a rush on the part of nations and companies to exploit the mineral resources that become accessible as the ice retreats.

It is a similar picture with broader issues of sustainability. The destruction of the world's rain forests is one of the sources of climate change, since that process releases large amounts of Co_2, and the capability the forests have to absorb it is being eroded. That destruction has consequences for many of the world's flora and fauna too. Here is another battle we are losing. The oceans are becoming increasingly acidified, with an accumulating threat to many of the life forms that exist in them. Species are disappearing. These wider ecological disasters overlap with other sources of stress and strain the world's ecosystems – such as the radical and continuing increase in the world's population. There are likely to be ten billion people on earth by 2050. Only about a century and a half ago, in 1850, there were less than one billion.

Why are we finding it so hard to take effective action against burgeoning climate change and environmental degradation? A cluster of reasons exists and it is important to see how powerful these are when added together: (1) There are powerful interests opposing the necessary policies – such as some of the most powerful fossil fuel companies. (2) These are global issues. Yet there is no effective system of global governance. Nations very often have different short-term interests from one another and collaboration over a longer time-span is especially difficult. Developing states are more vulnerable to the changing climate than the richer nations, who can spend more to defend themselves. Yet the richer countries are reluctant to transfer needed resources to the developing world on anything like the scale needed. (3) We are dependent upon the work of scientists to diagnose the risks we face and how severe they are. Most non-specialists – and especially the lay public – have no chance at all of mastering the detail of the scientific work involved. The climate change skeptics and those who say that proclamations of imminent ecological disaster are exaggerated step into this gap. (4) These are risks of which we have no previous historical experience, certainly on the scale they exist now. Many people find it hard, therefore, to treat them as real, since their worst consequences lie in the future. So far as climate change is concerned, the worry is that little will be done until there are cataclysmic disasters that can be unequivocally attributed to the warming of the world. Yet by this time it will be too late. For we have no way of getting the greenhouse gases out of the atmosphere once they are there. Some – like CO_2 in fact – will be there for centuries.

In trying to cope with these problems, more innovative policies will be needed than those that have dominated thinking up to the present. Moreover, a certain level of climate change is inevitable whatever happens now. As well as radically upgrading the struggle to reduce emissions, we shall have to think about how to improve environmental resilience and take proactive action to adapt. My preferred outlook for doing so is within the framework of what I call 'utopian realism'. A dose of utopianism is needed because policy thinking must go well beyond the parameters of the here and now if a more sustainable world economy is to be created. However, if not coupled to realistic modes of achieving them, such goals will remain mere aspirations.

This book is a significant contribution to the rethinking that will be necessary. It is global in scope, encompassing studies from a range of countries around the world. The work quite correctly focuses upon the need to bring the social and natural sciences much closer together. The task is a formidable one because the same holds true of different sectors of within these categories. For instance, economists have to work with sociologists, anthropologists and human geographers; climatologists have to collaborate with biologists and earth scientists. Younger researchers – such as represented in this volume – perhaps more able than their elders to think in radically new ways – should be in the vanguard.

Achieving greater sustainability must be grounded in the transformation of local practices, including in some of the poorest societies on earth. There causes for optimism alongside the entrenched problems. In the digital age, ideas can be directly discussed, and strategies developed, without the traditional limitations of time and space. The pace of change has accelerated, but so have the opportunities for positive innovation. To take the core example: the Internet barely existed 20 years ago. Now it is more or less universal, and embedded both in the largest global institutions and in the conduct of everyday life. Perhaps the same could happen with radical advances in renewable technologies? Perhaps the famous inertia of the fossil fuel industries, which seems to be condemning us to such a disturbing and dangerous future, could be less implacable than it seems? Think what happened with mobile phones in large parts of Africa. A whole stage of technological evolution was simply skipped.

We live today in what I like to call a high opportunity, high risk society. The biggest risks we face, as collective humanity, come not from nature but from ourselves. They derive from our newfound global interdependence and the fragility of the systems that are driving it. The level of technological innovation is so high that we do not know where it will lead. The very same research that could produce breakthroughs in genetic medicine, for example, might also inadvertently create lethal pathologies. A previously unknown virus, for instance, could produce a pandemic. As we saw in the financial crisis, the global electronic economy is inherently vulnerable and the effects of that crisis are as yet far from being fully resolved. We don't know in advance how the balance of opportunities and risks will pan out, simply because we have little past historical experience to go on. No previous generation has lived in a world as highly globalised as that of today. Some of the risks we face are existential. Yet the opportunity side of the equation is equally huge and we have to make it count.

London School of Economics Professor Lord Anthony Giddens
London

Acknowledgements

Many people have helped me in the production of this book. Its publication has been made possible through the dedication and commitment of a large number of individuals and organizations. Organizations supporting the project include first and foremost the International Council for Science (ICSU), the International Geographical Union (IGU) and the Stiftung für Technologie, Innovation und Forschung Thüringen (STIFT); their financial support has enabled young and senior researchers hailing from all continents to engage in fruitful exchange across disciplinary boundaries at the Transdisciplinary Intercultural Ecological Research for Sustainability (TIERS) conference held in Jena, Germany, in June 2012. This book is a selection of the papers presented there. From its inception to the final stages, the TIERS project has had the support of a large number of persons. I am especially grateful to Ron Abler, Heide Hackmann, Gordon McBean, Margaret Robertson and Adama Samassékou for having been a stalwart of support throughout the project.

The Friedrich Schiller University Jena provided the space for presenting and discussing the ambitious ideas of this project. I would like to thank the University's then rector Klaus Dicke and the then Minister for Economics, Labour and Technology in the Free State of Thuringia, Matthias Machnig for their engagement in this project.

Special thanks go to editors Ilse Evertse, Johan Grobler and Ilze Hugo who tirelessly and patiently worked to improve style and clarity of writing.

My thanks also go to everyone at Springer Publishing, namely Stefan Einarson, Mireille van Kan, Evelien Bakker and Bernadette Deelen-Mans, whose support and advice throughout the publishing phase has been invaluable.

Last but not least, a special thank you to Margrit Elsner, Dorothee Quade and Nadine Wassner who worked behind the scenes to ensure organizational and administrative processes of the project ran smoothly.

Acknowledgements

Contents

Introduction

Sustainability has been one of the keywords in the sciences, as well as in local, regional and global policies, since at least the United Nations Conference on Environment and Sustainable Development (also known as the Earth Summit) in Rio in 1992. It is also at the center of the latest – and probably one of the biggest – joint research initiative of the world's umbrella organizations for the natural and social sciences: Future Earth. With this initiative, led by the International Council for Science (ICSU) and the International Social Science Council (ISSC), sustainability is also at the heart of the Belmont Forum's activities, which is an association of some of the world's most important national research organizations, including those of the US, China, the UK, Germany, India, Brazil, Japan, France, Norway and others.

Although sustainability is of core importance for so many dimensions of human action, the concept lacks a convincing integration into the social and natural sciences. In addition, it is particularly important that our recognition of cultural dimension should be enhanced to achieve a shift towards sustainable lifestyles. We suggest that the social and natural sciences' contribution to achieving sustainability on a global scale may be to work towards a truly transdisciplinary approach: An approach that overcomes the merely additive logic of inter-disciplinary cooperation.

The adverse impacts of climate change, as well as earthquakes, hurricanes, and bush fires, are commonly reported as 'natural disasters'. It is, however, important to understand that there is a distinct social dimension to these events. Because they are very often primarily the outcome of inadequate knowledge and actions, they are above all social disasters. Basically, they are the unintended consequences of social actions and therefore as much a social as a natural problem.

These and other events demonstrate, however, that we need to better integrate natural and social-scientific research, as well as the insights from the cultural sciences and the humanities, in order to better understand the so-called natural disasters or ecological crises. This book, and the more encompassing initiative of the International Geographical Union for an UN International Year of Global Understanding, aims to make a significant contribution to this goal.

This book has two specific main objectives. First, it aims to document the latest theoretical developments in integrative research on the society-culture/nature interface within and outside the academic mainstream. Second, it aims to investigate the varied ways natural and social environmental sciences are applied in different cultural worlds.

These objectives have to be seen in their historical context. The theoretical foundations of ecological research and sustainability policies were developed at the end of the nineteenth century. They are largely based on investigations of living spaces, and the evolution and differentiation of varied life forms. It is important to realize that this perspective is embedded in the practical and theoretical European problem situations of their time. As a consequence of evolutionary imperialism – the dominant ideology at the time – the perspective lacks the appropriate social and cultural differentiation. The interim transformation of spatial and natural relations is based on the globalization process, which is so radical that we cannot continue trying to solve twenty-first century ecological problems with strategies derived from theories designed for nineteenth century problems.

This book offers new transdisciplinary research perspectives, in methodological and theoretical respects. The four parts offer insights into the twofold bio-physical and the socio-cultural global embeddedness of local living conditions on the basis of selected empirical studies from Latin America, Asia, Africa, Australia and Europe.

Part I: Integrated transdisciplinary research focuses on the critical evaluation of integrated research's current status in respect of developing a promising strategy for transdisciplinary problem solving. Taking into account the pace and magnitude of the environmental challenges that all life forms on Earth face, we need a systematic inventory and a careful appreciation of the existing research approaches, especially the ones that hold promise regarding overcoming clearly identified weaknesses, namely: (i) natural scientists and engineers' lack of appropriate understanding of the social world; (ii) social sciences and the humanities scholars' lack of understanding and integration of biophysical elements of; and (iii) the reductionist implications of systemic integration. It is imperative to identify research that goes beyond interdisciplinary frameworks towards genuine transdisciplinary research. To call for transdisciplinarity reinforces, rather than questions, the validity of disciplinary and interdisciplinary research, of course. But switching disciplinary perspectives alone will not yield an integrated view of the nature/society interface. To overcome disciplinary blind spots, we need a perspective that specifies and solves problem complexes regardless of the disciplinary interests and boundaries.

Part II: Knowledge encompasses contributions dealing with the basic conditions of everyday practices, the core 'elements' of global sustainability and transdisciplinary research. The pace and magnitude of climate change and other environmental issues rooted in human intervention in the natural world call for a systematic review and evaluation of the different research approaches in all cultural contexts. Promising approaches should be identified and evaluated. Besides consulting the leading figures in integrated research and in natural and social science sustainability research in the different scientific language communities and disciplines, it is essential to examine how scientific knowledge is integrated into everyday actions. Case studies in the different cultural realms exemplarily demonstrate the challenges of knowledge integration at the science-policy interface and policy-society interface as well as the complex nature of sustainability.

Everyday local human practice is based on powerful cultural needs, such as food, housing, ways of life, the requisites of leisure, mobility, communication and spirituality, which powerfully influence the states of global life-support systems. Part III: Society nature investigates approaches focusing on the society/nature interface to reveal the varying impacts of human practice on the global environment. In addition, understanding the impact of cultural interpretations is a prerequisite for achieving sustainable development. Therefore an investigation of the varied ways in which environmental sciences are applied in different parts of the world is essential, especially with specific attention paid to mitigating unsustainable practices and to the adaptation of the consequences of climate change at the local level.

These papers identify the impacts of cultural differences, as well as the role of cultural factors in the use of natural resources.

Part IV: Interfaces science policy introduces a systematic inventory and careful examination of existing national problem-solving strategies: Leading public figures in the different nation states are hypothetically the most experienced actors in sustainability. A comparative analysis from a cultural perspective of governmental environmental action units reveals cultural blind spots in political strategies for achieving sustainable development. In the politics of sustainability, the grand challenge is perhaps still that of changing human behaviour on the basis of scientific insights. Traditionally established customs, habits and everyday routines very often impede the application of scientific insights to unsustainable practices and policies. Young scientists and policy developers may be more open than their elders to the changes needed for the adoption of sustainable ways of life. The consequences for the adaptation of sustainable development standards are also discussed in this chapter.

Jena, Germany Benno Werlen

Part I
Integrated Transdisciplinary Research – New Approaches

From Local to Global Sustainability: Transdisciplinary Integrated Research in the Digital Age

Benno Werlen

1 Introduction

The theoretical foundations of ecological research and sustainability policies were developed at the end of the nineteenth century and were largely based on biological and geographical investigations of living spaces, and the evolution and differentiation of varied life forms. Both the biological and geographical disciplinary perspectives are embedded in the practical and theoretical European problem situations of their time. The two scientific investigations – one into life general and the other into human life on our planet – were linked to the social and spatial formations we currently regard as historical. Given the current problem constellations, we need to rethink these concepts. Systematically and historically, we can learn a lot from these biological and geographical scientific traditions: the positive outcomes and, even more importantly, the highly problematic implications.

The interim transformation of the spatial and natural relations is based on the globalization process. This process is so radical that we cannot continue to try and solve twenty-first century ecological problems with problem-solving strategies derived from theories designed for nineteenth century problems. To overcome the problematic implications of this constellation, we must trace back its historical roots. Since the theoretical horizons of the biology and geography disciplines are embedded in the practical European problem situations of their time, the theoretical foundation of the sustainability concept's original elaboration lies at the beginning of the eighteenth century's abuse of forests for early forms of industrial production in Saxony, central Germany.

The transformation of the spatial and natural relationships that have occurred in the interim is based on technical innovations aimed at mastering the spatiality of

B. Werlen (✉)
Department of Geography, Friedrich Schiller University Jena, Löbdergraben 32, 07743 Jena, Germany
e-mail: benno.werlen@uni-jena.de

B. Werlen (ed.), *Global Sustainability*, DOI 10.1007/978-3-319-16477-9_1

human existence (Werlen 2010, 324ff; 2013) and at empowering the transformation of contemporary natural living contexts. This process of innovation was first accelerated in the nineteenth century with the appearance of carbon-based engines, which is part of the current ecological crisis. The second wave of acceleration was a result of electronic communication innovations, which led to radical implemented changes in the spatial conditions for human actions. We must therefore reconsider the two key environmental policy concepts for current times and for our future. Under these globalized and globalizing conditions, it is indicated that we should move from a primarily space-centered regional perspective to a primarily practice-centered global perspective.

2 From Local to Global Sustainability

Today, global sustainability is needed, and global sustainability requires global understanding. To think globally and act locally, we require a better understanding of our local, daily activities' impacts on a global level. Reaching true global understanding is a prerequisite for a more sustainable planet through everyday practices. Our shared future on earth depends on successfully establishing sustainable everyday actions. This is where the local and global become one. To grasp this, we need a widespread awareness of how everyday actions create the challenges that impact humanity on a global scale. This includes our capacity to connect actions and thoughts that may seem disconnected across time and space.

Global environmental change research has produced unambiguous scientific insights into earth system processes. Nevertheless, these findings are globally too rarely translated into effective policies that can help prevent the worst consequences of global change. It seems that we not only need to deepen our knowledge of socio-cultural contexts, but must also improve the level of society's acceptance of scientific knowledge. The reasons for this are not only clear in the underrepresentation of the social sciences in environmental and sustainability research, but also in an even more radical – and very often widely accepted – diagnosis of the current situation: the social sciences and the humanities' lack of interest.

The reasons for this unsatisfactory situation lie very deep and can certainly not merely be overcome through the integration of the social sciences and the humanities into the existing methodology of environmental research, which are dominated by natural scientific research methodology and methods. This dominance is very often the reason for large parts of social sciences and the humanities' lack of interest in these two fields of investigation. Their role cannot just be to transmit natural scientific insights into environmental and sustainability research "into society," whatever that may entail. This constellation, with its highly problematic methodological background, can probably only be solved if we succeed in overcoming its shortcomings. These shortcomings are based in the afore-mentioned historical roots of environmental research, and has certainly nothing to do with any lack of competence in the social sciences and the humanities. Rather, these shortcomings are a result of the still dominant conceptualization of environmental research.

The theoretical foundations of ecological research were largely developed at the end of the nineteenth century. Ernst Haeckel's (1866, 1878/1879) biological investigations into living spaces and the conceptualization of human geography by Friedrich Ratzel (1882, 1901) formed the basis of the elaboration of ecology. In both scientific traditions, living spaces were seen as a kind of directory for selective evolution and the subsequent differentiation of varied life forms.

It is important to acknowledge that living spaces have much in common with the Newtonian container space elaborated in the context of the classical mechanics and optics (Newton 1704). We can conclude that the historical basis of ecological thinking is rooted in a mechanistic worldview (Jaeger et al. 2001), directly derived from the basic assumption of the Newtonian container space and its causal power. The mechanical view of all process implies that nature as well as society are controllable and maneuverable by the same means and same intervention type. Consequently, this mechanistic worldview requires – at least implicitly – that we neglect socio-cultural differences and the reflexivity potential of human practices. Therefore, it largely ignores the social scientific insights into the ways social realities are produced and reproduced, including the social context of the transformation of biophysical realities.

It is also important to understand that Hans Carl von Carlowitz's (1713) development of sustainability is also – at least indirectly – linked to the Newtonian container space. He developed the basis for sustainable ways of production in the mining and forestry contexts at the end of the seventeenth and beginning of the eighteenth centuries. In Germany and in many other places in continental Europe, most of social and economic life is characteristic of regional societies and their typical spatio-temporal constellations. Over time, stability has been based on the domination of local traditions. Transportation and communication's technical standards have led to narrow spatial limitations. The predominance of walking and the limited significance of writing restrict social and cultural expressions to the local and regional levels. Additionally, technological development has led to production processes tied to natural conditions. Economies and the (problematic) consequences of the transformation of nature for human purposes are, as a result, bound to a regional rather than a global reach. Pre-modern and early modern ways of living largely correspond to the basic assumption and shape of the Newtonian container space (Werlen 2005).

The methodological approach conceived by the above-mentioned scholars – the pioneers of regional environmental and local sustainability concepts – is still largely valid (World Commission on Environment and Development 1987; Hauff 1987) in respect of most international sustainability politics, especially the UN environmental politics and the UN decade of education for sustainable development (www.unesco.org/education/desd). The natural and spatial (including their ecological components) remain the starting points of ecological investigations, preceding all human actions (Werlen and Weingarten 2003, 2005, 2013; Grober 2010; Grunwald and Kopfmüller 2012). In this context, it is very important to realize that sustainability refers not only to the Newtonian container and Haeckel's living space, but also to life forms with clear local and regional borders, and certainly does not include the global reach of the social and economic spheres.

Thus, the logic of mainstream research in this field starts conventionally from pre-given spatial constellations, and normally also forms the empirically observed problem situation in the biophysical world. Therefore – and at first glance absolutely logically – natural sciences, including the geosciences, are seen as disciplines with first-order competences. In addition, sustainability problems have very often been regarded as spatial problems that should be addressed by spatial strategies, or at least territorial problem-solving strategies.

These views can be understood as expressions, or consequences, of spatial ontology, rather than of pre-modern space ontology in the tradition of the Newtonian container and Haeckel's living space. This understanding is based on highly problematic geodeterminist and even dramatically destructive geopolitical ways of thinking, especially in European history in the nineteenth and first half of the twentieth centuries. We simply cannot deny that this logic of thinking, acting, and politicizing has the same historical, scientific, and epistemological basis as traditional geopolitics practiced by the European imperialistic expansion politics and the national socialist blood and soil doctrines.

With the acceptance of all human action upstream, pre-given container or living spaces imply two naturalistic reductions (Werlen and Weingarten 2005, 291). Because only extended material facts can be localized in such spaces, all meaning aspects are at first neglected and reduced to a material vehicle of the representation of meaning. The social dimensions evaporate, and meaningful human action is reduced to the biological aspect of simple behavior. The second naturalistic reduction lies in the reification of the social. The social becomes a natural object with the same qualities as biomaterial world entities. Entities of the social world are thus addressed as biomaterial properties. On the basis of these kinds of operations, biomaterial properties such as race, sex, etc. are regarded as social features, and the derived racist, sexist, etc. discourses are believed to be, or at least implicitly presented as, naturally or even natural-scientifically justified. We must overcome this intellectually, politically, and ethically highly problematic constellation and move towards a post-colonial geographical imagination (Said 1978; Gregory 1993, 2004; Jazeel 2012) or, better still, towards a post-colonial worldview, taking into account the new spatio-temporal constellations, meaningful social-cultural worlds, and generalized democratic principles.

On the whole, the pre-given status of nature and space implies normative standards that undermine cultural and social differences in an affirmative, democratic way. From this we can draw the hypothesis that the container/living space perspective also undermines the required acceptance of the suggested ecological policies based on these premises.

3 The Needs for Revision and Sound Integration

There is much evidence that human action induces most ecological problems. Scientific research methodologies and designs should therefore adapt the same logic: They should start from social-cultural practices and regard them as the causes of, or better, the reasons for ecological problems. The global reach that the consequences of human practices have means that we should focus on tracing back the global problem constellation to daily actions and routines, with their manifold cultural differentiations, which are produced and reproduced in a specific local context.

If human actions do indeed cause the current ecological problems, the reasons for these actions consequentially and largely lie outside the competence of the realm of natural science. The nature of and the human reasons for non-sustainable practices are increasingly understood, but gaining knowledge on how to change individual and social practices concerning sustainability remains a major challenge for healthy nature-society relationships, as well as in terms of designing environmental policies informed by sound science. But this is only one side of the problematic constellation.

On the other hand, social scientists, with their competence to analyze human practices, must pay more attention – also in terms of traditional geodeterminist thinking – to biophysical conditions and to material culture. However, practice-centered approaches so far largely excluded the natural world from the outset. Because the biophysical world does not have meaning in itself, the social sciences' actions and practice-centered approaches – such as those of Max Weber (1912, 1913, 1920) and others standing in the tradition of his way of reasoning – excluded the natural world from their investigations. This meant a kind of spatial amnesia, which paradoxically parallels the spatial obsession of biological living space and geographical (environmental) research. Both these methodologies obstructed the non-deterministic integration of the biophysical into socio-cultural research, which would provide access to a non-reductionist form of social scientific ecological that would accept the importance of subjective meaning and interpretation in natural living conditions' transformation processes. This methodological constellation calls for a better understanding of material facts' significance and human actors' corporeality as the mediating biophysical part that constitutes the social realities that the newer practices-centered approaches address in the social sciences (Bourdieu 1977; Giddens 1984; Gregory and Urry 1985; Beck 1986; Lefebvre 1974; Latour 1991) and in social geography (Harvey 2005; Werlen 1993, 2007, 2010; Schmid 2005; Lippuner 2005; Gäbler 2015).

The historically established double blindness – regarding the social-cultural concerning the natural sciences and regarding the biophysical concerning most of the social and cultural sciences – have led to a widespread absence of non-reductionist social science and humanities insights into sustainability research and into global change issues such as the politics of climate change (Beck 2009; Giddens 2009; Stehr and Storch 2009; Urry 2011), which were until very recently excluded from social scientific research. This can be seen as the second challenge for nature-society and the science-policy interfaces. In short, the natural sciences do not have a differentiated view of the causes of and reasons for human-induced ecological problems, and mainstream social scientific approaches suffer largely from a near absence of biophysical world expertise concerning social practices.

The exclusion of a meaningful construction of the socio-cultural world and social-cultural meaning attribution to the biophysical world on the one hand (the naturalistic perspective) and the exclusion of the embodied actor as a medium of human practices from the biophysical world on the other hand (the culturalist perspective) is constitutive for the gap that has to be overcome by integrated ecological research. However, general systems theory is, to date, the most prominent approach to overcome this gap by means of a theoretical framework. This approach is applied in the natural and the social sciences as well in ecological research. It integrates biophysical and socio-economic systems on the same ontological level. However, the ways biophysical and socio-economic facts exist differ: Biophysical facts can be characterized as existing in a realm of materiality and (causal) determination, whereas socio-economic facts reside in a realm of contextuality, meaning, and path dependency. The two cannot be treated as if they were integrated into a single system governed by the same kinds of functional relationships; recognizing their distinctive logics is a prerequisite for successfully addressing socio-cultural realities and ecological challenges, or the dilemmas that human action produces. Understanding cultural interpretations' impacts is a prerequisite for achieving sustainable development to reach global sustainability.

The current status of integrated research broadly lacks an ontologically sound and promising strategy for integrated problem-solving and an acceptable consideration of the power of cultural schemas concerning interpretations of natural living conditions. We can conclude that we are facing four weaknesses – a lack of:

- appropriate understanding of the social world by natural scientists and engineers;
- understanding and integration of biophysical elements by social sciences and humanities;
- ontological differentiation resulting in the reductionist implications of systemic integration;
- sensitivity regarding the power of cultural differences when dealing with the natural.

4 New Realities, New Challenges

As briefly noted, new production and communication technologies are reshaping the geographies of the economic and cultural; time-space compression occurs through innovations in transportation and communication (McHale 1969; Harvey 1989; Rosa 2013), while resource and energy use is expanded. All these processes are interrelated, and transform daily life all over the globe. Daily, we are experience that globalization has brought – and is bringing – far-flung places and people into ever-closer contact. New kinds of supranational communities are emerging at an accelerating pace. At the same time, these trends do not efface the local. Globalization is also associated with a marked reaffirmation of places and regions as distinctive forums of human action. In one way or another, all human actions remain regionally and locally contextualized, even if their consequences have global impacts.

The acceleration of globalization is bringing about a new world order. This involves both the integration of natural-human ecosystems all over the world into a unified planetary environment and the steady emergence of an integrated global socio-economic reality. At the same time – perhaps as a result of these processes – global environmental change is occurring at the same, continuously increasing acceleration (Rosa 2013). The extent to which global environmental change is human induced was again documented in the latest IPCC report (IPCC 2014) on climate change mitigation. This report makes it even clearer that there is no alternative but to deepen our knowledge of socio-cultural contexts, to improve social and cultural acceptance of scientific knowledge, and to take culturally differentiated paths to global sustainability.

Although the recognition of global processes' interrelatedness is increasing, parochial discourses have become more forceful on the global stage, often in ways that seem to provoke discord rather than foster global understanding. Many of these discourses postulate a parochial or nationalist interpretation of global processes, or even territorial conflicts as 'solutions' for the most pressing issue of the day, instead of favoring interpretations based on global understanding as guidelines for local and regional actions and measures. In this regard, we also need to deepen our understanding of the new global realities so as to productively address the emerging interconnected challenges.

Although it is clear that local and global issues are intimately related, attempts to address the issue of global climate change have never focused adequately on the question of scale (Marston 2000; Brenner 2001; Beck 2006). In our view, this is one of the central issues of environmental politics and entails two aspects. The first is the specific spatial constellations of politics and global challenges; the second refers to the regional bases of action and the problem's planetary reach.

The first scale problem is that science develops planetary models, but decision-making bodies are primarily involved in finding solutions that have a national scope. The second scale-related problem affects the democratic legitimization of environmental politics in general, and climate politics in particular, beyond national borders. To become fully politically potent, the global orientation of political action and goal-setting must ultimately achieve national endorsement. To reach this sustainably, the first condition is for individuals to develop a global consciousness of their living situations. International institutions such as the UN and political networks should encourage and establish the development of global understanding in the everyday lives of the majority. In short, we need a stronger and broader awareness of the why and how of the link between the global and the local.

A fully binding decision-making body with global jurisdiction can certainly not be established in the timeframe available to humanity's most pressing problems. Global necessities require swift yet considered political action and goal-setting. We must promote global justice, while integrating local interests and concerns. In short, since it is unlikely that a global authority will be established in the near future, the most sustainable solution is to downscale planetary mitigation and adaptation programs to personal levels of decision-making in different everyday constellations.

5 From Interdisciplinary to Transdisciplinary

Given the described everyday and scientific conditions, we must acknowledge that societies and cultures determine the ways we live with and shape our natural living contexts. We must address the ways we live in an increasingly globalized world and the transformation of nature from the perspective of global sustainability for the sake of future generations.

To achieve the above, science must overcome the identified weakness in the field of integrated research in both natural science and social science, as well as in the humanities and even beyond general systems approaches. Natural and social scientific knowledge, as well as insights from the humanities, must be integrated by respecting the ontological differences of the various research realms.

Calling for transdisciplinarity is, first, to ask for a clear differentiation between interdisciplinary and transdisciplinary research methodologies. Attempts to establish interdisciplinary research cooperation clearly show that instead of promoting scientific cooperation and integration, the (paradoxical) opposite has occurred – the profiling of disciplinary boundaries. The disciplinary organization of modern science can, to a certain extent, be compared to nation states as a form of political organization.

Each discipline and each nation state proclaim their competence regarding and responsibility for the attributed or claimed sphere. If the problem constellation surpasses the capacity of one of them, the interaction between them is normally put on the agenda in the form of interdisciplinary or international cooperation. This tendency is a currently one. Since each nation state gains a certain profit from cooperation, each scientific discipline also gains. Consequently, disciplinary boundaries or identities are enforced rather than dismantled, and cooperation is only possible on the basis of a scientific division of labor and with specific – very often even incompatible – perspectives, methodologies, and methods. In most cases, the dilemma of incompatibility is solved by choosing to accept the logic of traditional environmental research, by maintaining the primacy of the natural and, as a result, the primacy of a natural scientific perspective, and a natural (scientific) research unit.

A typical example of such a 'solution' would be investigations into socio-cultural aspects of river catchment areas. The research unit of integrative research is natural, the topic is socio-cultural, and integration will only be possible on the basis of (geo)deterministic explanations of the socio-cultural by natural (living) space (river catchment area). Switching disciplinary perspectives will therefore not yield an integrated view that respects the ontological differences between the socio-cultural and the biophysical realms.

And if the additive logic of interdisciplinary cooperation does not lead to effective integration, this is not due to a lack of ability or goodwill. The lack of integration is inherently linked to the organizational logic of scientific research. While this logic is very effective for many purposes, it certainly does not aid integration.

To overcome disciplinary blind spots, we need a perspective that specifies and solves problem complexes independent of disciplinary interests and boundaries (Mittelstrass 2001). One possibility is to focus on problem constellations of general concern in the way human actions produce these human-induced problem constellations – such as global warming or establishing global sustainability – and by taking the integration of the natural and the social into account concerning the way these actions integrate them.

Therefore, a genuine transdisciplinary perspective starts with the logic of everyday actions rather than with traditional scientific disciplines. Learning how human actions produce ecological problems is the first step. The second step relies on seeking appropriate science-based analyses of the different aspects of human actions (socio-cultural, mental, biophysical, etc.). The third step does suggest solutions through the logic of problematic human actions, which provides suggestions on how to do things differently in the given everyday context, or in a context that must be newly established. In this way, topic-focused cooperation and not discipline-focused cooperation should become feasible between scientists and humanities scholars without reductionism and without one research sector inappropriately dominating the other.

The proposed form of transdisciplinary research perspective is marked by the logic of everyday human actions and practices, proceeding from the socio-cultural to the biophysical, and not according to the logic of a scientific, disciplinary division of labor. Therefore, we first require insights into the problematic context of action. We should then seek to answer the question of what kind of scientific stock of knowledge should be used to change harmful practices. It is not the disciplinary order determined by the analytical 'decomposition' that defines a problem as being a chemical, physical, sociological, historical, geographical, etc. one; instead, the practical problem provides the organizing principle of the use of knowledge, moving from the everyday to the science level and not the other way around. As a result, transdisciplinary perspectives should focus on actual embodied individual human activities performed by every person, every day, everywhere in the world. Finally, complex environmental problems should be addressed as social problems. This perspective will enable us to address society-nature and science-policy interfaces in new ways (Werlen and Weingarten 2004).

6 Building Bridges

According to the perspective suggested in this chapter, the predominance of the natural over the social should first be replaced by problem production logic, from the socio-cultural to the biophysical. To build a bridge between knowledge and action, knowledge should be presented in the logic of action, and not in the logic of nature.

Second, it is imperative that science should bridge the gap between global problems and national, regional, and local behavior, as well as decision-making. Thereafter, effective solutions should be prepared. According to this perspective, effective solutions based on bottom-up decisions and actions are at least as likely to be effective as top-down measures. This bottom-up strategy of scientific methodology aims to gain insights into the behavior of individuals, households, and firms that make the majority of decisions that collectively cause human-induced global change. In the process, differing perspectives and insights from everyday experiences should be integrated with research results from the natural and social sciences, as well as the humanities, into a geographical locality-based framework. The third bridging step addresses the world of sciences and the everyday worlds in their manifold constitutions.

Actions and thoughts that may seem disconnected in space and time are often fundamentally linked, and global understanding enables people to make such connections. Many people know about the need for sustainability, but few make the corresponding decisions. Consequently, the main scientific goal should be to promote global understanding so that actions and decisions yield sustainable outcomes, every day, all over the world.

7 Global Understanding: The IGU Initiative for an International Year

To help establish the necessary conditions for such global awareness, the IGU Initiative for the 2016 International Year of Global Understanding (IYGU) proposes a new geographical worldview that takes into account the specific spatial features of current living conditions. The outline of this initiative first illustrates the direction of the development of newly designed research methodologies, of subsequent research programs, and of everyday policies. Second, it seeks to make visible the implication of such an orientation and the three interfaces: between local and global, between the natural and social sciences and the humanities, and between the sciences and everyday life/policies.

The IYGU Program: Levels and Interfaces

Local || Global
Global understanding

Social || Natural
Transdisciplinary research

Science || Everyday
Integration of different types of knowledge
Bottom-up policies based on everyday decisions

The IGU initiative starts with the hypothesis that humanity's grand challenges (Reid et al. 2010) are self-made and affect the whole planet. A widespread awareness of how everyday actions create the challenges that impact humanity is a prerequisite to finding solutions. This initiative addresses the why and how of people's and peoples' distinct responses to similar ecological dilemmas. It seeks to improve conditions for the policies and strategies required to deal with these difficult situations.

First, the IYGU wants to make the global understandable through local everyday actions. True global understanding requires awareness of the global implications of everyday living and will demonstrate how the socio-cultural and biophysical spheres are connected in different life-worlds. The IYGU seeks to facilitate understanding of global processes, to encourage everyone to make daily decisions in light of the global challenges, and to contribute to bottom-up initiatives that connect individual, local actions to global sustainability.

To overcome the mentioned shortcomings of the currently dominant ecological approaches, the IYGU program, at a first level, ignores the differentiation between the natural and social, and starts with basic human practices such as working, housing, eating, etc. that encompass both realms in an integrated way.

Socio-cultural	Human actions as everyday practices	Natural
Cultural values	**Eating/drinking//surviving**	Food systems
Risk	**Working/housing//urbanizing**	Climate change
Cultural diversity	**Moving/staying//belonging**	Biodiversity
Schemes of interpretation	**Wasting/recycling//preserving**	Soil degradation
Glocalized lifestyles	**Communicating/networking//interacting**	Sea level
	Sports/entertaining//recreating	

The reconciliation of the global and the local, as well as the natural and social, is as much a part of global understanding as the reconciliation of science and everyday lives. To reach this goal, the IYGU wants to encourage scientists and citizens to effect change towards sustainable social actions, habits, and routines on the basis of a new geographical imagination of the world, one in which the practices of geography making are at the core, rather than spaces, countries, or continents. Social practices are at stake, not predominantly living spaces or ecotopes.

The International Social Science Council (ISSC), in collaboration with its member organizations, the international umbrella organization for philosophy and the humanities (CIPSH), and the ICSU, the global umbrella organization of all natural scientific disciplines, strongly supports this initiative.

The objectives of the International Year of Global Understanding will complement those of the Future Earth initiative by the International Council of Science (ICSU) and the International Social Science Council (ISSC):

The International Year of Global Understanding has five outreach objectives:

1. Establish a comprehensive global understanding and awareness of the natural and cultural embeddedness of all human action.
2. Contribute to changing ecologically harmful habits by designing culturally differentiated, exemplary alternative models of core everyday (best) practices.
3. Enhance awareness of the individual capacity and responsibility for everyday decision-making.
4. Produce school teaching modules and study guidelines to be implemented at all educational levels – from primary to tertiary and post-experience education.
5. Serve as a catalyst for transdisciplinary cooperation and improve the culturally differentiated transfer of scientific insights into social practice.

This way, the IGU initiative hopes to yield deep but actionable insights into the ways all peoples can live more sustainably on and with the earth, as well as with one another. It includes action sets that will establish scientific research initiatives and improve public awareness of key human life issues. Furthermore, the initiative intends to boost the scientific community's enthusiasm for transdisciplinary research.

References

Beck, U. (1986). *Risikogesellschaft. Auf dem Weg in eine andere Moderne*. Frankfurt: Suhrkamp.

Beck, U. (2006). *The cosmopolitan vision*. Cambridge: Polity Press.

Beck, U. (2009). *World at risk*. Cambridge: Polity Press.

Bourdieu, P. (1977). *Outline of a theory of practice*. Cambridge: Cambridge University Press.

Brenner, N. (2001). The limits to scale? Methodological reflections on scalar structuration. *Progress in Human Geography, 25*(4), 591–614.

Gäbler, K. (2015). *Gesellschaftlicher Klimawandel. Eine Sozialgeographie der ökologischen Transformation* (Sozialgeographische Bibliothek, Vol. 17). Stuttgart: Franz Steiner Verlag.

Giddens, A. (1984). *The constitution of society*. Cambridge: Polity Press.

Giddens, A. (2009). *The politics of climate change*. Cambridge: Polity Press.

Gregory, D. (1993). *Geographical imaginations*. London: Blackwell.

Gregory, D. (2004). *The colonial present. Afghanistan, Palestine, Iraq*. London: Blackwell.

Gregory, D., & Urry, J. (1985). *Social relations and spatial structures*. London: Macmillan.

Grober, U. (2010). *Die Entdeckung der Nachhaltigkeit*. München: Kunstmann.

Grunwald, A., & Kopfmüller, J. (2012). *Nachhaltigkeit*. Frankfurt: Campus Verlag.

Haeckel, E. (1866). *Generelle Morphologie der Organismen. Allgemeine Grundzüge der organischen Formen-Wissenschaft, mechanisch begründet durch die von Charles Darwin reformierte Descendenz-Theorie* (Allgemeine Entwickelungsgeschichte der Organismen, Vol. 2). Berlin: Georg Reimer.

Haeckel, E. (1878/1879). *Gesammelte populäre Vorträge aus dem Gebiete der Entwicklungslehre*. Bonn: Strauß.

Harvey, D. (1989). *The condition of postmodernity: An enquiry into the origins of cultural change*. Oxford: Blackwell.

Harvey, D. (2005). *Spaces of neoliberalization: Towards a theory of uneven geographical development*. Stuttgart: Steiner Verlag.

Hauff, V. (1987). *Unsere gemeinsame Zukunft – Der Brundtland-Bericht der Weltkommission für Umwelt und Entwicklung*. Greven: Eggenkamp.

IPCC. (2014). Summary for policymakers. In O. Edenhofer, R. Pichs-Madruga, Y. Sokona, E. Farahani, S. Kadner, K. Seyboth, A. Adler, I. Baum, S. Brunner, P. Eickemeier, B. Kriemann, J. Savolainen, S. Schlömer, C. von Stechow, T. Zwickel, & J. C. Minx (Eds.), *Climate change 2014, mitigation of climate change. Contribution of Working Group III to the fifth assessment report of the Intergovernmental Panel on Climate Change* (pp. 1–31). Cambridge/New York: Cambridge University Press.

Jaeger, C., Renn, O., Rosa, E. A., & Webler, T. (2001). *Risk, uncertainty, and rational action*. London: Earthscan Publications.

Jazeel, T. (2012). Postcolonialism: Orientalism and the geographical imagination. *Primary Geography, 97*(1), 4.

Latour, B. (1991). *Nous n'avons jamais été modernes. Essai d'anthropologie symétrique*. Paris: La Découverte.

Lefebvre, H. (1974). La production de l'espace. Paris: Éd. Anthropos

Lippuner, R. (2005). *Raum, Systeme, Praktiken. Zum Verhältnis von Alltag, Wissenschaft und Geographie*. Stuttgart: Steiner Verlag.

Marston, S. (2000). The social construction of scale. *Progress in Human Geography, 24*(2), 219–242.

McHale, J. (1969). *The Future of the Future*. New York: George Braziller Inc.

Mittelstrass, J. (2001). *Wissen und Grenzen. Philosophische Studien*. Frankfurt: Suhrkamp.

Newton, I. (1704). *Opticks: Or, treatise of the reflection, refraction, inflections and colours of light*. London: Printed for Sam. Smith, and Benj. Walford, Printers to the Royal Society, at the Prince's Arms in St. Paul's Church-yard.

Ratzel, F. (1882). *Anthropogeographie. Grundzüge der Anwendung der Erdkunde auf die Geschichte*. Stuttgart: J. Engelhorn.

Ratzel, F. (1901). *Der Lebensraum. Eine biogeographische Studie*. Tübingen: H. Laupp.

Reid, W. V., Chen, D., Goldfarb, E., Hackmann, H., Lee, Y. T., Mokhele, K., Ostrom, E., et al. (2010). Earth system science for global sustainability: Grand challenges. *Science, 330* (6006), 916–917.

Rosa, H. (2013). *Social acceleration: A new theory of modernity*. New York: Columbia University Press.

Said, E. W. (1978). *Orientalism*. London: Routledge & Kegan Paul.

Schmid, C. (2005). *Stadt, Raum und Gesellschaft. Henri Lefebvre und die Theorie der Produktion des Raumes*. Stuttgart: Steiner.

Stehr, N., & von Storch, H. (2009). *Climate and society: Climate as resource, climate as risk*. Singapore: World Scientific Publishing Company.

Urry, J. (2011). *Climate change and society*. Cambridge: Polity Press.

von Carlowitz, H. C. (1713). *Sylvicultura Oeconomica, oder haußwirthliche Nachricht und Naturmäßige Anweisung zur wilden Baum-Zucht*. Leipzig: Braun.

Weber, M. (1913). Über einige Kategorien der verstehenden Soziologie. *LOGOS: Internationale Zeitschrift für Philosophie der Kultur, 4*(3), 253–294.

Weber, M. (1920). *Wirtschaft und Gesellschaft*. Tübingen: Mohr Siebeck.

Weber, M. ([1912]1988). Geschäftsbericht und Diskussionsreden auf den deutschen soziologischen Tagungen (1910). In M. Weber (Ed.), *Gesammelte Aufsätze zur Soziologie und Sozialpolitik* (pp. 431–491). Tübingen: Mohr Siebeck.

Werlen, B. (1993). *Society, action and space: An alternative human geography*. London: Routledge.

Werlen, B. (2005). Andere Zeiten – Andere Räume? Zur Geographie der Globalisierung. In M. Ott & E. Uhl (Eds.), *Denken des Raums in Zeiten der Globalisierung* (pp. 57–72). Münster: Lit Verlag.
Werlen, B. (2007). Sozialgeographie alltäglicher Regionalisierungen. Globalisierung, Region und Regionalisierung. Stuttgart: Franz Steiner
Werlen, B. (2010). *Gesellschaftliche Räumlichkeit* (Konstruktion geographischer Wirklichkeiten, Vol. 2). Stuttgart: Steiner Verlag.
Werlen, B. (2013). Gesellschaft und Raum: Gesellschaftliche Raumverhältnisse. Grundlagen und Perspektiven einer sozialwissenschaftlichen Geographie. *Erwägen Wissen Ethik, 24*(1), 3–16.
Werlen, B., & Weingarten, M. (2003). Zum forschungsintegrativen Gehalt der (Sozial-) Geographie. In P. Meusburger (Ed.), *Humanökologie. Ansätze zur Überwindung der Natur-Kultur-Dichotomie* (pp. 197–216). Stuttgart: Steiner.
Werlen, B., & Weingarten, M. (2004). Tun, Handeln, Strukturieren – Gesellschaft, Struktur und Raum. In P. Weingarten (Ed.), *Strukturierung von Raum und Landschaft. Konzepte in Ökologie und der Theorie gesellschaftlicher Naturverhältnisse* (pp. 177–222). Münster: Westfälisches Dampfboot.
Werlen, B., & Weingarten, M. (2005). Integrative Forschung und „Anthropogeographie". In M. Weingarten (Ed.), *Strukturierung von Raum und Landschaft. Konzepte in Ökologie und der Theorie gesellschaftlicher Naturverhältnisse* (pp. 314–333). Münster: Westfälisches Dampfboot.
Werlen, B., & Weingarten, M. (2013). Alltägliches Geographie-Machen, Regionalisierung und die Grenzen des Materialismus-Paradigmas. In O. Brand, S. Dörhöfer, & P. Eser (Eds.), *Die konflikthafte Konstitution der Region: Kultur, Politik, Ökonomie* (pp. 74–91). Münster: Westfälisches Dampfboot.
World Commission on Environment and Development. (1987). *Our common future*. Oxford: Oxford University Press.

Challenges for Transdisciplinary Research

Joske F.G. Bunders, Arnout E. Bunders, and Marjolein B.M. Zweekhorst

1 Introduction

To make our human processes more sustainable, we need more knowledge and new inventions; yet neither science, nor technology, nor political will are sufficient: in many cases, one of the hardest nuts to crack is culture. Here, culture should be understood in the broadest sense of the word: as 'the way we do things around here.' The 'reigning regime' proves a formidable obstacle in the way of change – whether motivated by well-proven best practice standards or by religious considerations, by fashion (e.g., the connotation of words like 'natural' or 'chemical') or by rational risk management. (For an overview of the literature, see: Weber and Hemmelskamp 2005.)

Instruments to promote changes in culture are hard to come by. There are no methods with which to measure success in a sufficiently standardized way to draw conclusions from a range of case studies. However, one issue is highlighted in all the available literature: a top-down approach does not work. The active involvement of societal groups, or of representatives of those people for whom new methods are proposed, is needed as early as possible in the development process – even at the stage of setting the research agenda.[1] This requires cultural change at many levels, both inside and outside academia.

This article provides an overview of transdisciplinary research in the context of international efforts to achieve more sustainable processes; it takes a cultural

[1] For example, participation in research agenda setting was developed on the basis of the responsive methodology and the ILA approach, which operationalizes the partnership concept of Arnstein (Broerse et al. 2010; Abma et al. 2009; Caron-Flinterman 2005).

J.F.G. Bunders (✉) • A.E. Bunders • M.B.M. Zweekhorst
Athena Institute, Faculty of Earth and Life Sciences, VU University, De Boelelaan 1085,
1081 HV Amsterdam, The Netherlands
e-mail: j.g.f.bunders-aelen@vu.nl

B. Werlen (ed.), *Global Sustainability*, DOI 10.1007/978-3-319-16477-9_2

perspective. Transdisciplinary research hinges on societal involvement and is embedded in sustainability and research cultures. Culture has a number of different implications for transdisciplinary research, which range from cultures of different actors' communication in multistakeholder processes to the way different approaches to transdisciplinary research have emerged in different national and regional contexts (Bunders et al. 2010; Lang et al. 2012). This article will first outline transdisciplinary research's position in the context of sustainability cultures. Second, it describes transdisciplinary research in the context of the academic research culture, emphasizing the guiding principles and diverse approaches. Next, it highlights the need for methodologies and quality control. The challenges facing transdisciplinary research are then reviewed, on the basis of the interactive learning and action (ILA) example. The biotechnology and food example will be used to illustrate the issues raised.

2 Sustainability Cultures

Much has changed since the pioneering Brundtland Report (1987): today, few scientists, businessmen, or governments would deny the need for change towards the more sustainable use of energy, raw materials, land, water, and other resources. Public opinion does not lag behind in this, as virtually every public opinion barometer indicates (See, for example: Do Well Do Good 2012; Marist Poll 2009; The Yale project on Climate Change Communication 2012; European Commission 2009). But, if everybody agrees that we need to change our ways, why do we make such little progress?

Unsustainability – or, rather, the slow pace towards sustainability – is one of those problems that has been described as 'wicked problems' (see Text Box 1). One of the hallmarks of this type of problem is that there is no agreement on the definition of the problem. This is also true of sustainability: regardless of the differences in description, as soon as actual examples come into play, there are many different opinions of what *sustainable* means in this or that specific context. If we cannot even agree on what the problem is, how can we work together towards a solution?

This problem is exacerbated by the fact that, even if all parties were to agree to use English as the *lingua franca*, there are still many 'dialects': businessmen and scientists, civil servants, and laymen do not speak the same lingo. In a sense, this is a truism, but probably everyone with practical experience has at some stage been surprised not only by the depth of the rifts, but even more by the lack of awareness of the extent of this communication problem and its consequences.

The search for sustainability has been described as the search for "cultures of sustainability," because of the global mindset changes that are required (Kagan 2010, 1094). Many dialogue and cooperation efforts end in deadlock early on and, more often than not, there is no clarity whether this is due to genuine differences in opinion or interests, or to differences in the language and frame of reference. In our opinion, transdisciplinary research can contribute to sustainability cultures, because it can be applied in the absence of a clear and agreed problem definition and provides a comprehensive set of tools to overcome communication problems.

2.1 *The Nature of Culture*

The UNESCO Universal Declaration of Cultural Diversity (2001a) defines culture as:

> . . .the set of distinctive spiritual, material, intellectual and emotional features of society or a social group, and that it encompasses, in addition to art and literature, lifestyles, ways of living together, value systems, traditions and beliefs.

This has implications for:

> . . .contemporary debates about identity, social cohesion, and the development of a knowledge-based economy. (UNESCO 2001a, b)

According to Nurse (2006), building on the work of Williams (1981), there are at least four contested definitions of culture, which are in turn based on a 'cultured person's,' state of mind, the development of this interest as in 'cultural activities,' the arts, cultural heritage, and fourthly:

> . . . a whole way of life', 'a signifying system' through which a social order is communicated, reproduced, experienced and explored. (Nurse 2006, 36)

This article focuses on this fourth meaning of culture. In the literature on environmental sustainability, culture is regarded as the central pillar of sustainable development:

> Culture should be viewed not just as an additional pillar of sustainable development along with environmental, economic and social objectives because peoples' identities, signifying systems, cosmologies and epistemic frameworks shape how the environment is viewed and lived in. (Nurse 2006, 37)

Culture can be considered the elephant in the room for transdisciplinary research, which addresses complex societal problems and contributes to the development of more sustainable human processes.

2.2 *Cultural Forces Against Sustainability*

In the context of promoting change toward more sustainable industry and lifestyles, it has become increasingly apparent that knowledge dissemination, although a necessary precondition, is not enough. It has not been enough to keep people from smoking and it will certainly not be enough to make the public change their lifestyles, or to make businesses abandon their trusted but wasteful business models.

There are mutually reinforcing trends that undermine the role science can play to support sustainability cultures:

- Declining trust in science;
- Declining trust in the media, the foremost channel to disseminate scientific knowledge;
- Declining trust in regulators' ability to keep scientists in check.

We will only briefly discuss these trends here, since there is a readily available wide body of literature on this subject. However, they are relevant because they reflect on the mindset within which current efforts towards sustainability are developing.

In the past decade, surveys have demonstrated a growing mistrust in science, in the scientific information that the media present to the public, and in the entanglement of power, business interests, and science. In 2000, a broad analysis of available data (notably the results of polls and surveys from all continents) from the European Commission's Institute for Prospective Technological Studies could still claim that all the talk about the "declining trust in science" (Jensen 2000, 15) was largely a myth. However, the report found that the public do not trust governments, scientists working for governments, and the media. At roughly the same time, a report from the UK Office of Science and Technology and the Welcome Trust (Science and the public 2000, 5), demonstrated "ambivalence about politicians' motives for supporting science" as well as "concern about Government's ability to control science." More recently, in its 3-yearly monitor of public trust in biomedical science in the UK, the Welcome Trust found "A mismatch [...] between reliance on the media to be kept up-to-date and very mixed views about the accuracy of media reporting (and a lack of trust in journalists more broadly)" (Wellcome Trust Monitor 2010, 3).

Science has increasingly come under fire. The EC reports that, although there is "widespread agreement that science and technology make our lives healthier, easier and more comfortable," the share of Europeans with a general trust in science has declined from 78 to 66 % since 2005 (European Commission 2011, 500). The study also establishes that "Europeans trust science more or less at [sic] the same extent as citizens of the United States and Canada" (European Commission 2011, 452). In the US, Gauchat, following Mooney (2005), found a significant decline in trust in science among those who used to rely on science most:

> Conservatives began the period [1974 to 2010] with the highest trust in science, relative to liberals and moderates, and ended the period with the lowest. (Gauchat 2012, 167)

In the context of moving towards more sustainable industry, this seems particularly worrying, since the majority of entrepreneurs are conservatives. In addition, the decline in trust seems related to the growing number of publications and media debates about environmental issues (Gauchat 2012, 179; see also Leiserowitz et al. 2010).

2.3 Cultural Changes in Support of Sustainability

Some trends are, however, positive and support the development of cultures of sustainability. First, among major policy-level actors, such as governments and international bodies, there is increasing recognition that the one-sided dissemination of knowledge is not enough. The words 'dialogue' and 'participation' pop up in many reports and white papers. An example is the review of public trust in science that the UK Office of Science and Technology and the Wellcome Trust (Wellcome Trust Monitor 1 2010) undertook. The Wellcome Trust notes that "[i]t is important that the public debate on science issues is inclusive, and that no member of the public feels disenfranchised" (Sainsbury and Dexter 2000, foreword). Another example is a report from the Council of Europe (2008). Meanwhile, the EC notes that "an important minority (29 %)" of the public favors a more participatory approach to decision making on scientific issues (European Commission 2011, 466).

But how would one go about involving, or even consulting, the public? Clearly, opinion polls – which rely on information gathered from sources that the public increasingly mistrusts – will not work. Nor is it feasible to engage in a participatory process with millions. Here, other trends may point the way forward: there is increasing public trust in scientists working for independent institutions such as universities (Jensen 2000, 8). Furthermore, there is increasing trust in the scientific information that nongovernmental organizations (NGOs), consumer organizations, and other independent sources distribute (see, e.g., Edelman 2012; Jensen 2000).

2.4 The Cultural Dimensions of Persistent Problems

Persistent problems, such as unsustainability, are complex in a way that differs from complex monodisciplinary problems. Such problems are often called *wicked problems* and the questions related to climate change and sustainability have even been given the label *super-wicked problems* (see Text Box 1). Wicked problems have a number of key characteristics: (1) they cannot be understood from a single perspective (multiperspective); (2) they do not fit into disciplines, or into distinct policy domains (multifactor); (3) no single entity (multiactor) can devise and implement their solutions; and (4) features embedded in the system (multilevel) constrain solutions.

Text Box 1: Wicked Problems and Super-Wicked Problems

Wicked problems

> *Some problems are so complex that you have to be highly intelligent and well informed just to be undecided about them.* – Laurence J. Peter (as quoted in Platt 1993, 78)

Many issues relating to the detrimental effects of nonsustainable economic activities can be described as *wicked problems*, a term that C. West Churchman coined in 1967 to describe social planning problems that are complex, difficult to describe, resistant to logical analysis, contradictory, and continuously changing. This notion was later generalized by others (Rittel and Webber 1973; Conklin 2005). Our description concentrates on six features:

1. *You don't understand the problem until you have developed a solution*
   If we can formulate the problem by tracing it to its source, then we have also formulated a solution. In other words, the problem cannot be defined until a solution has been found. The formulation of a wicked problem is the problem.
2. *Every wicked problem is essentially novel and unique*
   Despite long lists of similarities between a current problem and previous ones, there will often be an additional distinguishing property that is of overriding importance. Consequently, every solution will have to be custom designed.
3. *A wicked problem has a no stopping rule*
   Since there is no definitive description of 'the problem,' there cannot be a state of affairs called 'the solution.' The problem-solving process ends when you run out of resources, not when some predefined solution is reached.
4. *Solutions for a wicked problem are not right or wrong*
   Solutions may be 'good enough,' 'better,' or 'worse' in the eyes of some, but others, who are equally equipped, interested, and/or entitled to judge, are likely to have a differing opinion. Each actor judges from his or her own societal context and according to his or her personal values, interests, and ideological preferences.
5. *Every solution for a wicked problem is a 'one-shot operation'*
   Given that every wicked problem is unique and novel, there is no opportunity to learn by trial-and-error. Every attempt is consequential, leaving traces that cannot be undone. Every attempt to correct the undesired consequences creates another set of wicked problems.
6. *A wicked problem has no given alternative solutions*
   There may be no solutions, or there may be a host of potential solutions and probably still many other solutions nobody even thought of. Thus, it is

(continued)

Text Box 1 (continued)

a matter of creativity to devise potential solutions and a matter of judgment to determine which solutions are viable.

Super-wicked problems

A recent subcategory of wicked problems are the 'super-wicked problems' inspired by the types of problems that are the subject of climate change discussions (Levin et al. 2007). Super-wicked problems have the following features in addition to those above:

1. *Time is running out*

   Time is running out to find and test solutions. The characteristics of the problem set deadlines that limit the choice of solutions, make the problem urgent, and put pressure on the need to find effective solutions before some irreversible tipping point is passed.
2. *There is no central authority*

   The problem is too large and complicated to fall within the authority of a single entity. This feature delays and complicates the definition of the problem, the negotiations about responsibilities, solutions, and the implementation of solutions.
3. *Those seeking to solve the problem are also causing it*

   In the words attributed to Einstein, "One can never solve problems at the same level of thinking that created them." Especially those who have caused a problem, will find it difficult to change their old way of thinking and move to a new kind of thinking. They may also have vested interests that hamper an adequate response and the finding of new options.
4. *Hyperbolic discounting*

   Given two similar rewards, humans show a preference for an immediate reward, even if the reward that comes later has much more value. In other words, people discount the value of the later reward. The extent to which this occurs depends – predictably – on the perceived difference between the rewards, as well as on the length of the delay but, oddly enough, also on the immediacy of the first reward. The relationship between the discount factor and the time-span until the first reward is described by a hyperbole, hence the term 'hyperbolic discounting.' Thus, not only are partial solutions that work immediately favored over more structural solutions, but public opinion is also likely to change in favor of the quick fix as time progresses.

Thus, to work towards a solution, we need to cross boundaries between science, policy, business, and civil society. However, within these different worlds, there are many cultural differences that stand in the way of cooperation towards a solution. Hence, various new forms of cooperation are necessary between groups of people who are usually separate, including:

- Cooperation between different departments or sectors within one organization (e.g., the ministry of the environment, cooperation on quality of life);
- Cooperation between different partners within one sector (e.g., academia, the private sector, and horticultural government); and
- Cooperation between state, market, and civil society actors on cross-sectoral problems (e.g., water, health, and poverty).

2.5 *Lessons from Intersectoral Cooperation*

We use the term *intersectoral cooperation* to describe the process of "bringing actors from the state, market, and civil society together to achieve mutual understanding on an issue and to negotiate and implement mutually agreeable plans for tackling the issue once it is identified" (Kalegaonkar and Brown 2000, 2). Many case studies (e.g., Kalegaonkar and Brown 2000) provide insights into the challenges of intersectoral cooperation, thus identifying the processes and conditions for successful cooperation. Lessons and toolkits have been derived from these case studies (see Table 1).

Furthermore, a great many toolkits are now readily available to set up and steer such intersectoral or transdisciplinary cooperation projects (see, e.g., Brown et al. 2012). Yet, contrary to what the term toolkit suggests, most of themlack a truly practical approach. Scant work has gone into the ways in which tools and rules can be implemented in practice.

> Dispersed literature on these approaches and a variety of empirical projects applying them make it difficult for interested researchers and practitioners to review and become familiar with key components and design principles of how to *do* transdisciplinary sustainability research. (Lang et al. 2012, 25)

In our experience, the main obstacle to a sustainability culture is the compartmentalisation of our society and of our institutions in particular, each with its own goals and incentive structures. Scientists and policy-makers work within their respective institutional settings and require space to deviate from the norm. Scientists are held accountable for producing scientific work, as determined by criteria developed in the context of monodisciplinary research. The public, who may or may not understand the issues involved or the consequences of certain choices, hold politicians accountable. Business has its own best practices that should be applied, if only to fulfil its legal obligations towards its shareholders and banks.

In such settings, the prescriptions of available toolkits are very hard to implement, since they tend to start from 'a problem' and work towards 'a solution.' When there is no shared frame of reference, a shared definition of the problem is also often lacking. Furthermore, as stated above, there is no single 'right' solution to persistent problems, at least not from a scientific perspective. Confronted with such hurdles, many ad hoc methods have been developed (Lang et al. 2012). As they are primarily concerned with the solution of a particular local problem, these methods are less

Table 1 Lessons from intersectoral cooperation

Lesson # 1 Intersectoral cooperation is most appropriate when:	Previous mono-sectoral problem-solving been unsatisfactory, particularly to high power actors.
	Problem solving requires information and resources held in more than one sector.
	Past relations among key actors do not fore-close cooperation.
Lesson #2 Starting intersectoral cooperation requires:	Conveners with the credibility to bring all the relevant parties together.
	Framing problems to emphasize the need for and mutual gains from multi-sectoral participation.
	Balancing power differences among the parties at the outset.
	Investing in relationship-building as well as problem-solving.
Lesson #3 Managing intersectoral cooperation requires organizational arrangements that enable parties to:	Develop shared plans that identify roles, responsibilities, and resources.
	Foster mutual influence in decision-making.
	Manage conflicts in implementation.
	Seek and identify mutual gains.
Lesson #4 Challenges to intersectoral cooperation include:	Cooptation that reduces contribution of party differences.
	Inequitable distributions of costs and benefits.
	Initial experiences that shape future cooperation.
Lesson #5 Positive outcomes of intersectoral cooperation include:	Innovative solution to intractable problems.
	Catalytic social change.
	Sustainable social change.
	Creation of multisectoral social capital and new capacity for joint local action.

Source: Kalegaonkar and Brown (2000, 20)

concerned with standards that would allow results to be aggregated, a body of transdisciplinary knowledge to be established, and scientific criteria, such as reproducibility, complied with. Thus, we need to go beyond toolkits and establish a methodology that comprises, in addition to tools, standards that enable us to compare and aggregate results and quality criteria that ensure recognition within the scientific community.

3 Research Cultures and Transdisciplinary Research

It has almost become standard practice to demand that research proposals demonstrate societal relevance and use transdisciplinary research methods, or at least a multistakeholder approach. This practice seems to suggest that we have a well-established methodology for organizing the required processes. But do we? We believe that it is time to assess transdisciplinary approaches and to learn from best practices in order to find out under which conditions they can be applied, and what they mean in the context of a change towards sustainability. We agree with Lang and colleagues' view that "the literature is fragmented and dispersed, without providing good guidance to interested researchers and practitioners on what can be learned from the different approaches and what needs to be considered when planning and carrying out transdisciplinary sustainability research" (Lang et al. 2012, 2). While the authors subsequently provide an excellent overview of the 'dispersed' literature, we would like to add that there is an increasing urgency to get a better grip on the possibilities and limitations, as well as to reflect critically on the principles, concepts, and core methods of transdisciplinary research. This is particularly warranted since transdisciplinary research appears to be a very demanding and challenging endeavor.

3.1 What Is Transdisciplinary Research?

Since Jantsch coined the term in 1972, many authors have proposed definitions of 'transdisciplinary'. We mention a few:

> [T]ransdisciplinarity refers to knowledge or research that frees itself of its specialised or disciplinary boundaries, that defines and solves its problems independently of disciplines, relating these problems to extra-scientific developments. (Mittelstraß 1992, 250, translation ours)

> A new form of learning and problem-solving involving co-operation between different parts of society and science in order to meet complex challenges of society. Transdisciplinary research starts from tangible, real-world problems. Solutions are devised in collaboration with multiple stakeholders. (Thompson Klein et al. 2001,7)

> Transdisciplinary research [. . .] denotes interdisciplinary cooperation, involving not only scientists but also practitioners from beyond the realm of science (for example the users) in the research work. (Defila and Di Giulio 1999, 13)

Even from these examples, it can be seen that the definition has changed over time as a result of a shift in emphasis in the different features. (For a more in-depth discussion, see Bunders et al. (2010)). An analysis of the current definitions of transdisciplinarity identified four such features (Pohl and Hirsch Hadorn 2007, 70):

(a) The transcendence and integration of disciplinary paradigms
(b) Participatory research
(c) The focus on real-world problems
(d) The search for unity of knowledge beyond disciplines.

The meaning of transdisciplinarity shifts with the relative weight given to each of these features. For example, the understanding as promoted by Nicolescu (1996) mainly emphasizes the search for unity of knowledge beyond disciplines, while other definitions combine several features. In the past few years, the participatory approach, namely the involvement of stakeholders, has gained emphasis to the point where, in our opinion, it has become transdisciplinary research's defining feature. This shift generally came about as a result of external pressure, notably from policy-makers, rather than as a result of developing scientific insights (Bunders et al. 2010).

3.2 *Shared Principles of Transdisciplinary Research*

In the scholarly literature, the core of transdisciplinary research is often presented as a shared set of principles (see Text Box 2). Principles differ from theories, methods, tools, and conditions because they refer to the researcher. The researcher is said to perform genuine transdisciplinary research when he or she acknowledges and acts in accordance with these principles' intention. A set of principles describes the intentions that guide the researcher in the choice of design, methods, tools, and their sequence – collectively known as the approach. This is consistent with the widespread convention of labeling specific realizations of transdisciplinary research as 'approaches.'

Text Box 2: Shared Principles of Transdisciplinary Approaches (Bunders et al. 2010)

- Joint process initiated outside academia (government, industry, public, NGOs), or by scientists on an ill-defined, societally relevant, real-world problem
- Joint problem definition (including definition of system boundaries)
- A method-based analysis of the complexity of a system (actor analysis, causal analysis, system analysis)
- Mutual learning enhanced in focus groups, round tables, expert sessions, stakeholder dialogues, etc.)
- The construction of 'robust orientations' for the development of outcomes.

3.3 *Types of Transdisciplinary Research*

Given that the type of approach is more important than choices of tools in determining variations in transdisciplinary research, the variety of approaches within transdisciplinary research will be analysed in more detail. From the perspective of knowledge democracy, which is characterized by not just dominant actors, but also nondominant actors having free access to relevant knowledge,[2] we can distinguish

[2] For an explanation of 'knowledge democracy,' see, e.g., Bunders et al. (2010).

Table 2 Four research styles regarding two dimensions: the relationship between the degree of input of lay knowledge and the degree of focus on nondominant actors (Bunders et al. 2010)

	Degree of focus on involvement of nondominant actors	
Degree of lay knowledge input	Low	High
Low	Self-referential knowledge production	Knowledge dissemination (transmission)
	Monodisciplinary, interdisciplinary, and multidisciplinary research	Creating access to knowledge
High	Mutual learning for knowledge production between scientists and societal actors	Knowledge co-creation between scientists and societal actors with a specific focus on the involvement of nondominant actors

two important dimensions in approaches: the degree of lay groups' knowledge input included in a specific transdisciplinary project, and the degree to which nondominant actors are explicitly involved in the decision-making process regarding policies or research agendas. Four research types, based on these two dimensions, are distinguished in Table 2. From these two dimensions, we have developed a typology of research from the knowledge democracy perspective.

3.3.1 Type 1: Self-Referential Knowledge Production

This type of research – characteristic of monodisciplinary, multidisciplinary, and interdisciplinary academic research – takes into account questionnaires and polls from the stakeholder groups in order to enrich the academic expert's view on the issue. However, there is little active engagement of stakeholder groups and there is no intention to share academic knowledge with the non-academic lay groups. In this style of knowledge production, knowledge democracy goes hand in hand with accountability and transparency. Society has the right to know what is done in the ivory tower of research and researchers have a duty to make information transparent. Democratic societies are interested in which experiments are undertaken, whether the costs are justified, and whether there might be some unexpected side effects. Inappropriate use of this type of production results in a lack of transparency and accountability.

3.3.2 Type 2: Knowledge Dissemination

This type of research involves a process in which knowledge, developed largely in the absence of lay knowledge, is transferred to the wider public. This type is characteristic of health promotion and disease prevention activities in which improvements in lifestyle and behavior are encouraged. It involves a unidirectional interaction between scientists and nonscientists, excluding mutual learning and joint problem definition. In this type of knowledge production, knowledge democracy implies being responsible for providing balanced information. The responsibility rests with those who disseminate knowledge. They should ensure that they are not blindly reproducing the views of specific pressure groups, such as those who uncritically promote biotechnology, nanotechnology, or medicine as panacea. Dissemination in knowledge democracy means providing society with state-of-the-art information on the pros, cons, and uncertainties of new technologies in order to enable society to make well-informed decisions. Inappropriate use of this type includes providing biased information, manipulation, and the coercion of societal actors.

3.3.3 Type 3: Mutual Learning for Knowledge Production Between Scientists and Societal Actors

This type of research is described in the transdisciplinary case study (TCS) approach by Scholz et al. (2006). The approach comprises multiphase projects (2–4 years) to support societal decision-makers and the public in a joint analysis with academic researchers to tackle complex multistakeholder problems at the regional level. The transdisciplinary process generally starts with collaboration with a legitimized decision-maker, such as a community mayor, the owner of an urban site for potential property development, or the head of a local environmental agency. In this type of collaboration, relevant stakeholder groups' involvement is essential. Mutual learning among the various actors involved is the key objective, but actors have different societal roles in this process. Thus, for example, the TCS approach is conceived as the preparation for, not the replacement of, a subsequent democratic decision-making process in which all members of society should have a say about the concrete actions to be taken. This is an appropriate approach if relevant information on the issue is available among the participating actors and can be utilized in the transdisciplinary process, and if the decision-makers involved continue to stay in power. The approach faces challenges if nondominant actors are not included in the transdisciplinary process because of a tight project schedule (and hence insufficient time to allow for participation of various stakeholders), because the stakeholders have not been identified, or because the subsequent decision-making process is not organized in an inclusive way. When nondominant actors are excluded, fierce resistance can be expected during implementation.

3.3.4 Type 4: Knowledge Co-creation Between Scientists and Societal Actors with a Specific Focus on Nondominant Actors

The interactive learning and action (ILA) approach, which contains cyclical, multiphase programs stretching over a longer time period represents this type (Bunders et al. 1999; Zweekhorst et al. 2001; Regeer 2009, 117). Here, transdisciplinary researchers help dominant and nondominant actors engage in a joint exploration and analysis of the social problem they choose to solve. Various tools facilitate by the co-creation of insights and solutions in the midst of academic and non-academic expertise. This approach represents one of the competencies learned in the process. This knowledge production style is most appropriate if the decision-making process and its implementation depend on a number of different actors within academia, civil society, governmental agencies, and the private sector. This also applies if the knowledge needed to address the issue is distributed among these actors. Here, the challenge for knowledge democracy is to find a balance between the knowledge of those who are in power and that of those who are marginalized, since most existing mechanisms tend to exclude nondominant actors.

4 The Culture of Transdisciplinary Research

Many transdisciplinary approaches have a different name, because they are embedded in local scientific, cultural, and political practices that differ by country. The fundamental embedding of transdisciplinary research in society leads to different routes, in which windows of opportunity and development challenges are taken into account. Although we distinguish the shared characteristics of transdisciplinary research as it is practised, there are also many differences that stem from different cultures and different development routes. As transdisciplinary research has grown out of practical – often local – needs and in very different contexts, the methodologies and conceptual frameworks that pioneers have developed are known under a number of terms. For example, the term *participatory development* (now often called *participatory governance*) was initially used loosely, generally in the US to describe an emerging methodology involving citizens in the solution of societal problems with the specific aim of helping the underprivileged.[3] Currently, it is not only a well-known methodology, but also a well-established field of academic research, underpinned by a wide body of literature.[4]

Another example: the *constructive technology assessment* (CTA) framework was developed in the Netherlands and Denmark (Schot and Rip 1997; Bunders et al. 2010). CTA stimulates the creation of an infrastructure through which users, producers, and others whom a technology will affect can regularly interact with each other and exchange information (Smits and Leyten 1991). Learning processes

[3] In the 1990s, many experiments with a participatory approach were undertaken with an explicit emancipatory goal in mind. Mayoux (1995, p. 235) for instance, mentions "meeting the demands of poor women in the South."

[4] The publication *Participatory governance and the millennium development goals (MDGs)* (United Nations 2008) aims to "identify and examine various tools, innovations, best practices and possible areas of cooperation needed to strengthen the capacities of governments and civil society groups for a participatory approach to governance and to the implementation of the MDGs" (United Nations 2008, iii).

between various actors in research, diffusion, and application are enhanced to create a more sophisticated understanding of the problems and of possible (technological) options (Schot and Rip 1997), while reflexivity is fostered to avoid falling back on old positions and preconceived opinions. Important tools include interviews, focus groups, dialogue workshops, and consensus conferences. CTA facilitates the development of a societal agenda for technology research and development (R&D).

Clearly, both participatory development and CTA, as well as a host of other methodologies, are examples of transdisciplinary activities in which participation is a defining factor. Furthermore, they largely employ comparable methods to involve stakeholders: interviews, focus groups, dialogue workshops, and consensus conferences (Regeer and Bunders 2003; Regeer et al. 2009). However, the lack of common ground between all these approaches makes it almost impossible to aggregate results, to evaluate methods, and to establish quality control standards. Thus, we argue that a shared transdisciplinary research framework needs to be developed. This should be one that furthers result comparisons and offers policy-makers and civil society transdisciplinary quality controls.

4.1 Methodologies, Standards, and Quality Control

A shared framework for the various transdisciplinary methodologies, under whatever name they are practised, should focus on common standards, methodologies, and quality controls. The field of transdisciplinary research needs to be established as a sound scientific discipline, answerable to high scientific standards, if it is to fulfil its role in engaging industry and civil society in the scientific debate, in reinforcing public trust in science and, ultimately, contributing to attaining shared goals. Furthermore, we believe there is a clear need in transdisciplinary research to agree on standards. At present, given the great variety in methods and implementation, it is impossible to evaluate methods' effectiveness across the boundaries of various contexts, countries, and subcultures. Data on the effectiveness of the various approaches is hard to aggregate, if only because of the widely differing terminology.[5] Finally, in our view, the debate on scientific quality standards, particularly as they relate to participatory research, should not stand in the way of establishing quality criteria and data gathering standards.

What Is Quality? There are two perspectives on the quality of transdisciplinary research: the societal perspective and the scientific perspective. From a societal perspective, quality research should yield results that are recognizable to societal actors and applicable in practice. From the scientific point of view, results should be publishable in peer reviewed journals. Scientific research and research proposals

[5] Even common terms such as 'sector' and 'intersectoral' are used in a significantly differing way in the USA and western Europe.

are increasingly being evaluated from the perspective of societal relevance. In many countries, special bodies have been created to assist funding bodies with this evaluation.[6]

In scientific circles, interdisciplinary and multidisciplinary research have long suffered from the image of not delivering publishable results. In the current '*publish or perish*' scientific world, this is still a formidable obstacle to transdisciplinary research. Nevertheless, scientists with established reputations have often advocated and undertaken such innovative collaborations. There is also the fundamental need to build a body of knowledge on transdisciplinarity. This requires publication in scientific journals, rather than reports that address only a few people. Thus, a transdisciplinary research project should create a *win-win* situation, i.e., be advantageous for society and science.

We need innovative strategies to ensure valid and reproducible results. These should be related to different aspects of validity:

- *Internal validity* (is the data consistent?) can be achieved by triangulation:
  - The use of different, complementary, methods
  - The use of different data and data-sources
  - Assuring data saturation.
- *External validity*: is the data significant?
- *Statistical validity*: can the results be generalized?
- *Conceptual validity*: does it help theory building?

4.2 The Roles of Universities and Publicly Funded Research Institutions

Why should universities and research institutes facilitate and promote transdisciplinary research? After all, some critics argue that the core business of such institutions is the creation of knowledge, not societal reform. The knowledge created should be fairly relevant to society and should certainly be available to all, but thereafter it is up to others (politicians, businesspeople, NGOs, the public) to make use of it.

We disagree with such critics on the following grounds: If society is faced with a complicated problem, it is reasonable to turn to science for solutions. If the problem does not fit within established fields such as physics or biology, but turns out to be something like 'getting scientists to work together' or 'involving the public in the scientific debate,' there is no reason for science to turn its back on what is a very

[6] In the Netherlands, the ERiC (Evaluating Research in Context) institute, part of the government body that funds most innovative research, is tasked with defining evaluation criteria for and evaluate scientific research regarding its societal relevance.

real problem. On the contrary, science should adapt to accommodate such real-world problems and make them the subject of scientific endeavor. Furthermore, universities and government-funded science institutes have a particular responsibility in this regard: They are funded by the public, and the public rightly expects them to pay attention to societal problems. The public trusts universities and independent institutions far more then they trust the government, the media, and business. This places a special responsibility on the shoulders of scientists, particularly publicly funded scientists.

5 Interactive Learning and Action: The Challenges Facing Transdisciplinary Research

Important ‘problem fields’ for transdisciplinary cooperation seeking to solve persistent societal problems are:

- No shared understanding of the problem (discussed in the introduction)
- Stakeholders are involved too late in the process
- No shared view among stakeholders on a path towards a solution
- Stakeholders operate from their own frames of reference.

We next discuss these problems and, based on the example of the interactive learning and action methodology, provide our vision of possible solutions.

In our efforts to establish a generalized approach to transdisciplinary research, we developed ILA as a broad framework – not a rigid method. The approach includes practical tools, developed in many projects across the world and applied in practice by our institute, as well as monitoring and evaluation methods. All of these result in reflection and adaptation.[7] ILA aims to produce results that are both relevant to society and scientifically valid with regard to the four criteria of validity: internal, external, statistical, and conceptual. We draw attention to four *crucial elements* in transdisciplinary research to summarize this methodology: participation, knowledge integration, reflection, and the emergent design of solutions.

5.1 Participation

Although stakeholder participation is now generally regarded as a necessity, a vast number of case studies demonstrate that stakeholders are often only involved after a sizeable amount of preliminary work. In such cases, the problem, stakeholders, and possible solutions may have already been defined. Experience of successful projects

[7] The practical tools we used are described in a number of publications, for example: De Cock Buning et al. 2008; Regeer et al. 2011; Moens 2010.

demonstrates the necessity of involving all stakeholders as early as possible in a project. Indeed, stakeholders should be defined by the problem and not, as is often the case, by the more or less accidental network of contacts of institutions.

When involving stakeholders, the right conditions have to be created for constructive ideas exchange. To achieve this, and depending on the group's composition, a series of methods and techniques have to be applied to avoid some group processes' pitfalls. Habermas (1971) described the three basic attitudes needed for a constructive analytical discussion: (a) conscious avoidance of strategic power games, (b) respect for the position of the other party, and (c) being prepared, in principle, to question one's own assumptions. According to Habermas, such a 'power-free' dialogue enables a depolarisation of viewpoints and a synthesis of ideas. By subjecting participants' comments to rational enquiry by repeatedly asking 'why?,' the basic assumptions underlying the comments are revealed, as well as any shared (or conflicting) values and norms. From this, we obtain insights into the underlying value systems and their relationships to one another by using communicative action theory.

5.2 *Knowledge Integration*

Given the different frames of reference and differences in opinion, integrating the knowledge of the various stakeholders is not easy. Some stakeholders' knowledge will be implicit, neither written down (codified) nor put into words. For scientists or others used to 'book knowledge,' it may be hard to accept that a gut feeling may be rationally based on a sound body of practical experience. Thus, the challenge is to integrate this knowledge in a way that respects, rather than lowers, certain quality criteria. In past decades, methods have been developed to guide and steer this process (Zweekhorst et al. 2001).

5.3 *Reflection*[8]

We use the term reflection in its psychological sense and point to several similar processes (or actions):

- Reflection on one's own frame of reference (embedded ways of thinking, ways of framing the problem field, ways of talking about it, ways of devising solutions, etc.),
- Reflection on other stakeholders' frames of reference as they emerge during an interaction,

[8] In the literature, what we call 'reflection' and 'emergent design' are often called 'action research' (Regeer and Bunders 2009). We prefer the terms we use here in order to make a clear distinction.

- Reflection on the process of interaction itself.

In the context of persistent problems, such reflection is crucial, because any solution will invariably require a change in perspective from all stakeholders, as well as changes in behavior and routines. In short, it is a rejection of the conundrum that problems cannot be solved by those who made them.

5.4 Emergent Design

The second important feedback mechanism comes from putting parts of the perceived solution into experimental practice. Data from these experiments is then evaluated, which may lead to a change in plans and give rise to a new action point agenda. This leads to an iterative – or, rather, a spiral – design process in which the final design emerges after a number of iterations (see Fig. 1). There is no natural end to this process other than a general feeling among the participants that 'this is about as far as we can go'.

The four crucial elements of transdisciplinary research discussed above provide ways to overcome the hurdles we outlined before. The lack of a shared definition of the problem at the outset should be addressed as part of the transdisciplinary process, leading to a collaborative effort to formulate a shared definition.

Stakeholders should be involved as soon as possible and the choice of who is invited should not depend on chance or accidental contacts but on analyses of the problem. Vastly differing frames of reference and thinking can be brought to fruitful interaction through reflection. Agreement on a path towards a solution is

Fig. 1 Action research spiral (From: Kemmis and McTaggart 1988, 154)

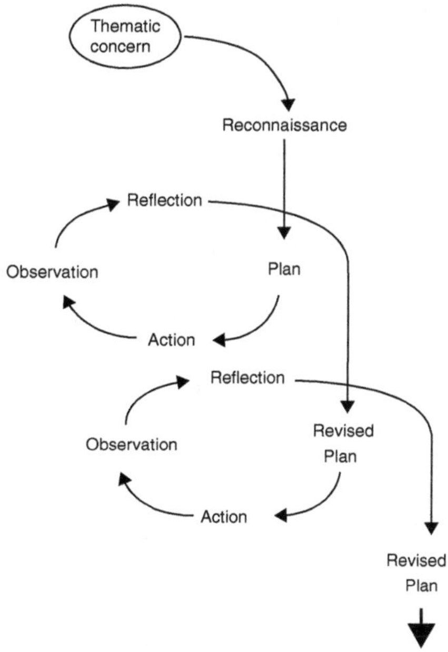

not necessary at the outset because, through various cycles of action-evaluation-reflection, some kind of solution will emerge.

Much of this approach is already practiced in numerous innovative projects across the world (for a Bangladeshi example, see Zweekhorst et al. (2002)). However, we fear that the implementation of these principles is often very ad hoc, and lacks standards and quality controls. Therefore, the implementation does not sufficiently contribute to knowledge growth. In our view, transdisciplinary research needs quality standards, especially for transparency. Most of all, it needs reproducible results. Only then can it help address super-wicked problems such as climate change and sustainability.

6 Biotechnology and Food: A Case Study

As an example of our approach, we provide a case study in which a research agenda was developed for the use of biotechnology in food production taking into account the concerns and views of all stakeholders, including civil society.[9]

6.1 Biotechnology: Who Decides?

Society seems to have a love-hate relationship with certain scientific innovations, such as nuclear energy and biotechnology (De Cock Buning et al. 2008). In the early 1990s, with the lamentable consequences of food shortages vividly shown on TV screens almost every day, biotechnology was hailed as the solution to the world's food problems. Since then, public opinion has changed, with some societal actors focusing on Faustian scenarios in which 'escaped genes' destroy creation once and for all (De Cock Buning et al. 2008, 7).

A great deal of research has been done on public responses to biotechnology in the past decade. These developments are a striking example of Hajer's argument (2003) that, in postmodern society, decision-making power no longer lies with the classical-modern institutions. Instead, civil society has organized itself to become a shrewd, transnational, and significant but elusive actor. British opponents of genetically modified food products, for example, reached a broad network of supporters by means of the media and internet, which quickly led to the collapse of the market for this type of product. In 2000, the introduction of genetically modified maize seed (Starlink™) in the US failed after consumers in Japan and Europe refused to accept this technological innovation (De Cock Buning et al. 2008). The US government, which had approved the product, was unable to resolve the problem (Kettle 2002). This example illustrates 'traditional' decision-making processes' loss

[9] The following text is a summary of the introductory chapter of De Cock Buning et al. (2008, 9–18).

of power, legitimacy, and efficacy. Although civil society exercised considerable influence on biotechnology's development and acceptance, in this case, it took the form of obstructive actions and countervailing power. This has the effect of maintaining or even reinforcing the established parties' polarized positions.

In response, political, scientific, and commercial organizations' attitudes have changed. Whereas the relationship between technology and society was previously approached from a top-down or technology-push perspective (how can we make an already developed product acceptable?), scholars currently assume a strong exchange or even interweaving of technological and societal developments. This new perspective is very important for issues relating to the management of scientific activities (particularly research programming). It implies a different method with which to relate societal perspectives to technological developments, and a new phase in these efforts. It is not so much a case of removing society's concerns about technology, but of analyzing these concerns and then applying this analysis to research policy contents. One way of achieving this is to formulate a societal research and action agenda in an interactive process.

6.2 Informed Discussion

The formulation of a societal research and action agenda requires an informed discussion between stakeholders that affords them the opportunity to voice their aspirations, anxieties, and irrational exuberance. This discussion needs to critically address societal concerns, including truths and beliefs. For some, such a discussion will be interpreted as window dressing, i.e., seeking to make decisions already taken more palatable. Consequently, such a discussion – led by publicly paid scientists from universities and other knowledge institutions – should be scrupulous in its methods, transparent, and open in the sense of refraining from a priori limitations. In his research into the British debate on biotechnology and food, Irwin (2001) describes how the then Minister for Science determined the direction of the debate by defining precisely which questions would be discussed.

Our analysis shows the significance of institutional location, preframing, and predefining the issue (bioscience, biotechnology) for a discussion's outcome and effects. The objective of our approach is to articulate the unarticulated societal agenda, which requires some distance from the institutionalized policy-making bodies to achieve this. Our approach is primarily focused on the content of the discussion, uncovering the fundamental reasons behind the concerns that have been validated in the group process, focusing on the 'why?' behind citizens and other stakeholders' emotions and comments. The objective is a comprehensive overview of concerns about biotechnology and the underlying issues, rather than a list of separate and unrelated items.

An important conclusion from studies of deliberative policy making is that, in spite of principles such as equality between participants, some participants nonetheless have an advantage:

> The dice, however, may in many ways be loaded in favour of those with superior resources, including through better access to customized knowledge and rhetorical resources. (Hagendijk and Kallerud 2005, 174)

Society lacks rhetorical resources because opinions, feelings, and knowledge about biotechnology are not articulated. In line with a great deal of *science, technology, and society* (STS) research, our approach assumes that society can play a role in the discussion, deliberation, and decision making on research and policies, provided that the working methods facilitate the articulation of opinions that are held but not expressed. In research into a public debate on biotechnology and food in the UK, Irwin (2001) observed that:

> Public groups expressed well-developed views on these topics (despite their initial unfamiliarity), once they had been given the opportunity to reflect on and discuss them both inside and outside the workshop. (Irwin 2001, 12)

6.3 From Consultation to Construction

While the ideas and opinions of important stakeholders (government, industry, NGOs, the scientific community) have largely been shaped and even institutionalized, the body of ideas that society holds is generally underarticulated. Although the institutional agenda includes inputs from society (via NGOs and ministries), we hypothesized that these inputs may differ considerably from the societal agenda. In order to go beyond well-known opinions, professional societal spokesmen were not consulted in this research. Instead, a societal agenda was constructed with citizens who were interested but not professionally involved. The construction of a societal agenda might provide options to overcome the vested polarization between the institutional agendas. A starting point for elaborating this issue can be found in the work of Grin and van de Graaf (1994; van de Graaf and Grin 2001), who indicate that societal actors' actions are based on their frames of meaning (communicative action theory), i.e., the sets of assumptions that guide behavior and that operate at different levels (a multilayered set of assumptions). These range from levels concerned with assessing solutions and defining problems, to more fundamental levels of underlying theory and value systems, as illustrated in Table 3.

Our approach focuses on the multiple patterns of preferences, value systems, and problem definitions. Communicative action theory teaches us that societal groups should be questioned in terms of their assessments of solutions, expressions of concerns (problem definitions), and underlying value systems and world views. Thus, the development of a societal agenda that transcends polarized viewpoints is subject to a number of criteria:

- The problems and dilemmas that people experience, not the formal policy agenda, are the starting point.
- Relevant issues should be approached integrally (e.g., food safety instead of the safety of genetically modified food).

Table 3 Layers within communicative action theory

Levels	Notion from communicative action theory	Order
1. Technical	Assess solutions	First order of argumentation: specific situation
2. Situational	Problem definition (policy problems, practical problems)	
3. System	Background theories (value systems, perceptions)	Second order of argumentation: value systems and world views
4. Societal-rational perspective	Final preferences (e.g., on the preferred structure of society)	

Adapted De Cock Buning et al. (2008, 12) based on Grin and van de Graaf (2001)

- The participants will better elaborate the problem definition if intuitive assumptions are made more explicit through 'why?' questions.
- The participants are addressed as persons rather than as stakeholders.

6.4 *Ten Steps Tools in Transdisciplinary Research*

The practical use of 'tools' in the process of discussion is depicted as a cycle with ten discrete steps (Fig. 1). It starts with one dedicated institution or group undertaking the initial preparation and thereafter involves stakeholders at the institutional level: policy-makers, scientists, business representatives, NGOs, and other established societal groups. In Step D, the set of stakeholders is further enlarged by involving the public. From Step G, the number of people involved decreases, returning to the institutional level in order to formulate better conclusions and advice (Step J). The process is circular: research will be undertaken, new developments will take place, and after evaluation the process should be repeated if necessary.

Step 1: Report Analysis During the survey of literature from a diverse range of entities (advisory councils, industry, government, politics, research), three application areas for biotechnology were distinguished: healthcare, agriculture, and the environment. In a number of documents, cross-cutting issues such as food, developing countries, and society's knowledge of and attitude towards technology were mentioned.

The initial analysis of the documents resulted in a review that summarized the explicit views regarding 'opportunities' and 'threats' (Roelofsen 2003, 5–13), while excluding the political landscape, for instance, the positions taken by the different institutions and 'authors.' This review is a necessary instrumental activity to prepare for the next step, in which we focus on the genealogy of accepted argumentation lines.

After analyzing a number of documents, the number of new viewpoints found in a document (using the focus described above) diminished asymptotically. Thus, we feel confident that our initial analysis was exhaustive – not with regard to the number of relevant documents, but with regard to the number of different viewpoints. This review was subsequently analyzed with regard to the underlying values that are in line or conflict with specific biotechnology applications. Table 4 illustrates a part of the working document that enabled us to obtain a clear picture of how viewpoints articulated in the formal, institutional discourse (central column) relate to the implicit values behind opportunities (left-hand column) and threats (right-hand column) (De Cock Buning et al. 2008, 21).

This analysis of seven fields demonstrated considerable agreement on several well-known values, such as 'safeguarding human health' and the 'equality principle.' Simultaneously, some potential conflict areas also emerged. Technological options trigger value dilemmas, where different values directly oppose one another. The 'solidarity' concept contradicts 'own responsibility' in the discussion that links genes and behavior. Dilemmas such as these hamper the societal embedding of biotechnology by typically triggering conflicting behavior.

Step 2: Removing Contextual Restraints (Questionnaire) In addition to these second-order insights, as determined by communicative action theory (see Table 3), the literature study also produced a first-order list of potential issues for the societal agenda. However, it must be noted that each policy document was limited by its contextualization. In other words, advisory board experts limited their expert contribution to the context of the advisory body's field (environment, agriculture, health, etc.), which constrained the policy document's topic (biotechnology, biomedical innovation, etc.). A specific member sitting in an environmental board may want to prioritize research on biomedical biotechnology rather than environmental issues, but this expert judgment is omitted as not relevant for the board's field, nor for the policy document. The aim of the next step was to investigate the existing policy expertise without contextual constraints. The result of these individual expert views on the innovative biotechnology landscape was seen as one of the important grounds to focus this project on biotechnology in general. Using questionnaires, we arrived at the conclusions depicted in Figs. 2 and 3.

Step 3: Follow-Up (Semistructured Interviews) At the institutional level, questionnaires were followed by interviews to do justice to the different stakeholder perspectives. The subsequent analysis of these interviews reconstructed the sphere of conflict in which each stakeholder placed biotechnology's technological developments. This analysis emphasizes the positions stakeholders adopt in the value discussions in order to provide a comparative overview of the existing value debates, thus illustrating the spectra of the spheres of conflict.

In addition to this second-order analysis, the in-depth interviews also provided new first-order insights, since all informants were also asked to comment on the preliminary agenda. The aim was to discover points missing from the agenda and to gain insight into the different stakeholders' priorities. The key question was: *What*

Table 4 Example of one analysis of viewpoints from the institutional discourse on implicit value orientations

Opportunities	Viewpoints articulated in the formal, institutional discourse	Threats
Well-being	Food	
Initiating self respect	Biotechnology will have enormous consequences in the area of nutrition. Knowledge of genetically valuable characteristics can lead to improvements in production techniques and introducing (manipulated) genes can improve product quality. In combination with identifying an individual's personal genetic profile, it may well be possible to indicate which nutritional components are most suitable for him or her	
Well-being	Functional foods and nutriceuticals	Accountability
Health	A particular development is that food products – so-called 'functional foods' – will come onto the market claiming to protect and promote health. Food supplements with a medical significance will also be developed ('nutriceuticals'). This blurs the distinction between food and medicine. It is important that not only the quality of these products should be tested, but also their effectiveness. In addition, a monitoring system is required to identify and evaluate the negative consequences of these new types of food products. The question is: who will be responsible for the quality and the effects of food which is genetically modified? Will it be the farmer, the pharmacologist, the doctor? In the interplay between lifestyle and functional foods, new kinds of responsibility issues arise. The choice that the consumer has to make between several diets that match several of his or her personal genetic qualities may induce some stress	Responsibility for choice of lifestyle
Personal benefits	In the end, the main question is whether the consumer is prepared to be genetically screened, and how far he or she is prepared to take risks with nutrition	Well-being
	With functional foods, it is possible for the consumer to have an overdose of 'healthy' substances. This can have undesirable consequences, particularly if products contain high concentrations of physiologically active components	Health
	The symbolic function of food may change from 'food as nourishment' to 'food to remain healthy'. At present, genetically modified products are the object of considerable societal resistance. Consumers make choices on the basis of food's functional value and will therefore first have to be convinced of these products' usefulness	Medicalization, narcissism Autonomy

Adapted from De Cock Buning et al. (2008, 21)

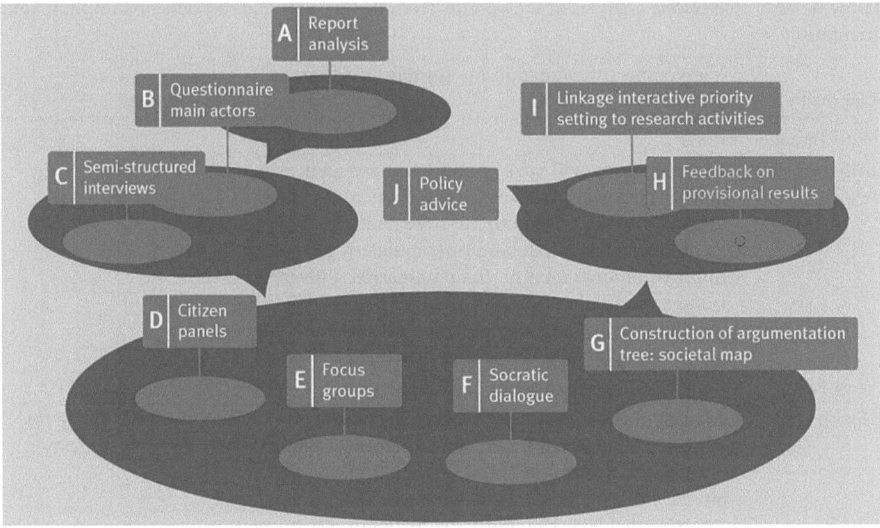

Fig. 2 The ten-step cycle in Transdisciplinary research (De Cock Buning et al. 2008, 85)

Fig. 3 Weighted importance of research subjects (most important subjects at the *top* of each column) (De Cock Buning et al. 2008). *Dark grey* natural science questions, *Light grey* humanities questions, *white* legal or administrative questions

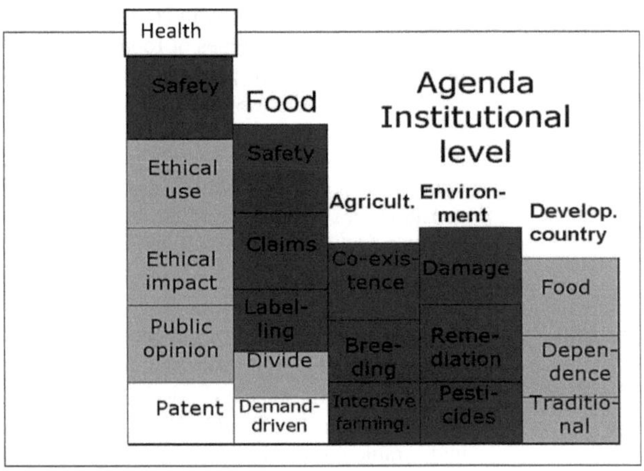

Dark grey = natural science questions
Light grey = humanities questions
White = legal or administrative questions

should be at the top of the stakeholder agenda, and why? From a comparison of the interviews, a number of top issues were distilled: the issues that different stakeholders with differing perspectives stressed.

The Next Steps: Focus Groups and Argumentation Trees Our approach is not aimed at a quantitative description of opinions. On the contrary, it is a qualitative study of the underlying reasons why people hold a particular opinion. Every effort is made to map a complete spectrum of the lines of argumentation within the framework of biotechnological innovations. Group interviews, such as focus groups, citizen panels and Socratic dialogues, are research methods that are suited to helping people express concerns, problems, and solutions, and also involve the elaboration of underlying values (Text Box 3).

Text Box 3: Society in this Study

During Step 2 in this project, 'society' was questioned about the personal viewpoints held by its members regarding the challenges and restrictions arising from biological innovations relating to food. Who is this society? And how can one determine the diversity of views relating to biotechnology and food? Because this project is about articulating the societal agenda, a quantitative study of the opinions of a representative selection of all individuals (a large-scale enquiry) is not appropriate. A quantitative approach supposes that the possible agenda issues are known and have been articulated. A qualitative study was therefore carried out among a diverse group of interested 'citizens'.

When compiling an inventory of the societal spectrum, participants were sought who were interested in the subject, but were not pressure group lobbyists. In other words, participants who were motivated to consider the subject, but had not yet adopted any clear position; they were, as it were, still trying to make up their minds. Also, every care was taken not to mix experts and non experts due to experts' tendency to lecture other participants, and the tendency of non experts to adopt the views that experts expressed. Finally, the decision was taken to keep the groups fairly small and to focus the discussion on the themes emerging from the institutional level. It was observed that, in a public setting – for example, a public debate – a farmer will tend to speak on behalf of all farmers, while the same farmer takes part in a small focus group as a 'person'.

Such a group interview is made up of three stages. First, the participants are facilitated in a process during which they can clarify their views and determine their position in a debate. Next, the moderator facilitates an inventory of these positions. Finally, the group structures the problem in a validated way. To achieve these three objectives in the space of time of a single session, a focus group approach was chosen. The group formulated the focal points within a problem via a strictly managed question-and-answer session. Using visualization techniques, the interrelations and prioritizing of the aspects discussed were mapped on flip-overs. A summary and a construction of the discussion – in the form of an argumentation tree (see Figs. 4 and 5) – were sent to the participants after the session for their authorization. In the first stage, people with comparable roles and perspectives were grouped together: laymen, civil servants, businessmen, scientists, representatives of NGOs, etc. In a later stage, these perspectives were brought together when the various argumentation trees were integrated (Regeer and Bunders 2009).

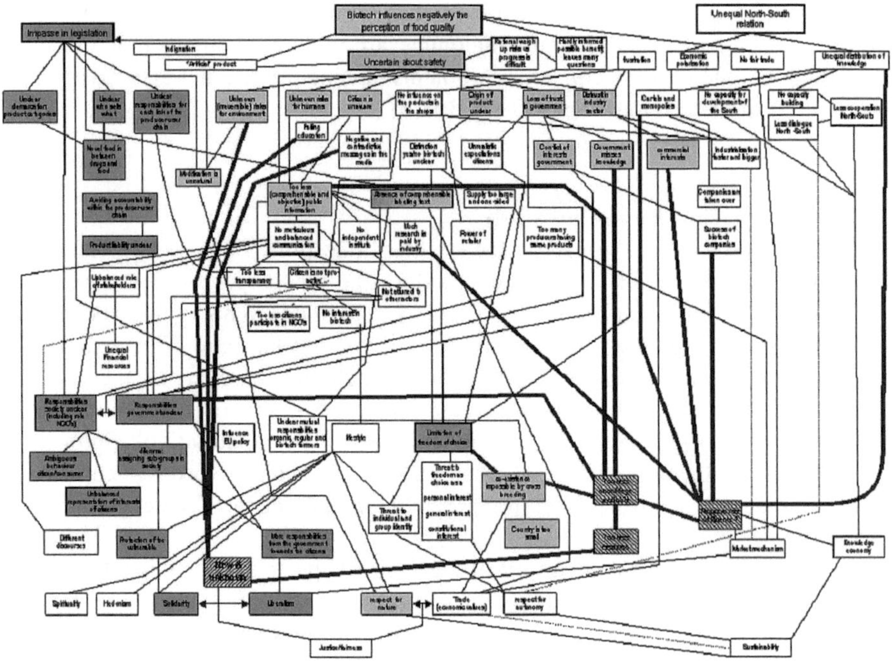

Fig. 4 Different views usually spring from different lines of argumentation based on different underlying causes. The various lines of argumentation can be summarized in an integrated argumentation tree, as depicted here. (The large original is in *color*, which makes the various themes stand out more clearly.) The object of such a drawing is to show the intertwined relationships of the arguments. The more symptomatic aspects are at the *top* and the more basic aspects at the *bottom*. This results in a comprehensible cohesion between the themes

An argumentation tree is a diagram that lists a concrete problem at the top. The increasingly deeper causes of the problem are shown below. By asking – in each box – *why is this a problem*?, one descends via a line to a lower box that provides the answer to the question. Conversely, if you start below, with each upward line, the statement must be: 'If the statement in the lower box is correct, then a consequence of this is shown in the higher box.' In this way, one can move from the top to the bottom of the argumentation tree using *why*? questions, and vice versa. This almost has the same structure as the four levels of the theory of communicative action, with the more concrete and situation-specific questions at the top ('insecure about food safety,' 'inequality between North and South') and the underlying values and dilemmas at the bottom. These last issues are very general in nature, for example, 'societal security', 'liberalism', 'sustainability', or 'respect for autonomy'. These underlying, second-order arguments are generally common to different societal roles, which means the argumentation trees of different group meetings overlap at the deeper, second-order arguments. This makes it possible to connect the argumentation trees to one another.

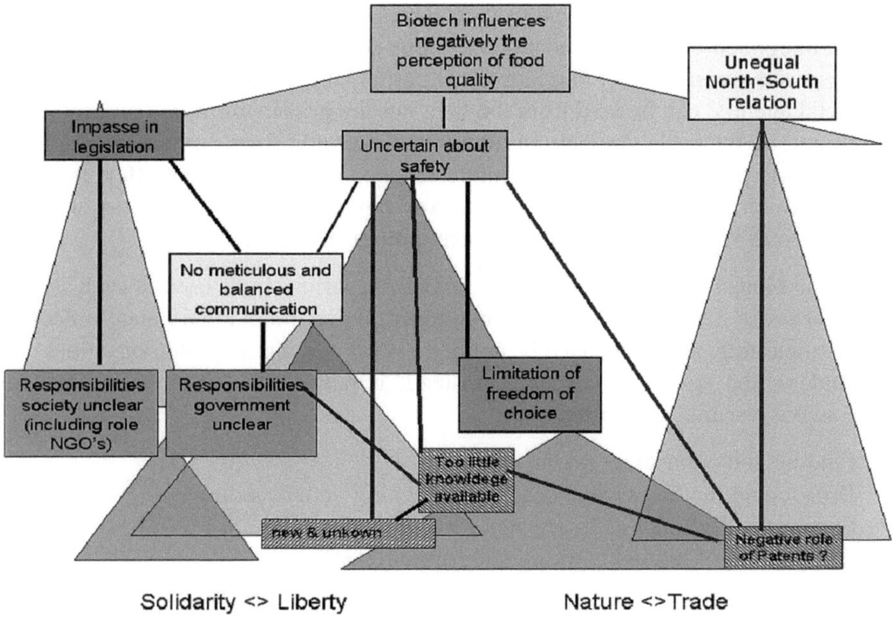

Fig. 5 Scaleddown version of the analysed argumentation tree

The integrated and interconnected argumentation tree (see Figs. 4 and 5) therefore provided a reconstruction of the complex cohesion between arguments relating to the societal aspects of biotechnology. It must be noted that the same applies to the argumentation trees in this study as to van de Graaf and Grin's (2001) theories of communicative action when they state:

> We expressly do not suggest there is any hierarchy between the different layers, and neither do we believe that the underlying insights and preferences are firmly rooted. We are more likely to be of the opinion that there may be a consistency between the layers which comes about not only because these insights and preferences in a specific context give a reason for a particular definition of motives and concerns, but also because the reverse is true: these insights and preferences acquire significance through the course of progress because they are validated in different contexts. (van de Graaf and Grin 2001, 210, translated by the authors)

The argumentation tree should also not be understood as a single group of actors' single, clear theory of communicative action, but as a complete spectrum of argumentation lines. One group of actors will feel their perspectives are covered on the left, whilst others may feel their perspectives are covered on the right, thus enabling fruitful dialogue between people with different views on the matter.

Societal Agenda We based our questioning of the group interview participants on Habermas's assumption of the argumentative rationale built into communication. By means of persistent questioning of the reasons behind statements, we arrived at a level of shared values that apply to a particular study group. The total argumentation tree is a logical construct of statements and deeper causes that arise from the rational analysis, which participants themselves carried out in the group interviews. In the integrated tree, the rational connections between the boxes is important, not the personal preferences of any individual participant regarding specific boxes. No 'societal agenda' can be read from the tree, nor the priority of important problems. However, on the basis of the group interviews, we indicate the extent to which the analysis and problem perception of interested, non involved citizen differs from the way in which policy-makers, politicians, key figures, and institutional-level stakeholders construct the discussion on biotechnology and food.

In the biotechnology report, we discussed the various tools used in each step of the process in depth. Although not mentioned above, tools to stimulate reflection and co-learning were extensively applied. However, the methodology does not depend on the specific tools used; instead, it purports to be a framework for interactive research concerned with:

- Finding and connecting people,
- Broadening the knowledge base and enriching understanding,
- Systematic and joint analyses, and
- The process of mutual learning and frame reflection.

7 Conclusion

We know that even extensive media campaigns are not effective in making people change their lifestyles towards a more sustainable one. Clearly, a much more fundamental approach is needed. This approach should involve all parties of society and respect all angles and perspectives. For this to work, the existing boundaries between science, policy, business, and civil society and citizens should be breached, since all these groups of actors need to work together. Fundamentally, transdisciplinary research is defined by the participation of a variety of stakeholders in solving real-world problems. Transdisciplinary research is therefore a key methodology for the future. In the transdisciplinary research process, the researcher becomes a knowledge creation facilitator between scientists and societal actors, with a specific focus on involving nondominant actors whose voices are not generally heard but whose involvement is critical for success or progress.

With growing public distrust of the media, politicians, and governments, scientists – particularly, publicly funded scientists – stand out as relatively trustworthy. Furthermore, in the eyes of the public, scientists are the preferred communicators of scientific knowledge. Thus, researchers are in a good position to facilitate such emergent change processes.

After the lessons from many practical implementations of transdisciplinary research, as illustrated by the biotechnology example above, the need for a comprehensive practical toolkit that can deal with highly complex, culturally embedded problems becomes apparent, as does the need for the development of even more robust processes. These should include quality controls, transparency standards, and – even more importantly – reproducible results.

8 Sources

This article comprises a review of transdisciplinary research, drawing heavily on two previous publications involving the authors. The sections on the nature of transdisciplinary research are derived from Bunders et al. (2010), and the biotechnology example originates from De Cock Buning et al. (2008). We have cited page numbers from these texts when referring to specific figures and elements in the text.

Acknowledgements The authors acknowledge the editorial contributions of Sarah Cummings, in particular in the first section on culture, with gratitude.

References

Abma, T. A., Nierse, C. J., et al. (2009). Patients as partners in responsive research: Methodological notions for collaborations in mixed research teams. *Qualitative Health Research, 19*(3), 401–415.

Broerse, J. E. W., Zweekhorst, M. B. M., van Rensen, A. J., & de Haan, M. J. (2010). Involving burn survivors in agenda setting on burn research: An added value? *Burns, 36*(2), 217–231.

Brown, V. A., Harris, J. A., & Russell, J. Y. (2012). *Tackling wicked problems: Through the transdisciplinary imagination*. London: Earthscan.

Brundtland, G. H. (Chairman). (1987). Report of the world commission on environment and development: Our common future, UN document

Bunders, J. F. G., Broerse, J. E. W., & Zweekhorst, M. B. M. (1999). The triple helix enriched with the user perspective: A view from Bangladesh. *Journal of Technology Transfer, 24*(2–3), 235–246.

Bunders, J. F. G., Broerse, J. E. W., Keil, F., Phol, C., Scholtz, R. W., & Zweekhorst, M. B. M. (2010). How can transdisciplinary research contribute to knowledge democracy? In R. In 'tVeld (Ed.), *Knowledge democracy, consequences for science, politics and media* (pp. 125–152). Dordrecht: Springer.

Caron-Flinterman, F. (2005). *A new voice in science. Patient participation in decision-making on biomedical research*. Zutphen: Wohrmann Print Service.

Churchman, C. W. (1967). Guest editorial: Wicked problems. *Management Science, 14*(4), 141–142.

Conklin, J. (2005). *Dialogue mapping: Building shared understanding of wicked problems*. Chichester: Wiley.

Council of Europe. (2008). *White paper on intercultural dialogue*. http://www.coe.int/t/dg4/intercultural/whitepaper_interculturaldialogue_2_EN.asp. Accessed 20 May 2013.

De Cock Buning, T. J., Regeer, B. G., & Bunders, J. F. G. (2008). *Biotechnology and food*. The Hague: The Netherlands Advisory Council for Spatial Planning, Nature and the Environment (RMNO).

Defila, R., & Di Guilio, A. (1999). Evaluating trandisciplinary research. *PANORAMA Special Issue, 1*, 1–28. http://www.ikaoe.unibe.ch/forschung/ip/Specialissue.Pano.1.99.pdf. Accessed 15 Jan 2012.

Do Well Do Good. (2012). *Second annual public opinion survey on sustainability*. http://dowelldogood.net/wp-content/uploads/2011/11/Second-Annual-Report-on-Sustainability-FINAL.pdf. Accessed 5 Apr 2013.

Edelman. (2012). *Edelman trust barometer 2012, executive summary*. http://de.scribd.com/doc/79026497/2012-Edelman-Trust-Barometer-Executive-Summary. Accessed 12 Nov 2012.

European Commission. (2009). *Flash Eurobarometer 256 (Europeans' attitudes towards the issue of sustainable consumption and production)*. Brussels: The GALLUP Organisation. http://ec.europa.eu/public_opinion/flash/fl_256_en.pdf. Accessed 4 April 2013.
European Commission. (2011). *Innovation union competitiveness report 2011*, Directorate-General for Research and Innovation.
Gauchat, G. (2012). Politicization of science in the public sphere: A study of public trust in the United States, 1974 to 2010. *American Sociological Review, 77*(2), 167–187.
Grin, J., & van de Graaf, H. (1994). Handelingstheorieën en beïnvloeding in netwerken: Ongelijksoortige rationaliteiten en congruente betekenissen. *Beleidswetenschap, 8*(4), 349–366.
Habermas, J. (1971). *Theorie und Praxis*. Frankfurt: Suhrkamp.
Hagendijk, R., & Kallerud, E. (2005). Changing conceptions and practices of governance in science and technology in Europe: A framework for analysis. In R. Hagendijk, P. Healey, M. Horst, & A. Irwin (Eds.), *Science, technology and governance in Europe: Challenges of public engagement* (Vol. 1, Annex 2, pp. 155–180). Online report available at: http://curis.ku.dk/ws/files/34380488/STAGE_Final_Report.pdf
Hajer, M. A. (2003). Policy without polity? Policy analysis and the institutional void. *Policy Sciences, 36*(2), 175–195.
Irwin, A. (2001). Constructing the scientific citizen: Science and democracy in the biosciences. *Public Understanding of Science, 10*(1), 1–18.
Jantsch, E. (1972). *Technological planning and social futures*. Chichester: Wiley.
Jensen, P. (2000). *Public trust in scientific information*. Sevilla: Institute for Prospective Technological Studies, Joint Research Center, European Commission.
Kagan, S. (2010). Cultures of sustainability and the aesthetics of the pattern that connects. *Futures, 42*(10), 1094–1101.
Kalegaonkar, A., & Brown, L. D. (2000). *Intersectoral cooperation: Lessons for practice*. Boston: IDR.
Kemmis, S., & McTaggart, R. (1988). *The action research planner*. Geelong: Deakin University.
Kettle, D. F. (2002). *The transformation of governance. Public administration for twenty-first century America*. Baltimore: John Hopkins University Press.
Lang, D., Wiek, A., Bergmann, M., Stauffacher, M., Martens, P., Moll, P., et al. (2012). Transdisciplinary research in sustainability science: Practice, principles, and challenges. *Sustainability Science, 7*(1 Supplement), 25–43.
Leiserowitz, A., Maibach, E., & Roser-Renouf, C. (2010). *Climate change in the American mind: Americans' global warming beliefs and attitudes in January 2010*. Yale University and George Mason University. New Haven: Yale Project on Climate Change. http://environment.yale.edu/uploads/AmericansGlobalWarmingBeliefs2010.pdf. Accessed 20 May 2013.
Levin, K., Cashore, B., Bernstein, S., & Auld, G. (2007). *Playing it forward: Path dependency, progressive incrementalism, and the 'super wicked' problem of global climate change*, International Studies Association Convention Chicago.
Mayoux, L. (1995). Beyond naivety: Women, gender inequality and participatory development. *Development and Change, 26*(2), 235–258.
Mittelstraß, J. (1992). Auf dem Weg zur Transdisziplinarität. *GAIA, 1*(5), 250.
Moens, N. (2010). *Innovation in sectoral governance and development with ICT in agriculture, education and health*. Amsterdam: VU University Press.
Mooney, C. (2005). The Republican war on science. *New York Times*
Nicolescu, B. (1996). *Levels of complexity and levels of reality*. In: *The emergence of complexity in mathematics, physics, chemistry, and biology*. Proceedings of the Plenary Session of the Pontifical Academy of Sciences, Casina Pio IV, Vatican, 27–31 October 1992. Pontificia Academia Scientiarum Editions, Vatican City (distributed by Princeton University Press).
Nurse, K. (2006). *Culture as the fourth pillar of sustainable development*. London: Commonwealth Secretariat. http://www.fao.org/sard/common/ecg/2785/en/Cultureas4thPillarSD.pdf. Accessed 19 Nov 2012.

Platt, S. (Ed.). (1993). *Respectfully quoted: A dictionary of quotations*. Washington, DC: Library of Congress.

Pohl, C., & Hirsch Hadorn, G. (2007). *Principles for designing transdisciplinary research, proposed by the swiss academies of arts and sciences*. München: Oekom.

Poll, M. (2009). Threat of global warming to grow, say Americans. http://maristpoll.marist.edu/threat-of-global-warming-to-grow-say-americans/. Accessed 4 Apr 2013.

Regeer, B. J. (2009). *Making the invisible visible*. Oisterwijk: Boxpress.

Regeer, B. J., & Bunders, J. F. G. (2003). The epistemology of transdisciplinary research: From knowledge integration to communities of practice. *Interdisciplinary Environmental Review, 5* (2), 98–118.

Regeer, B. J., & Bunders, J. F. G. (2009). *Knowledge co-creation: Interaction between science and society. A transdisciplinary approach to complex societal issues*. Den Haag: RMNO/COS.

Regeer, B. J., Hoes, A. C., Van Amstel-Saane, M., Caron-Flinterman, J. F., & Bunders, J. F. G. (2009). Six guiding principles for evaluating mode-2 strategies for sustainable development. *American Journal of Evaluation, 30*(4), 515–537.

Regeer, B. J., Mager, S., & van Oorschouw, Y. (2011). *Licence to grow, innovating sustainable development by connecting values*. Amsterdam: VU University Press.

Rittel, H. J., & Webber, M. (1973). Dilemmas in a general theory of planning. *Policy Sciences, 4* (2), 155–169. doi:10.1007/BF01405730.

Roelofsen, A. (2003). *Een maatschappelijke agenda biotechnologie, verslag eerste en tweede fase, maart – september 2003*. Vrije Universiteit Amsterdam, Biologie en Samenleving (report of internship).

Sainsbury, D., & Dexter, M., (2000). Foreword. In The UK Office of Science and Technology and the Wellcome Trust to Science and the Public (Eds.), *Science and the public. A review of science communication and public attitudes to science in Britain*. http://www.wellcome.ac.uk/stellent/groups/corporatesite/@msh_peda/documents/web_document/wtd003419.pdf. Accessed 19 Nov 2012.

Scholz, R. W., Lang, D. J., Wiek, A., Walter, A. I., & Stauffacher, M. (2006). Transdisciplinary case studies as a means of sustainability learning; historical framework and theory. *International Journal of Sustainability in Higher Education, 7*(3), 226–251.

Schot, J. W., & Rip, A. (1997). The past and the future of constructive technology assessment. *Technological Forecasting and Social Change, 54*(2–3), 251–268.

Smits, R., & Leyten, A. (1991). *Technology assessment: Waakhond of speurhond? Naar een integraal technologiebeleid*. Zeist: Kerckebosch.

The UK Office of Science and Technology and the Wellcome Trust (Eds.). (2000). *Science and the public, a review of science communication and public attitudes to science in Britain*. http://www.wellcome.ac.uk/stellent/groups/corporatesite/@msh_peda/documents/web_document/wtd003419.pdf. Accessed 19 Nov 2012.

The Yale project on Climate Change Communication. (2012). *Public support for climate and energy policies in September 2012*. http://environment.yale.edu/climate/publications/Policy-Support-September-2012/. Accessed 4 Apr 2013.

Thompson Klein, J., Grossenbacher-Mansuy, W., Häberli, R., & Bill, A. (Eds.). (2001). *Transdisciplinarity: Joint problem solving among science, technology, and society. An effective way for managing complexity*. Basel: Birkhauser.

UN Educational, Scientific and Cultural Organisation. (2001a, November 2). *UNESCO universal declaration on cultural diversity*. http://www.unhcr.org/refworld/docid/435cbcd64.html Accessed 6 Nov 2012.

UNESCO. (2001b). *UNESCO universal declaration on cultural diversity*. http://portal.unesco.org/en/ev.php-URL_ID=13179&URL_DO=DO_TOPIC&URL_SECTION=201.html. Accessed 21 May 2013.

United Nations. (2008). *Participatory governance and the millennium development goals (MDGs)*. http://unpan1.un.org/intradoc/groups/public/documents/UN/UNPAN028359.pdf. Accessed 27 Nov 2012.

van de Graaf, H., & Grin, J. (2001). Variëteit in rationaliteit en de legitimiteit van beleid. In T. Abma & R. In't Veld (Eds.), *Handboek Beleidswetenschap* (pp. 208–221). Amsterdam: Boom.

Weber, M., & Hemmelskamp, J. (Eds.). (2005). *Towards environmental innovation systems*. Heidelberg: Springer.

Wellcome Trust Monitor 1: Key Points. (2010). London: The Wellcome Trust. http://www.wellcome.ac.uk/stellent/groups/corporatesite/@msh_grants/documents/web_document/wtx058869.pdf. Accessed 19 Nov 2012.

Williams, R. (1981). *Culture*. Glasgow: Fontana Paperbacks.

Zweekhorst, M. B. M., Broerse, J. E. W., & Bunders, J. F. G. (2001). Innovations for sustainable development: The need for transdisciplinary knowledge integration. *Interdisciplinary Environmental Review, 3*(1), 75–94.

Zweekhorst, M. B. M., Broerse, J. E. W., & Bunders, J. F. G. (2002). Institutionalising a transdisciplinary approach to technology development in a Bangladeshi NGO. *Interdisciplinary Environmental Review, 4*(1), 43–65.

Narratives for a Sustainable Future: Vision and Motivation for Collective Action

Ilan Chabay

1 Introduction

On August 28, 1963, on the steps of the Lincoln Monument in Washington, DC, Martin Luther King Jr. delivered an important and deeply moving speech in which he used the phrase "I have a dream" repeatedly. That speech was a defining moment in the American civil rights movement. The phrase became an emotionally powerful symbol of the aspiration for justice and equality for blacks in the United States. The phrase itself became the kernel of a narrative that was repeated, explained, and retold in many forms as lessons, songs, and images that helped motivate, unite, and animate people to act collectively to urge the passage of civil rights legislation and change patterns of societal behavior. This suggests the following question: What visionary narratives – expressed as words, videos, paintings, dances, sculptures, or songs – will inspire and motivate us to act collectively to address the most urgent and critical issues facing humanity today?

The scope and magnitude of the highly complex and interrelated challenges are the most critical issues currently facing humanity living in the midst of rapid and accelerating change in physical, biological, and societal conditions on multiple spatial and temporal levels (see, e.g., ICSU Earth System Science for Global Sustainability: *The Grand Challenges* 2010 and Future Earth 2025 Vision). Owing to the vital importance and urgency of the global change issues, global, as well as temporally and spatially specific, responses are needed from humanity. Anticipating, characterizing, understanding, and learning how to avoid, mitigate, or adapt to these changes requires approaches that fit the complex conditions and issues. While we certainly need existing and new knowledge and a deeper understanding of the complexities, it is important to note that knowledge alone is not sufficient to guide and catalyze the necessary transformation of the global society.

I. Chabay (✉)
Institute for Advanced Sustainability Studies, 130 Berliner Straße, Potsdam 14467, Germany
e-mail: ilan.chabay@iass-potsdam.de

B. Werlen (ed.), *Global Sustainability*, DOI 10.1007/978-3-319-16477-9_3

We also need scientifically sound and culturally relevant visionary narratives for action to inspire us and motivate collective decisions on policy and action at all scales and levels.

In this article, we will consider what role narratives – especially in the sense of affective communication – can and do play in motivating transformative steps toward sustainability, why substantive collaborations between those engaged in science, art, technology, and humanities are needed to create new visions for and reflections on sustainability, and how the impacts and outcomes of different forms of narrative expression can be assessed to improve our understanding of how knowledge, learning, and societal change are linked (Blackmore et al. 2011).

2 Complexity, Models, and Uncertainty

The Earth system is highly complex, as many of the component systems are multiply connected and non-linearly coupled to one another. This includes the human social, technical, and economic systems, which are not only complex components in themselves, but also interact with components in the biophysical sphere. Component systems, such as the cryosphere or atmosphere, are often studied and modeled as dynamic systems with static, non-interacting input from other components, such as anthropogenic effects. This is normal in the reductionist scientific tradition, but a holistic, systems approach is essential when addressing the global system (Liu et al. 2015), even when it is examined on a more localized spatial and temporal scale. Our total accumulated knowledge of the global system and its local aspects is vast, but it is also fragmented, incomplete, constantly developing, and changing, as insights, methods, and conditions change. In many important respects, our knowledge is, to varying degrees, ambiguous, uncertain, and contested and will remain so.

This complexity and inherent uncertainty cannot be dismissed or denied by simplistic thinking. However, complexity can be managed to some degree through the use of models and scenarios. As forms of approximation that limit the degree of complexity considered in order to categorize, explain, and predict the behavior of physical, biological, social, and economic phenomena and systems, models can provide both a range of options in scenarios for decision making and insights into the potential consequences of making certain choices and tradeoffs (Morgan and Morrison 1999). Analytical and computational models are increasingly used in attempts to understand complex issues pertaining to global change and sustainability on a range of temporal and spatial scales and to develop scenarios for decision making and action (see, e.g., Pidgeon and Fischoff 2011). Models used for decision making for governance at different levels and over a range of spatial and temporal scales include global system dynamics models (e.g., of atmospheric circulation with chemical interactions, see Seinfeld et al. 1998) and agent-based models (e.g., of the interaction between ecological systems and governance processes, as in Janssen and Ostrom 2006). Such models constitute an essential set of tools with which scientists can systematically analyze complex issues in order to generate insights into and to develop scenarios for specific conditions and purposes.

As important as the tools, methods, and knowledge are to researchers, the latter are only one of the various categories of stakeholders – private citizens, members of corporations, businesses, governmental and non-governmental organizations, and professional and trade organizations – that have to make meaningful use of the knowledge if it is to help move society toward sustainability. Stakeholders at all levels who make decisions and take action need to understand that the output of model calculations indicates options within the limitations of the models and data, rather than providing a singular answer or solution. Thus, stakeholders are faced with the cognitive demand of having to understand the output of the model, while keeping the limitations of the model and data in mind. Equally important, but hitherto usually neglected, is that scientists and other stakeholders need to jointly consider the questions, concerns, and priorities of the stakeholders impacted by the research, and have to choose which to incorporate in what form into models and scenarios.

In most instances, decision making is further complicated by the conflicts surrounding the tradeoffs between options for action due to stakeholders' divergent beliefs and values. Stakeholders are only able, and perhaps willing, to make informed decisions and undertake related collective actions if they can make sense of information from within their own context of cultural, beliefs, and values. They should be able to trust the "messenger" conveying the narrative through a form of direct or indirect relationship (Siegrist et al. 2005). The knowledge about *global* issues has to be salient, credible, and legitimate in the eyes of *local* stakeholders (Webler and Tuler 2001). That is, the knowledge in itself is not sufficient to be accepted and acted upon – the context and culture in each locality or stakeholder community are crucial and are indeed a knowledge resource that should be part of the larger knowledge system (Tàbara and Chabay 2012). The connections to closely held and culturally specific values and beliefs are key to stakeholders' engagement in and commitment to action on the issues.

Therefore, it is essential not only to generate and communicate information content, i.e. the knowledge itself, but also to identify, acknowledge, and address – either explicitly or implicitly – stakeholders' beliefs and values. In doing so, diverse groups of stakeholders at different levels may be inspired and motivated to join in collective decisions on policy and action that move humanity toward a sustainable future. Narratives are essential for this.

3 Narratives, Identities, and Ideologies

Narrative kernels that convey messages and visions of sustainability can be expressed and elaborated upon in many forms and can be crafted to fit different media, cultures, and local contexts. Narratives that, in addition to content, either implicitly or overtly convey their values and identity are central to communication

(Wynne 1992). Effective narratives primarily establish links to recipients' sense of identity and ideology through affective rather than cognitive means, and, thus, play a crucial role in what is accepted, internalized, and remembered. While different conditions and locations affect the manifestation of global challenges, responses to those challenges reflect not only the differences in biophysical conditions, but, importantly, also derive from other dynamic contexts, such as cultural perspectives, political processes, power relationships, and economic realities (Van Kerkhoff and Lebel 2006). The latter aspects are generally reflected in the ideologies and identities of different sections of communities and are thus at least partial determinants of accepted and reinforced narrative kernels.

In this article, the term "narrative" or "narrative kernel" is considered a fundamental element of expression shaped by critical links to personal, social, and political identity. Thus, the use of the word (according to Humpty Dumpty in Lewis Carroll's *Through the Looking Glass* (Gutenberg project edition 2013) "means just what I choose it to mean"), does not refer to stories themselves in the Aristotelian sense – with a beginning, middle, and end – but to kernels of ideas that underlie the many ways in which they are expressed. The "I have a dream" phrase mentioned at the beginning of this article illustrates this. Stories, games, visual art, social media, dance, and music embellish narrative kernels and clothe them in accordance with particular contexts and cultures. The narratives of most interest in this article are those that manifest in different forms to express visions of socially innovative trajectories toward a sustainable future and the landscapes of that future. In this sense, the narrative kernels that humanity uses are compact, emotionally communicative messages for mobilizing collective action toward sustainable practices in accordance with current knowledge. In addition, these narrative kernels are based on an ethical foundation that fosters equity across economic, cultural, spatial, and temporal domains. They may also reflect the current or past practices in a way that stimulates awareness of the need for change and highlight perspectives on societal transformations.

Compelling narrative expressions, in which knowledge has been selected, distilled, and framed in culturally recognizable ways, can be used to establish or strengthen links between stakeholders and facilitate a process leading to meaningful decisions and collective action toward sustainability. For example, a central narrative in sustainability is the concept of the "Anthropocene" (Crutzen and Stoermer 2000) which was derived from the interpretation of observational data on and sophisticated models of atmospheric chemistry and physics. The idea itself represents a tremendous change in how we perceive human "agency" vis-à-vis the Earth system. Recognition (or rejection) of the scale of human intervention's impacts on the Earth system has strong emotional resonances for many people. Governing, managing, engineering, stewarding, navigating, or denying the Anthropocene are profoundly different views of the issues and challenges, and may thus open up or restrict policy measures and personal decisions accordingly. Individuals' positive engagement in finding a sense of purpose and agency for actions in response to the critical challenges they perceive involves issues of agency and responsibility for action on both the individual and collective scales.

Fig. 1 Example of an anti-Agenda 21 campaign in the U.S. (Newman 2012)

While it is important to focus on positive narratives for sustainability, it is also worthwhile acknowledging, and learning from, the nature of negative or oppositional narratives. Narratives that are intentionally, or sometimes inadvertently, attached to an idea or vision often obscure or directly oppose it (discussed more broadly in Oreskes and Conway 2010). The recent passage of laws in Kansas, New Hampshire, and Tennessee, USA that oppose any measures meeting the voluntary guidelines of Agenda 21 (originally adopted by the UN Conference on Environment and Development (UNCED) in Rio de Janeiro, Brazil in 1992) is a striking example of the power of ideological narratives that oppose sustainability. Tea-party and libertarian activists in the USA have vigorously promoted the anti-Agenda 21 efforts in several states on the grounds that it is an "attack on private property rights, individual liberty, and national sovereignty" (Newman 2012) and is part of an "insidious, covert" UN effort to ensure its global governance. The ad shown above was posted on a libertarian blog and illustrates the link between the American sense of individual freedom and ownership of a car, home, and even a teddy bear, which Agenda 21 is presumed to imperil (Fig. 1).

The key point is that the anti-Agenda 21 campaign does not address the scientific evidence of urgent global change issues and their local impacts, but uses a narrative of individual freedom and rights of ownership to oppose voluntary adaptive measures, ostensibly because they stem from an "outside" organization. This closes the door on addressing issues that affect the future of the particular state (not to mention other states and nations) regarding global change, including, for example, the agricultural and commercial sectors (as well as the global export of crops) that have helped build and sustain Kansas.

This narrative completely lacks evidence and arguments against the urgent need to act in order to mitigate and adapt in the face of global changes, at which Agenda 21 is directed. Direct challenges of the narrative by means of evidence of the need for action on sustainability are not likely to influence those who adhere to the anti-Agenda 21 view. A paper pointedly entitled "I will see it when I believe it" (Yariv 2002) incorporates the problem of existing beliefs that oppose new or belief-challenging information in a model of an agent utility function.

We need to understand the forms and nature of positive and negative narratives to learn which narratives influence which stakeholders at any given time. This can catalyze the creation of effective narratives and allow assessment of their outcomes, which will allow us to foster more sustainable practices and policies. Moreover, we need to determine the dynamics of narratives in different communities and what causes them to change in impact and importance. The field of marketing has developed great expertise in crafting messages to appeal to specific groups and cultures, including making use of traditional and emerging social media, from which lessons can be learned to gain an understanding of customers and tailoring messages to suit them (e.g., Rewerts and Hanf 2006). However, the lessons we learn should pay careful attention to ethical issues to ensure that when we craft messages and narratives they represent diverse voices with different values, economics, and power, and are based on clear and accessible information. If constructive, collective action on sustainability is to emerge, it is of the utmost importance that the relevant stakeholders also consider these narratives trustworthy, credible, and legitimate within their own context. The point is not that one narrative should suffice for all stakeholders and all issues, but that multiple narratives can each have an impact on some of the local or regional stakeholder groups with regard to specific practices or policies.

4 Creating New Narratives of Sustainability

Many artists and groups have expressed their views of sustainability and related issues (Grant 2012). Some use natural and social sciences' interpretations as the impetus for artistic expression, while others reflect on their personal concerns and aspirations. Owing to the complexity of the issues and the cultural diversity of the stakeholders, a multiplicity of narratives and expressions are needed to engage with and represent different segments of society. This suggests a process of developing new collaborations and on-going workshops that, in various ways, bring together different people whose experiences and expertise lie in diverse endeavors and fields, including art, design, engineering, information and communications technology (ICT), law, journalism, philosophy, natural and social sciences, marketing, and history. The key to making this productive for the participants and for outcomes in society is to realize that it is a mutual learning process to which many can contribute their expertise and learn from it. In most instances, careful selection and preparation of the participants prior to the workshops, as well as facilitation of the group process, are necessary to ensure that everybody derives value from the workshops. The purpose of the workshops is not necessarily to create new narratives in order to "sell" the science, but to first form working relationships through face-to-face contact, which helps build the trust needed for effective collaboration, brainstorming ideas, and discussing common challenges. The diverse participants can then start collaborating – the actual hard work – by means of electronic forms via ICT, as well as in face-to-face meetings. Clearly, ICT can also play an important dual role as a medium to express the collaboration's narrative output, and as the means of recording the collaborators' processes.

A significant challenge in developing effective collaborations across such diverse domains is to allow each person to contribute the most critical knowledge from their domain of expertise without favoring one domain over another. Since the urgency and identification of sustainability and global change issues are still largely based on natural science, as articulated in the planetary boundary concept (Rockström et al. 2009), the content of many natural scientists', engineers', and technologists' dialogues focus on natural science or technology. The social sciences, humanities, and the arts have been underrepresented in research and dialogues, although that is changing toward greater inclusion in addressing global change challenges through, for example, the establishment of Future Earth (http://www.futureearth.org/) to coordinate solution-focused transdisciplinary research on global challenges. The collaborative process and relationships need to be facilitated and nurtured in order to keep the focus on creating effective narratives that draw on diverse perspectives and insights from all field. Simultaneously, if there is no process to select those with relatively similar styles and a propensity for collaboration across unusual boundaries, and the process is not actively facilitated, the wealth of knowledge available from multiple sources and the complexity of the issues could easily overwhelm the collaborators and overburden the effort, resulting in frustration or uninspiring didactic expositions.

Both live and digital expressions of narratives can contribute to the process of engaging stakeholders. The immediacy of live performances and the heightened sense of engagement due to social feedback during an event, as well as the impact of real objects, are very powerful influences on people's emotional states and memory. Moreover, the memorable quality of the narratives and the ease with which they are communicated can inspire the persistence of their visions. The electronic capture of live events and real objects, as well as native electronic expressions, is essential because this can help reach a much wider audience and can be structured to actively engage people, including in game formats. The increasing pervasiveness of ICT, including cell phone access to most places in the world, offers the opportunity to adapt and compare narratives appropriate in one cultural and contextual form to another. For example, images with a strong regional, local, or cultural focus might convey the urgency and saliency of climate change by highlighting social impacts of open, nearly ice-free seas on people living in the Arctic regions, or barren, dried former grasslands on those living in sub-Saharan Africa.

Another endogenous approach to generating narratives in stakeholder communities is to develop narratives from citizen science and ICT experiments by using the real-time monitoring of individual exposure to pollutants and conditions – such as noise and CO_2 – in daily living via sensors and GPS-equipped cellphones (e.g., the European Commission EveryAware project http://www.everyaware.eu/). Each participant's spatially-resolved exposure over time can be collected and the aggregate input from all the participants can be mapped and displayed. Such a display of near real-time exposure can then be complemented by the participants' feedback on their circumstances when monitoring and reactions to the aggregate or individual information. Common responses and shared perceptions may evolve into or trigger the expression of new narratives of reflection or future visions.

5 Assessing the Impact and Outcomes of Narratives for (and Against) Sustainability

It is not only necessary to develop narratives that reflect our current concerns about and visions of a sustainable future, but also ones that help people understand what works, under which conditions, and for whom. Both quantitative and qualitative methods are needed to better assess the outcomes of existing and new narratives on stakeholders' attitudes, decisions, and actions, as well on the dynamics regarding the diffusion and evolution of narratives.

Methods for assessment include interviews and surveys before and after individuals' exposure to particular narrative expressions, for example, before and after a performance, a dialogue, or a visit to an exhibition. Protocols for interviews and surveys will need to be adapted or developed to gain information on the immediate responses to narratives in whatever form they are expressed, as well as on longer-term changes in attitudes and practices related to sustainability. With the participants' explicit permission video and audio recordings can also be used to gain insight into the verbal, socially mediated, and non-verbal responses (e.g., facial and gestural reactions recorded when specific images or words are presented) during events or exhibitions. For example, in unpublished video-recorded studies of science café sessions that took place in Stockholm, Sweden in 2007, the author observed frequent reactions to a particular point in a presentation that resulted in brief whispered conversations with neighbors, often before an individual raised a question. In the videos, more questions were raised during the event than were self-reported in post-event surveys; the video thus provided evidence of the extent and frequency with which social interactions modulated the dialogues between the participants.

Monitoring participants' Twitter feeds during events may also provide insight into their immediate responses to specific images, narrative framings, and information in presentations and dialogues. More elaborate methods, such as eye-tracking and brain-imaging techniques, including functional magnetic resonance imaging (fMRI), electroencephalography (EEG), and near-infrared (NIR) optical tomography, could be used to observe responses to images, sounds, and words in video narrative presentations. That data could then be used to investigate the correlation between the responses and emotional state, the immediate and long-term memory (recall) of the input and response, and the relationship between the subjects' retained memory and their expressed attitudes towards and practices regarding sustainability.

It is generally very difficult to track down and follow up on individuals after a particular event in order to assess changes in attitudes and practice. However, depending on cost, time, and sampling limitations due to permission and privacy, this can still be done. ICT can be very helpful both in actively following up on and questioning those who have access to an electronic form of narrative expression, as well as in "passively" collecting information on clicks on and specific forms of response to web sites.

A variety of ICT methods have been developed and employed that can be used to analyze the impact and outcomes of exposure to narratives. These include semantic analysis to uncover implicit and explicit narratives, and data mining with dynamic and geo-specific network analysis (e.g., Chavalarias et al. 2011). Computer and mobile app games are being designed to challenge players regarding the decisions and tradeoffs of sustainability challenges in a variety of contexts, while recording and analyzing players' choices. These games can be used as research tools by examining the correlations between the players' game input and their self-reported attitudes, community, or cultural identities. The objective is to bring qualitative and quantitative lenses into play to understand the immediate and long-term socially mediated (social learning and diffusion) impact of sustainability narratives.

6 Conclusion

Humanity in its myriad communities and contexts needs the inspiration and motivation that comes from new narrative visions of sustainable futures. Humanity needs to share dreams of a more sustainable future, as well as feel and be able to contribute to it in locally appropriate forms. This can be catalyzed through an open and continual process of collaboration between stakeholders with a wide range of expertise and experience, which can result in powerful expressions in different cultures and contexts. These collaborations are a mutual learning process between disciplines that will be informed by careful analysis of the impact and outcomes of the wide range of narrative expressions. Developing and deploying scientific tools and methods when examining the impacts and outcomes of affective discourse, as well as of knowledge and cognitive discourse, will create avenues for research to examine critical aspects of our development as a species, which is essential to our survival on Earth.

Acknowledgements I would like to thank Professor Benno Werlen for the opportunity to engage in a stimulating workshop on transdisciplinary intercultural ecological research for sustainability (TIERS) in Jena, Germany in June 2012, which contributed to the ideas discussed in this article. I am also very grateful for conversations with Dr. David Tàbara, who enriched my thoughts on narratives.

References

Blackmore, C., Chabay, I., Collins, K., Gutscher, H., Lotz-Sisitka, H., McCauley, S., Niles, D., Pfeiffer, E., et al. (2011). *Knowledge, Learning and Societal Change (KLSC) – Finding paths to a sustainable future*. Science plan for International Human Dimensions Programme on Global Environmental Change (IHDP). Research alliance. http://www.proclim.ch/4dcgi/klsc/en/News?2070. Accessed 8 Nov 2012.

Chavalarias, D., Coinet J.-P., Cornilleau, L., Duong T. K., Mogoutov, A., Roth, C., et al. (2011). *Stream of media issues: Approaches for monitoring world food security*. http://pulseweb.cortext.net/static/files/wp1.pdf. Accessed 24 Jan 2012.

Crutzen, P. J., & Stoermer, E. F. (2000). The "Anthropocene". *International Geosphere-Biosphere Programme (IGBP) Newsletter, 41*, 17–18.

Grant, D. (2012). *'Sustainability' has become a growing focus of artists' (and art schools') attention*. Huffington post blog. http://www.huffingtonpost.com/daniel-grant/sustainability-and-art-schools_b_1417562.html. Accessed 3 Dec 2012.

Future Earth (2014) *Future earth strategic research agenda 2014*. Paris: International Council for Science (ICSU). http://www.futureearth.org/sites/default/files/strategic_research_agenda_2014.pdf

ICSU. (2010). *Earth system science for global sustainability: The grand challenges International Council for Science, Paris*. http://www.icsu.org/publications/reports-and-reviews/grand-challenges/GrandChallenges_Oct2010.pdf. Accessed 5 Feb 2012.

Janssen, M. A., & Ostrom, E. (2006). Empirically based, agent-based models. *Ecology and Society, 11*(2), 37.

Liu, J., Mooney, H., Hull, V., Davis, S. J., Gaskell, J., Hertel, T., Lubchenco, J., Seto, K.C., Gleick, P., Kremen, C., & Li, S. (2015). Systems integration for global sustainability. *Science, 347*, 1258832. doi:10.1126/science.1258832.

Morgan, M. M., & Morrison, M. (1999). *Models as mediators. Perspectives on natural and social science*. Cambridge: Cambridge University Press.

Newman, A. (2012). Kansas lawmakers seek resolution against "Insidious" UN Agenda 21. *The New American*. Blog. www.thenewamerican.com/tech/environment/item/11318-kansas-lawmakers-seek-resolution-against-%E2%80%9Cinsidious%E2%80%9D-un-agenda-21. Accessed 10 Jan 2013.

Oreskes, N., & Conway, E. M. (2010). *Merchants of doubt*. New York: Bloomsbury Press.

Pidgeon, N., & Fischoff, B. (2011). The role of social and decision sciences in communicating uncertain climate risks. *Nature Climate Change, 1*, 35–41.

Rewerts, A. L., & Hanf, J. H. (2006). *Culture and values – their relevance for marketing strategies*. European Association of Agricultural Economists Seminar. http://purl.umn.edu/10113. Accessed 12 Feb 2012.

Rockström, J., Steffen, W., Noone, K., Persson, Å., Chapin, F. S. I., Lambin, E. F., et al. (2009). Planetary boundaries: Exploring the safe operating space for humanity. *Ecology and Society, 14*(2), 1–33.

Seinfeld, J. H., Pandis, S. N., & Noone, K. (1998). Atmospheric chemistry and physics: From air pollution to climate change. *Physics Today, 51*(10), 88.

Siegrist, M., Gutscher, H., & Earle, T. C. (2005). Perception of risk: The influence of general trust, and general confidence. *Journal of Risk Research, 8*(2), 145–156.
Tàbara, J. D., & Chabay, I. (2012). Coupling human information and knowledge systems with social–ecological systems change: Reframing research, education, and policy for sustainability. *Environmental Science and Policy, 28*, 71–81.
van Kerkhoff, L., & Lebel, L. (2006). Linking knowledge and action for sustainable development. *Annual Review of Environment and Resources, 31*, 445–477.
Webler, T., & Tuler, S. (2001). Public participation in watershed management planning: Views on process from people in the field. *Human Ecology Review, 8*(2), 29–39.
Wynne, B. (1992). Misunderstood misunderstanding: Social identities and public uptake of science. *Public Understanding of Science, 1*(3), 281–304.
Yariv, L. (2002). *I'll see it when i believe it – A simple model of cognitive consistency* (Cowles Foundation Discussion Paper Number 1352).

Green Capitalism, Sustainability, and Everyday Practice

Karsten Gäbler

1 Introduction

Today, more ado than ever is being made about appropriate ways of dealing with the global ecological crisis and of building sustainable futures.[1] In these debates, two opposed modes of thought can be distinguished. One type of diagnosis claims that the capitalist societies of the Global North will be able to initiate the necessary 'self-healing' processes without fundamentally revising their modern lifestyles. The trademark of this strategy is *green capitalism* – conceived as a reconciliation of economic growth and sustainable development. On the other side, we find a more critical approach that highlights capitalism's inner contradictions and calls for a transformation of the economic, political, and cultural framework in order to address global (environmental) problems. The fundamental characteristic of this second approach is its emphasis on *multiple* crises, instead of the superficial talk of a decoupled ecological crisis. It considers global environmental problems as the most prominent side-effect of capitalism's general crisis.

Despite their theoretical antagonism, I argue that both approaches share the same deficiency. Although everyday practice is – explicitly or implicitly – at the center of their arguments, they have inadequate concepts of human conduct. The disciples of green capitalism, for instance, hold that the *sustainability shift* ultimately consists of the challenge to change individual *behavior*. They emphasize the role of knowledge, rationality, and values in the process of modifying everyday resource use patterns. But why, then, do individuals and societies not simply change their

[1] In this article, I use terms such as *sustainability shift* or *socio-ecological transformation* as rough equivalents.

K. Gäbler (✉)
Department of Geography, Friedrich Schiller University Jena,
Löbdergraben 32, 07743 Jena, Germany
e-mail: karsten.gaebler@uni-jena.de

B. Werlen (ed.), *Global Sustainability*, DOI 10.1007/978-3-319-16477-9_4

everyday behavior, although it is widely known that current lifestyles bring about devastating consequences on a global scale? In light of the widespread hiatus between knowledge and practice – the notorious attitude-behavior gap – it does not seem as if the road to sustainability is particularly well paved with green 'enlightenment' programs.

The same problem, albeit in a more subtle form, holds true for capitalism's environmental critics. While emphasizing the necessity of a fundamental turnaround, i.e. a break with capitalist modes of production, these critics often fall back on a rhetoric of social macrostructures without recognizing the central role of everyday human activity in addressing (socio-ecological) transformation processes. While left-wing environmental positions stress the state's role (as a power relationship) and deliver deep insights into the state's prosperous marriage with capitalism, what people do in their everyday lives has received very little conceptual attention. It is therefore unsurprising that capitalism's perseverance in the face of its – obvious – contradictions has remained a mystery for many of these approaches.

I will now pursue two objectives: First, I provide a brief survey of some proponents of the above-mentioned approaches and outline their arguments and conceptual deficiencies. Second, I show that *theories of practice* offer useful insights into the nature of human conduct and might thus help us to understand the persistence of unsustainable activities. I argue that a practice theoretical approach involves all key aspects of the sustainability shift and offers a theoretically ambitious framework of its interpretation.

2 Green Capitalism and Everyday Behavior

I would like to state from the outset that talk of a single, unified *green capitalism approach* is fairly misleading. Just as there is no universal capitalism, there is also no homogenous *greening* of capitalism. As Tienhaara (2013) points out, alongside a myriad of other approaches, three particularly popular strategies try to reconcile economy and ecology. The *green new deal* approach primarily seeks to reform of the finance sector in order to prevent debt-based unsustainable patterns of energy and resource consumption, and to promote public and private investment in environmentally friendly technologies (Tienhaara 2013, 3; see Friedman 2007, 72). *Green stimulus* strategies, on the other hand, apply fiscal measures (e.g., taxation, or direct investment) to support the 'green' sectors of the economy and thus indirectly enhance resource use efficiency (Tienhaara 2013, 4). Finally, the centerpiece of *green economy* approaches is the attempt to price 'ecosystem services' and thus systematically include nature in economic calculations (Tienhaara 2013, 7).

Regarding these different approaches, it seems at first glance as if socio-ecological transformation was only a matter of (government) designing smart institutional frameworks. The protagonists of such rather technocratic, seemingly top-down, approaches are legislative bodies, tax authorities, departments of commerce or labor, and so forth.

However, concern with institutional measures and public policy cannot hide that the vision of a green capitalism in all three paradigms depends substantially on *individualist* thinking. Neither tax incentives nor high interest rates for green investment, price increases of environmentally 'unfriendly' goods, or job offers in the green industries bring about socio-ecological change. *Individual* market participants making the 'right' choices are required. The various strands of green capitalism – regardless of the concrete measures they propose – are organized around the idea that individual behavior and consumer-citizens' sense of responsibility are the crucial instances for achieving socio-ecological change. Steinberg therefore calls green capitalism *green liberalism*, which he conceives as "the idea that market forces combined with individuals all doing their part can save the planet" (Steinberg 2010, 8).

However, talk of green *liberalism* must not hide that the measures applied interpret the idea of individual freedom (of choice) in distinct ways. On the one hand, tenacious advocates of libertarian thought seek to establish a framework in which "individuals pursuing their own best interests could create a sustainable society" (Steinberg 2010, 11). Using a classic metaphor of economic liberalism, such an approach could also be understood as the greening of Adam Smith's *invisible hand*. In an appropriate setting, the libertarian reasoning is that the individual's freely-made choices bring about ecologically sound conditions.[2] According to this position, the main goal of environmental policy is to make free markets work and to internalize (ecological) externalities as far as possible.

On the other hand, strategies to establish green capitalism often also imply attempts to *directly influence* market participants' behavior and hence shape their choices and decisions. In recent years, ideas of 'soft' or 'libertarian' *paternalism* have emerged and have in some way become the *dernier cri* of green capitalism strategies (see Thaler and Sunstein 2003). At first sight, this appears surprising, since paternalism has a bad name in libertarian debates. In a strict sense, paternalism is conceived as "the interference of a state or an individual with another person, against their will, and defended or motivated by a claim that the person interfered with will be better off or protected from harm" (Dworkin 2014). Clearly, 'freedom-loving' disciples of neoliberal capitalism reject such reasoning outright. However, as Friedman (2007, 72) notes in his summary of some of the arguments in favor of green capitalism, "[a]n unusual situation like this calls for the ethic of stewardship." Thus, in the seemingly oxymoronic phrase *libertarian paternalism*, the *libertarian* serves to soften the authoritarian connotation by emphasizing the actors' agency, while at the same time claiming that subjects *ought* to make their choices in a certain way. This notion is perfectly encapsulated by the definition of Thaler and Sunstein (2003, 179), who consider libertarian paternalism "an approach that preserves freedom of choice but that authorizes both private and public institutions to steer people in directions that will promote their welfare."

[2] For instance, when an ecological rucksack of goods and services receives a proper price tag, the market will do the rest.

To summarize: Green capitalism is mainly based on the attempt to achieve sustainability through *individual* behavior, or individual behavior change. Both the orthodox market approach (internalize externalities and leave the rest to the market) and libertarian paternalism agree on the fundamental role of individual choice and, in the end, hold individuals *responsible* for socio-ecological change. Thus, most policy approaches developed in the context of green capitalism clearly draw implicitly or explicitly on – more or less sophisticated – *behavioral theories* of everyday conduct.

It is important to note here that different behavioral theories translate differently into concrete policy measures. Radical followers of market solutions – who would reject paternalist proposals – draw on fairly simple, rationalistic models of behavior. Their implicit economic anthropology is the classic model of the *economic man* (rational agent). Green economy approaches, for instance, use this thinking when they argue in favor of (changing) pricing mechanisms. Such strategies presuppose rational choices by fully informed individuals with clear preferences. The change in individual economic behavior (such as consumption patterns, for instance) is thus a function of changing price tags, or changing utility functions. The best way to instigate change is thus to set economic or material incentives.

Behavioral economics present a different approach. Following the idea of libertarian paternalism, behavioral economists suggest 'supporting' individuals by *nudging* them to make rational decisions (see Thaler and Sunstein 2008). The basic argument of the 'nudging theorists' is that suboptimal choices are widely observable, because individuals often do not possess the capacities to act rationally, and are thus prone to make – what ex post appear to be – mistakes (i.e. choices that do not improve well-being, or serve one's long-term interests). Behavioral economics thus suggests steering people's behavior towards rationality, albeit without eliminating freedom of choice.

It is important to recognize that nudging means more than just materially incentivizing individuals to behave in a certain way. Nudging, as Thaler and Sunstein (2008, 3) repeatedly point out, means to improve the 'choice architecture.' Through a deliberate design of situations of choice, the "psychological quirks" (Wilkinson 2013, 341) responsible for irrational behavior could be overcome and, ultimately, people's lives could be made "longer, healthier, and better" (Thaler and Sunstein 2008, 5).

The means applied in choice architecture range from the material arrangement of offered goods (taking the idea of architecture quite literally), to information campaigns and default rules (see Thaler and Sunstein 2008). The latter two have specifically become classic strategies in environmental contexts. Information campaigns seek to provide better information, and thus aim to improve individual knowledge of the consequences of one's decisions. The German Advisory Council on Global Change (WBGU), for instance, recommends 'supporting' changes in consumer behavior "through information measures such as product labelling, product standards and targeted communication policies" (WBGU 2011, 185). The underlying argument goes that given the right information at the right time in the right way, individuals are very likely to make proper choices. Default rules, on the other hand, are concerned with steering people's behavior by pre-defining procedures or courses of events. In this sense, *default* is what happens when people simply do nothing, i.e. when they do not explicitly articulate their (diverging) preferences and stick to the pre-defined option (Sunstein and Reisch 2014, 131). A standard example is electric energy supply. Setting 'green' electricity as the standard that must actively be opted out of often increases the number of people purchasing eco-power (Sunstein and Reisch 2014, 134 ff.). Default strategies thus build on the establishment of obstacles, requiring a deliberate decision, compared to more or less effortless routine.

Both the setting of material incentives and the nudging strategy center around the idea of rational action as the *normal* case. Whereas the former strategy expects rational decision making and seeks to exercise control by materially rewarding 'good' choices, the latter proactively intervenes in what is regarded as (potentially) irrational behavior and takes countermeasures.

Despite the fundamental role behavioral economics plays in public policy making, and especially so in the context of green capitalism, it must be emphasized that the various approaches have been challenged. Besides a wide range of normative questions about libertarian paternalism (its manipulative or coercive dimension, for instance; see Wilkinson 2013; Henderson 2014), three (interrelated) critical aspects can be specifically highlighted: first, a reductionist understanding of everyday human conduct; second, an overemphasis of individual capabilities to bring about socio-ecological change; and third, an incapacity to think outside the box, i.e. to question the systemic frame.

The theoretical *reductionisms* that come with most green capitalism approaches become apparent in the use of a rather simplistic *ABC model* (Shove 2010) of social and behavioral change. In short, the ABC model assumes that people's *values and attitudes* (the A) "drive the kinds of behaviour (the B) that individuals choose (the C) to adopt" (Shove 2010, 1274; see Shove et al. 2012, 142). Social change is thus solely thought of as a function of an individual's values, attitudes, and knowledge; public policy's task is to initiate and steer social change by affecting these determinants (Shove 2010, 1275). Material incentive strategies draw on this model, just as nudging approaches do. Although, at first glance, nudging seems to transcend simple models of economic behavior by acknowledging social context's role (see Sunstein and Reisch 2014, 128–130), it is nevertheless based on the search for behavior *drivers*, and takes rationality as its normative horizon, i.e. it seeks to "cultivat[e] the rational and reflexive aspects of the mind" (Pykett 2011, 220) in situations where people are inclined to act irrationally. However, such an outlook reduces the social sphere to a physicalistic world of *drivers* and *determinants*. It obscures the intrinsic logic and situatedness of our everyday conduct by introducing criteria that are foreign to it.[3] An understanding of (non)sustainable practice,

[3] The academic and political surprise at a *knowledge-behavior gap* serves as a succinct example here. Whereas 'lay people' do not wonder at inconsistencies in behavior, the gap becomes "mystifying if we suppose that values do (or should) translate into action" (Shove 2010, 1276).

though, will have to avoid the reductionist language of drivers and determinants, and develop a language of *complex* social realities instead.

The second difficulty that behavior-based approaches face relates to the view that individuals are responsible for and the central unit of socio-ecological change. Two strands of critique must be distinguished here: On the one hand, assigning environmental responsibility to individuals can be criticized from a *moral* perspective as a form of overburdening. Since personal environmental stewardship faces profoundly complex, in part unpredictable, systems (both environmental and socio-cultural/political-economical), individual actors are barely to blame for the ecological crisis (see Steinberg 2010, 12). Green capitalism's role model of an active citizen tends to obliterate the *structural* deficiencies that prevent individual initiative from bringing about the desired consequences.

On the other hand, individualism can be criticized for its theoretical insufficiencies. As Shove et al. (2012, 144) concisely note, rather than being arbitrarily changeable acts, everyday practices "are outcomes of complex, essentially emergent processes over which no single actor has control." The social and cultural context of practices (conventions and values, for instance) is not an external driver of behavior, but an integral part of practice itself (Shove 2010, 1279). A theoretical language that captures the fundamental sociality of our everyday conduct is needed. As I will argue later, practice theory offers such a language.

The third point of criticism relates to behavioral economics' limited capacity to challenge green capitalism's ideological basis. Since the approaches outlined above focus on the more or less technical problem of changing individual behavior, questions for the wider context of everyday conduct – fundamental economic and political institutions, shared thought patterns, and patterns of interpretation, for instance – are systematically omitted. Thus, behavioral economics – theoretically and empirically – remains within the system. It reproduces capitalism's economic anthropology and, simultaneously, suggests that minor adjustments are sufficient to overcome unsustainable modes of living. In other words, it is incapable of articulating criticism. In the next section, I introduce an approach that seeks to overcome this shortcoming.

3 Challenging the *Imperial Mode of Living*

As noted, there are a great number of objections against approaches such as green capitalism. Besides a general critique of capitalism, as expressed, for instance, by critical left projects such as the Occupy movement, an explicitly ecological critique of the dominant socio-political-economic system of our time has developed, or – in terms of the environmental aspects in Marx's writings (see Foster 2000) – is currently being *revitalized*. I now focus on the approach developed by the political scientists Ulrich Brand and Markus Wissen. They provide an implicit (ecological) critique by emphasizing capitalism's systemic contradictions and environmental myopia.

Brand and Wissen are particularly interesting for my account, not only because they address the problem of capitalism's tenacious adherence to unsustainable development,[4] but also because they partly use a language similar to that of the practice-theoretical approach I will discuss. Yet, importantly, the authors do not elucidate a theoretical concept of practice, thereby leaving a glaring gap at the very center of their theory.

The argument presented by Brand and Wissen is, in its widest sense, concerned with the nexus of state, capitalism, and nature (or nature-society relationships). It is based on the basic premise that, in order to understand the ecological crisis, we must acknowledge the interdependence of ecological and social issues (Brand and Wissen 2012, 556). However, in opposition to the standard view in current politics, the authors are skeptical of the state's capacities to reshape capitalist nature-society relationships and to implement serious socio-ecological transformation processes (Brand and Wissen 2012) (see Brand 2012a). Therefore, they propose to talk of a *multiple crisis* of capitalism, calling for a fundamental renegotiation of the roles of the state and the economy.

The core piece of Brand and Wissen's analysis is the notion of the *imperial mode of living* (Brand and Wissen 2012), by which the authors mean "dominant patterns of production, distribution, and consumption that are deeply rooted in the everyday practices of the upper and middle classes of the global North and increasingly in the emerging countries of the global South" (Brand and Wissen 2012, 548). In everyday life, the imperial mode of living expresses itself in things such as a certain standard package of consumer goods; the widespread availability of services; so-called minimum conditions for labor; the promise of progress, growth, and prosperity; and – most importantly – the demand that the state permanently guarantees all these conditions, for instance, through a relevant labor market policy, or by ensuring the supply of natural resources.

With its emphasis on production, distribution, and consumption, the *imperial mode of living* generally conceives of societies' relationships with nature as capitalist *patterns of matter and energy use*. Thus, from the outset, it is clear what separates the imperial mode of living from common notions of lifestyle, customs, and habitude: First, Brand and Wissen advocate a strong emphasis on structural conditions, or constraints of everyday resource use and its change (see Brand 2012b, 30). Drawing on the theoretical assumptions of *regulation theory*, they place the imperial mode of living in relation to specific regimes of accumulation, which – guaranteed by the state and social institutions – constitute coercions that cannot readily be avoided (Brand and Wissen 2012, 548). Second, the imperial mode of living also entails a normative position that contests the generalizability of Western ways of life. While classic notions of lifestyles are often confined to a mere reconstruction of individual choices, Brand and Wissen offer a global view that considers questions of structural inequality and injustice. Since the imperial mode of living generalizes resource usage patterns that – for ecological reasons,

---

[4] Even though they do not use the term *sustainability*.

at least – cannot be globalized, they conclude that, in the end, the "existing rules of the game" must be "call[ed] into question" (Brand and Wissen 2012, 554, 556).

It is worth noting the two central properties that Brand and Wissen attribute to the imperial mode of living. On the one hand, and unsurprisingly, the imperial mode of living is seen as reinforcing and advancing asymmetrical global power relationships. As the term *imperial* indicates, the *normal* lifestyle of people in the Global North is inextricably linked to the capacity to access the raw materials, cheap labor, and waste-sink capacities of the Global South (Brand and Wissen 2012, 555). The perpetuation of capitalist consumption patterns thus inevitably presupposes the acceptance of global inequalities and exploitation or, to put it bluntly, Western imperialism.

On the other hand, the imperial mode of living is also characterized as *hegemonic*. Broadly speaking, this aspect may be regarded as a tacit consensus on *the good life* ideal in capitalist societies (Brand and Wissen 2012, 549). By and large, there is usually only little doubt about what we are supposed to have or to experience in everyday life, even though there might be significant differences in the actual choices we make. However, it is crucial that *hegemonic* not only means that the current capitalist lifestyle is omnipresent and 'operates' behind people's backs, it also means that the imperial mode of living successfully conceals its nongeneralizability. According to Brand and Wissen (2012, 551), state institutions normalize and secure the capitalist mode of living (despite all its contradictions), thus promoting the specific (unsustainable, unjust, etc.) nature-society *and* global North-South relationships associated with it.

The irony of this lies in the 'supposed to have it' phenomenon[5] (see Wittmayer et al. 1994), which holds true even for those who are not in a position to share the affluent life considered *normal* in capitalist societies. Both the lower classes of the Global North and the lower and emerging middle classes of the Global South seem to share the ideal of a globally generalizable capitalist lifestyle, even though this is only possible on the basis of an (asymmetrical) exploitation of the workforce and natural resources. From Brand and Wissen's perspective, this might ultimately be the key paradox in need of explanation: How can a specific lifestyle that necessarily involves (global and intra-societal) inequalities be successfully disseminated globally?

To sum up this brief outline of an ecologically grounded critique of capitalism, it can be concluded that Brand and Wissen substantially broaden the debate about socio-ecological transformation processes and seek to overcome the oversimplifying technocratic approaches common to current politics. They show that capitalism in its present form relies upon and perpetuates global inequalities that are inherently incompatible with sustainable development. With their reference to people's living conditions and ideas about *the good life*, they add a further dimension to the inquiry of capitalism's spread and perseverance (besides macro-structural arguments such as David Harvey's idea of spatio-temporal fixes, for instance). In a way, Brand and Wissen's approach is one of the few critiques of

[5] That is, ideas about what, materially speaking, is a *normal* or *good* life.

capitalism that – prima facie, at least – takes the commonly neglected logics of everyday life seriously.

Yet, as instructive as Brand and Wissen's approach might be, it must be theoretically enhanced in order to deliver a truly convincing framework for an analysis of socio-ecological transformation processes. While Brand and Wissen's borrowings from and advancement of regulation theory provide a cogent account of the capitalist state's role in the ecological crisis, their inquiry into the nature of everyday practice is less convincing. The two authors repeatedly speak of the imperial mode of living's *deep-rootedness* in everyday practice (see Brand and Wissen 2012, 548–555). Furthermore, they likewise speak of "microstructures of daily life," in which the imperial mode of living is "hegemonically reproduced" (Brand and Wissen 2012, 555), or they claim that globalized liberal markets *inscribe* themselves in "everyday practices" (Brand 2012b, 30). But what, one might ask, does *deeply rooted* or *inscribed* mean? Are these descriptions just a synonym for "routine", or does the term "practice" mean more that that? Brand and Wissen provide little and indirect information about their understanding of *practice*. Besides the obvious association with activities, knowledge, and values (such as ideas about *the good life*), the only hint we get is that practices seem to be stable or inert and, therefore, they serve as a strong anchor of everyday life capitalist structures. However, to explain why so little change occurs, an elaborate concept of practice is necessary in order to elucidate not only the routine character of everyday conduct, but also its social embeddedness, and the role of the actors that carry social practices.

4 Interlude: Sustainability and Practice

Arguing that sustainability research must be anchored in practice theory might at first glance appear somewhat redundant – it seems all too obvious that any socio-ecological transformation process is bound to practice from the outset, because someone is urged to *do* things in a different manner. It is worth noting that we can observe a remarkable affinity between the sustainability concept and a practice-centered approach from the very beginnings of the use of sustainability. Sustainability, or *Nachhaltigkeit*, as introduced in German by Hans Carl von Carlowitz in the early eighteenth century (see Carlowitz 2000, 105–106), means, in its broadest sense, to *do* things in a particular way. As the head of the Royal Mining Office of the Kingdom of Saxony, Carlowitz was entrusted with managing forests in order to prevent a scarcity of timber that would have sent the economically crucial silver ore mining in Saxony into crisis (Grober 2012, 76 ff.). The guidelines he developed for the conservation and cultivation of timber were centered on the idea of taking forests' natural regeneration rates into account. The loss of a certain stock of trees (owing to harvesting or natural depletion) must be replaced in order to ensure a durable – in principle *indefinite* – use of a forest. Hence, the essential *rule of (sustainable) action* is "Do not use a resource beyond its regeneration rate." This reproducibility principle, which in Carlowitz's profession became known under the title *sustained yield forestry*, is also indicated by the term sustainability's etymological roots. *Sustainability* derives from the Latin *sustinere* (to keep up); sustainability thus means "being able to keep in being" or "being able to keep something in a certain state" (Grober 2012, 19). Irrespective of the particular character of the thing to be kept in a certain state, sustainability means to *do* things in a way that takes the past, present, and future use into account. This very core of sustainability is also contained in the term's contemporary use. Despite the semantic shift to *sustainable development*, the 1987 Brundtland Report, for instance, conceives sustainability in terms of long-term thinking and *actively* balancing the present against the future.

Two primary aspects of these practical foundations of sustainability must be specifically highlighted. The *first* relates to the ecological dimension. While sustainability in Carlowitz's technical sense denotes a *certain type* of nature-society relationships (as a result of sustainable practice), it must be emphasized that *all* kinds of nature-society relationships are established, reproduced, contested, and transformed in (everyday) practice. Any transformation of nature, or intervention in the material world – whether or not it follows institutionalized rules, as in forestry, and whether or not it successfully matches present and future demands – ultimately occurs through what can be conceived as social, supra-individual practice. In this sense, *sustainability* denotes a specific mode of practice, i.e. an attempt to arrange nature-society relationships to successfully deal with the natural environment's unavailability.[6] It is important to bear this in mind, because it shows that socio-ecological transformation processes can only be properly understood – and initiated – when we acknowledge the everyday logic of practice.

The *second* aspect leads us to the complexity, multidimensionality, and sociality of practice. Carlowitz's original use of *sustainability* serves as an example. As a practice, sustainable forest management is not limited to observing concrete, individual acts of harvesting or planting trees. Since practice inherently transcends the level of individual agency – thereby resembling the aforementioned *mode of living* – an appropriate understanding of sustainable forestry must necessarily involve the shared knowledges and cosmologies, the technical means and social positions, etc. that make related individual acts possible. To understand the (non) sustainable use of forests in Carlowitz's time, it is necessary to consider historical contexts such as the eighteenth century economic worldviews, the forestry technology applied, common concepts of nature and nature-society relationships, the symbolic value of possessing natural resources, and so on. To some extent, it can be said that practice-centered approaches prioritize complex, social practices over individual, atomistic acts, even though practice 'reveals' itself only in such concrete human conduct.

[6] *Unavailability of nature* means that society has only limited capacities to transform, use, or influence nature in the directions it wants, since nature (or the bio-physical world) follows its own logic.

5 Theories of Practice

Given the striking 'practical' nature of sustainability issues – and ecological issues in general – it is unsurprising that theories of practice provide a useful theoretical framework for analyzing nature-society relationships Over the past 35 years there was, and still is, much debate about praxeological approaches, culminating in the (seemingly inevitable) proclamation of a *practice turn* in social theory (see Schatzki et al. 2001). Since the late 1970s, scholars such as Pierre Bourdieu – to whom I will return later in detail – Anthony Giddens, and Theodore Schatzki (to mention just three of the most prominent names) have addressed questions of practice.[7] Their various oeuvres draw on theoretical and methodological considerations in ethnology, sociology, and philosophy, thus forming a fairly *heterogeneous* field of research rather than a unified theory of practice. All these divergent approaches share the general notion of practice as "a routinized type of behaviour which consists of several elements, interconnected to one other: forms of bodily activities, forms of mental activities, 'things' and their use, a background knowledge in the form of understanding, know-how, states of emotion and motivational knowledge" (Reckwitz 2002, 249).[8]

Practice theoretical approaches thus do not focus on subjective or individual acts, or on 'practice' as opposed to 'theory,' as is sometimes supposed, but on *types* of human activity and their socio-material contexts. These types bind together knowledge, bodies, social relations, material objects, etc., and constitute the focal point (or *essential unit*, to use a common expression in social theory) of the social world. Thus, theories of practice not only pose questions concerning human agency – the capacity to *act* – but also include questions of social order and its change. Practice is thus necessarily *social*, and practice theory allows for a systematic integration of the micro level and the macro level of the social.

While the general approach outlined above is widely shared by all theorists of practice, the specific concepts and ideas to which the term *practice* is applied differ (in some cases considerably) between authors. I next outline the practice approach as developed by Pierre Bourdieu.

[7] Their central works include Giddens (1979), (2012), Bourdieu (1995), (1990a), (1998), Bourdieu and Wacquant (1992), and Schatzki (1996).

[8] A shorter, but also widely shared understanding, considers practice as "embodied, materially mediated arrays of human activity centrally organized around shared practical understanding" (Schatzki 2001, 2).

6 Bourdieu's Praxeology

Pierre Bourdieu's praxeology centers on the concepts of *habitus* and *field*. In a very broad sense, these concepts address questions of agency, social reproduction, and transformation. Similar to Giddens's theory of structuration, Bourdieusian praxeology explains the tension between the seemingly coercive power of social structures on the one hand, and its construction in and through everyday practice on the other. In doing so, praxeology conceives of actors as neither completely free agents (which theories based on the assumption of rational choice seem to), nor as subjects determined by structures (as some strands of Marxist or structuralist theories tend to). Instead, it seeks to develop a language that can capture the social, cultural, and historical contextuality of everyday practice. In the following paragraphs, I address the meanings of Bourdieu's central concepts, habitus and field, in order to clarify how praxeology differs from the two approaches outlined earlier.

Habitus denotes Bourdieu's heuristic device to grasp the – seemingly paradoxical – nondetermined directedness of everyday practice.[9] As Bourdieu notes, "types of behaviour can be directed towards certain ends without being consciously directed to these ends, or determined by them" (Bourdieu 1990b, 9–10). This says that, on the one hand, everyday life follows certain rules that manifest in comparatively stable behavior types that a number of people share. Praxeology traces these behavior types back to past experiences that individuals incorporate, and thus provide the basis for their current "*perceptions, appreciations, and actions*" (Bourdieu 1995, 83; emphasis in original). In this sense, habitus denotes our *acquired* dispositions and the ability to reproduce the past. However, it must be emphasized that, on the other hand, habitus does not postulate the total predictability of social life on the basis of social experiences' formative role. Since habitus also denotes acquired *dispositions*, i.e. *tendencies* to perceive, appreciate, and act, there is nevertheless always the possibility of deviation, or variance, transformation, and disruption in everyday practice (which, in this view, is necessarily contingent).

These reproductive and generative/transformative aspects of habitus are also captured in its characterization as an opus operatum and a modus operandi. Bourdieu calls habitus an *opus operatum*, because it can be regarded as a result of the internalization of objective structures in the course of socialization (Bourdieu 1995, 81; see Bourdieu 1990a, 52 ff.). *Embodiment* is the key mechanism through which social experiences are appropriated and memorized as an opus operatum (see Bourdieu 1995, 87 ff.). Praxeological thinking contends that the lived body serves as a repository of past experiences (e.g., the learning processes of how to walk, eat,

[9] In Bourdieu's view, habitus and field are not 'existing' entities in the sense that everyday agents use them as a means to describe their social experiences. Instead, they are specific, sociological ways of comprehending the social world. Neither habitus nor field are thus simply observable objects. On the contrary, to take them 'for real' would, according to Bourdieu, conflate the logic of theory with the inherent logic of practice (see Bourdieu 1990a, 81, c).

or position oneself in the presence of other people), and at the same time represents our current access to the world. In this perspective, the body, "*enacts* the past" (Bourdieu 1990a, 73; emphasis in the original).

This pivotal role of embodied histories and biographies draws attention to two aspects of habitus that are of particular interest to questions of social and socio-ecological change. The first aspect relates to the capabilities to comprehend and access one's habitus. If the idea of embodiment or the *incorporation* of biographical experiences is taken seriously, habitus must be conceived of as being more than mere knowledge. To equate habitus with knowledge is to confuse conscious aspects of the mind with pre-conscious ones. While knowledge is amenable to reflection, habitus is not. Or, in Bourdieu's words, "[w]hat is 'learned by the body' is not something that one has, like knowledge that can be brandished, but something that one is" (Bourdieu 1990a, 73). Given the deep anchorage of dispositions in bodies, it is clear that they cannot be as easily modified as theories of behavior change seem to assume. Since dispositions and subsequent forms of practice draw on long-term *experiences*, mere information campaigns – addressing the reflexive parts of the mind – are unlikely to be sufficient to trigger behavior change.

The second aspect concerns the consequences of the durability of habitus. According to Bourdieu, our dispositions are inclined to be inert and relatively stable. For this reason, they may sometimes no longer fit the living conditions in which agents find themselves. In praxeology, this phenomenon is called *hysteresis*, or the delay of habitus (Bourdieu 1995, 83). The hysteresis effect idea contends that a rapid change in living conditions is very likely to result in adaptation problems, since the adjustment of everyday practice takes time and is not a matter of conscious decision, even though reflexive deliberation plays a role in changing habitus. Conceptualizing processes of social transition in terms of hysteresis draws attention to the actors' strategies to cope with new contexts and, thus, helps us understand seemingly irrational or unreasonable behavior as an effect of the inertia of habitus.

Whereas the aspects concerned with habitus as an opus operatum are encapsulated in the concepts of embodiment and hysteresis, those concepts concerned with habitus as a modus operandi can best be grasped through the notion of practical sense. Practical sense is one of the keys to understanding the difference between behavioral theories and practice theories. The notion of practical sense draws attention to our everyday conduct being largely based on *pre-reflexive* capacities, or tacit knowledge about "what is to be done in a given situation – what is called in sport a 'feel' for the game" (Bourdieu 1998, 25; see Bourdieu 1990b, 61). From the actor's perspective, it is unnecessary to deliberately decide what to do in each and every moment; in most everyday situations, actors simply act without consciously examining their options. Practical sense thus provides us with the strategies to cope with everyday life through well-adjusted, albeit always improvisatory routines (Bourdieu 1990b, 61).

The concept of habitus not only allows us to understand the persistence of practice, but also draws attention to the fundamental difference between social science and everyday life worlds. Whereas a plethora of approaches to collective behavior change – at least implicitly – urges individuals to act more rationally or self-consciously, Bourdieusian praxeology acknowledges the inherent logics of practice and reminds us not to mistake scientific rationality for everyday conduct: "the logic of practice," Bourdieu concludes, "is logical up to the point where to be logical would cease being practical" (Bourdieu, as cited in Wacquant 1992, 22–3).

To summarize: The praxeological concept of habitus seeks to link objective structures (e.g., living conditions) with the subjective reproduction and transformation of these structures. Treating habitus as an opus operatum and a modus operandi means highlighting the agents' acquired abilities to generate well-adjusted practices by means of practical sense. Yet, the habitus and practical sense concepts do not lead to a sociologically sophisticated picture of the social world. In addition, we need a concept that can incorporate practice's social context. In praxeology, the notion of (social) *fields* serves this purpose.

Thinking in terms of social fields involves two central aspects: a symbolic mapping of the social world and an agonistic, relational perspective. The first aspect might best be described with reference to sociological modernization theories. Like other theories of modernization, Bourdieu's praxeology claims that the social world is differentiated into various autonomous spheres of action – fields. Therefore, the process of modern societies becoming increasingly complex simply means that a growing number of distinct social fields are being developed. From the individual's perspective, 'being modern' means engaging in specific types of practice, each with their own "stakes and interests, which are irreducible to the stakes and interests specific to other fields" (Bourdieu 1993, 72).[10] Thus, fields are *social microcosms*. Classic fields are, for instance, economy, arts, religion, and science (see Bourdieu and Wacquant 1992, 97 f.). Each of these social spheres follows its own logic, and what counts in one sphere does not necessarily translate into other spheres; for instance, to gain a reputation as an artist and succeed in art, one must do things differently than one would do in the economic field, and religious practice follows different rules than scientific practice, and so on.

Thus, field theory highlights not only the striking differentiation of the modern social world (macro-perspective), but also the effects this has on the agents (micro-perspective). To enter a field, an agent must have a sense of what the field is about – he or she must accept and apply its logic. The term Bourdieu uses for this phenomenon is *illusio* – a (deliberately) false etymology from the Latin *ludere* (to play) (Bourdieu 1998, 76; see Bourdieu and Wacquant 1992, 98 f.). Illusio denotes that agents engaged in a specific field agree on the fundamental value of 'playing' or engaging in the field despite all their differences. And they need a feel for the game, or a field-specific habitus and practical sense to succeed in the field.

Insofar as Bourdieu's praxeology draws attention to the common denominator of practice, one might assume that fields are mainly about consensus. Quite the opposite is true, however. In line with Bourdieu's thoroughly agonistic vision of

[10] As an aside, this view is close to what Max Weber and Alfred Schütz refer to as *spheres of values* or *provinces of meaning*.

the social world, he envisions fields not only as spheres of meaning, but also as areas of *struggle* for resources and social positions. Ultimately, Bourdieu's social theory is based on the assumption that "to act in the social world means necessarily to engage in certain arenas of competition" (Peters 2011, 66). Agents compete – not necessarily consciously – for positions within a sphere, for resources such as field-specific capital, for the rules of the field, and for interpretational sovereignty over what the field is about. Competitors can be divided into *orthodox* agents, defending their status and the existing rules of the game, and *heterodox* agents (or 'heretics'), who challenge the field's order (Bourdieu 1993, 73). The result of these struggles is a relational space of positions, or a social topology, which reflects the power relationships in the field. Praxeological field theory thus draws attention to the inherent dynamics of social arenas and helps make the various strategies to preserve, or challenge, a field's topology visible.

To sum up this brief outline of Bourdieusian practice theory, we might say that praxeological thinking provides a useful account of the social nature of everyday conduct and thus helps illuminate behavioral economics' inability to explain, or even acknowledge, the knowledge-action gap. Further, Bourdieu's approach offers one way to theoretically fill the conceptual void in Brand and Wissen's critique of capitalism. In other words, praxeology allows us to widen the scope of observing seemingly individual behaviors, and to discover powerful social relationships and trajectories, where behavioral economics more or less only sees individuals in concrete situations of choice. According to practice theory, the decisions people make in their everyday lives are deeply anchored in embodied (social) experiences that cannot simply be overcome by providing additional information. The implications of the anchorage and subsequent inertia of habitus must – specifically in terms of the global ecological crisis – not be underestimated. As Leggewie and Welzer (2009, 11) persuasively note:

> After 250 years of superior power, economy, and technology, our self-image and habitus are bound to conditions that no longer exist. The time-lag between our perception and self-image on the one hand, and the 'globalized world's' speed of alteration on the other becomes obvious also in other respects of our existence – for instance with regard to the energy-, environmental- and climate-crises (own translation).

In the next and last part of this article, I briefly point out what follows from this diagnosis of the issues concerning the socio-ecological transformation processes.

7 Sustainability as Practice

Applying the practice paradigm to socio-ecological transformation processes has only become common in recent years. Broadly speaking, practice theory is primarily used to investigate *specific fields* of the socio-ecological transformation. Gram-Hanssen (2011) and Wilhite (2014), for instance, introduce practice theory to energy consumption research; Haluza-DeLay (2008) uses Bourdieu's praxeology to theoretically frame environmental movements; Sahakian and Wilhite (2014), as well as Røpke (2009), apply theories of practice to the study of everyday consumption; while Scott et al. (2012) enhance (sustainable) design theory with praxeological insights, and so on.

Whereas these thematically narrow strands of practice theoretical research have – deservedly – experienced a certain boom, a wider perspective on large-scale social change does not seem particularly fashionable today. Besides a few studies on the general nature of (ecological) practice and its modification (see Shove 2010; Shove and Walker 2010; Shove and Spurling 2013; Spaargaren 2011), questions of socio-cultural change and disruption in the context of the ecological crisis have only rarely been addressed using practice theory. Furthermore, most of the few studies concerned with the sustainability shift do not seem to recognize the heuristic value of Bourdieu's praxeology for examining social change.[11] Contrary to conventional wisdom in the social sciences, I argue that Bourdieu provides a cogent approach to examining the socio-cultural conditions and consequences of far-reaching change in a given society (see also Wacquant 2004). Based on his early experiences in the Algerian War, Bourdieu's *ethnosociology* "deal[s] squarely with cultural disjuncture, social disruption, and structural rupture at levels ranging from the individual to the societal, and in temporalities spanning the biographical to the epochal" (Wacquant 2004, 389). Thus, the concepts of practical sense, habitus, and field make an alternative language available to describe socio-ecological transformation processes. What, then, does it mean to conceive the sustainability shift in terms of Bourdieusian praxeology?

In a general sense, speaking of sustainability as practice does mean to acknowledge the routinized, unexamined, embodied character of the everyday activities that propel the nature-society metabolism. However, this seemingly straightforward premise exposes the profound difficulties of contemporary market-based attempts to steer socio-ecological transformation processes. In this view, a shift towards sustainable practices does not consist of cognitively driven, rational decisions to act differently, but involves the transformation of a whole set of inherited 'ecological' dispositions and the socio-material context in which they have developed. Using Bourdieu's vocabulary, we could say that the road to sustainability necessarily includes a *habituation* of sustainable behavior patterns against the inertia (*hysteresis*) of nonsustainable lifestyles and their socio-material preconditions. The outcome of this process can be called *ecological habitus* – the agents' capacity to generate sustainable practices. Following the praxeological premises outlined above, it should be clear that *ecological habitus* means more than *environmental attitudes*, *knowledge about one's ecological impact* or, as Daniel Goleman (2010) famously put it, *ecological intelligence*. *Ecological habitus* transcends the level of cognition and values concerning the environment; it is applied to the complex package of embodied dispositions in order to perform sustainable acts and obtain a *feel for the game* of sustainable living.

[11] I discuss some exceptions below.

While it does not (yet) belong to the core concepts of sustainability research, the notion of an ecological habitus developed here has been used elsewhere, albeit in a slightly different fashion. Debbie Kasper, for instance, uses the term in a broad sense to generally denote the ecological dimension of everyday acts. She argues that "[e]verybody, whether aware of it or not, lives in, depends on, affects, and is affected by their ecosystem and the wider biosphere" (Kasper 2009, 320). Kasper concludes that an ecological habitus concept must be *value neutral*, i.e. it must not focus on pro- or anti-environmental tendencies, but must consider *any* ecologically relevant outcome of human conduct (ibid.). Randolph Haluza-DeLay, in contrast, applies the heuristic of an *ecological habitus* explicitly to the "routinization of environmentally sound practices" and the "everyday environmentalisms" that can be currently observed (Haluza-DeLay 2008, 206, 207). Haluza-DeLay has in mind social or environmental movements that, to him, are a breeding ground for the development of ecological habitus (Haluza-DeLay 2008, 210). Finally, in *Ethics of Place* (Smith 2001), Mick Smith suggests an ecological ethics based on acknowledging the level of the local and the logics of practice. In respect of the moral consideration of nature, he argues that "[o]ur moral identity is best understood in terms of inspiring, developing, and sustaining an ethical *habitus* rather than a compliance with codes, rules, or conscious calculations of benefits and losses" (Smith 2001, 202). To put it more plainly, Smith argues that, in order to prevent ecological crises, our everyday ecological outlook must be radicalized and take the form of a 'second nature,' or an "ecological habitus," creating a "practical environmental sense" (Smith 2001, 204).

Clearly, all three approaches are based on Bourdieu's basic notion of habitus as a system of embodied dispositions and therefore share the same vantage point with my argument; however, they draw different conclusions.

Whereas Kasper's general idea of all practices' ecological impact helps point out the significance of everyday life in building sustainable futures, the weakness of her approach seems to be the emptiness of the attribute *ecological*. Kasper broadens the meaning of the *ecological habitus* to such a degree that it becomes hard to distinguish *ecological* from *non-ecological* habitus. If one takes her interpretation of quotidian life seriously, every habitus must ultimately be considered *ecological*, since undoubtedly all – embodied! – practices have relationships with nature and the material world.

At first sight, Haluza-DeLay's approach appears much more convincing. Yet, in my view, he empirically emphasizes domains too narrowly to fully exhaust the habitus concept's heuristic power. While one cannot but agree with Haluza-DeLay's view of environmental movements as sites of counternarratives and alternative ways of life, he runs the risk of losing track of the manifold changes happening in the *midst* of contemporary society with this focus (although changes at the margins may play a role). The widespread emergence of a whole new sphere of everyday environmental logic – in other contexts referred to as the *greening of society* – is not only crucial owing to its dimension (compared to 'marginal' phenomena such as environmental movements), but also because it produces its own social dynamics. It might be true that widespread cultural change almost always originates in a small avant-garde, but in order to fully attain the dynamics of the current socio-ecological transformation processes, it seems necessary to hypothesize the emergence of ecological habitus at the center of society.

The same problem ultimately applies to Smith's approach. While he convincingly (and quite close to Bourdieu here) argues for ecological habitus to be a form of *attunement* to locations (or environments), he first seems to confine the project of a place-based ecology to the realm of radical environmentalism (see Smith 2001, 25–26). This might have been due to the paucity of evidence of the substantial presence of 'green' issues in wider society at the time of publication (the earliest in 2001). However, as noted, to explain the current *sustainability wave* (see Woehrle 2010), one cannot only look for ecological habitus in radical pro-environmental circles.

As an interim conclusion, we could first propose that the heuristic device of an ecological habitus enables us to understand what a genuine sustainability shift must be about: If we are to durably keep the natural world in a certain state – roughly speaking, the state of reproducibility[12] – the pre-reflexive routines that govern our everyday resource use must be 'equipped' with an 'ecological bias.' By taking the impact of everyday practice seriously – virtually everything we do leaves an ecological footprint –, it seems particularly unlikely that a sustainability transformation exclusively based on knowledge and the demand for a permanent reflexive evaluation will succeed. Consider such mundane practices as housing, the use of private consumer goods, nutrition, waste disposal, personal mobility, etc. – we so rapidly become customized to a certain way of being-in-the-world that even the task of reflexively *imagining* a different mode of living seems difficult in many cases.[13]

The notion of an ecological habitus does not imply that socio-ecological change is impossible. Because habitus correlate with specific socio-cultural contexts, they are also likely to change when an environment changes, or when a specific context emerges. My central argument is that the increasing public relevance of sustainability issues in countries of the Global North goes hand in hand with the rise of a specific realm of ecological practice – an ecological field. This sphere, like other fields, has its own rules, a certain consensus among the agents about what is at stake, and a competition for dominant positions and capitals within the field.

Applying *field theory* to the sustainability shift might at first glance seem somewhat irritating. In Bourdieu's empirical works on fields there is a clear tendency to limit this perspective to spheres that are closely linked to established professions and institutions (see Lahire 2011, 29). As noted, in this strict sense,

[12] This, of course, is a disgraceful oversimplification of the debates about sustainability, but it will suffice for our purposes.

[13] Practice theory also shows that through these internalization processes, the perseverance of nonsustainable lifestyles is not only a question of existing material infrastructures, but also of mental and bodily 'infrastructures,' so to speak.

fields are realms of action equipped with relatively clear entry points (e.g., professional qualifications or exams), accurately defined roles, and so on.

However, I contend that extending field theory beyond the realm of professions helps make the converging point of the myriad of activities around sustainability and everyday environmentalism visible.[14] Agents such as environmental movements, politicians, natural and social scientists, consultants, 'eco-authors,' and 'ordinary people' try to define what is ecologically sound (or *sustainable*) behavior, even though they act in different professional and nonprofessional contexts and are thus, simultaneously, part of other fields. However, through their contributions, they create a sphere that is increasingly emancipating itself from other social life arenas. Incidentally, it does not matter that the agents engaged in the ecological field often disagree on the particular definitions and strategies, it is important that they *share* the belief (or illusio) that nature-society relationships need to be actively managed and that sustainability must be brought about rather than being a 'natural' state of affairs.

The specific capital (or commodity) with which the ecological field is concerned might be described as *ecological integrity*. By this I mean the result of an agent's ability to minimize his or her ecological footprint and not (directly or indirectly) negatively impact other beings' present and future living conditions. In other words, ecological integrity roughly means to adhere to the reproducibility principle suggested in the sustainability debate. With this field-theoretical framework in mind, we can also further specify the above-mentioned idea of an ecological habitus: Possessing a feel for the game means to accept the idea of and strive for ecological integrity.

It is crucial to see here that the dispositions, which I suggest we call *ecological habitus*, generate practices that *fit the field*, i.e. that match what a given cultural and historical situation considers environmentally friendly or sustainable. This does not mean that the 'actual' ecological outcomes of our practice, i.e. the physical/material dynamics that our bodily acts set in motion, are insignificant. Yet, as social constructivism convincingly argues, we only have symbolically *mediated* access to the world. The socio-cultural domain is thus the only source of the ideas and worldviews that organize our practices, and what sustainability (or environmental friendliness) 'is' must thus be regarded as subject to negotiation.

Such a perspective is far from relativist. Bourdieusian practice theory specifically urges us to notice the ambiguities and power-ladenness of our taken-for-granted views. Field-specific practice (like sustainable practice) is neither immutable nor arbitrary. It is the result of dynamic and contingent struggles within the fields but, is simultaneously fairly inert and stable through embodiment or habitualization processes.

For instance, take the two approaches to the ecological crisis I mentioned at the outset. The increasingly dominant (or *orthodox*) view that the 'green capitalists'

[14] In her work on emotional capitalism and emotional fields Eva Illouz develops a similar, inspiring approach, albeit in an entirely different context (see Illouz 2007).

grosso modo hold, claims that individual agents are ecologically significant and should change their lives. Approaches like that of Brand and Wissen, however, challenge this position and place more emphasis on structural aspects. They often try to unmask the individualization narrative as a means of capitalist domination: Individual efforts to live an ecologically sound life, they conclude, ultimately serve to distract from capitalism's structural flaws.

We find different everyday life strategies to cope with the ecological crisis that correspond to these two – at first sight fairly academic – positions. In keeping with the green capitalism perspective, there are (on the one hand) widely observable attempts to achieve sustainability through 'greener' consumption. The rationale goes that if we were to substitute the consumption of environmentally more benign goods for our pre-ecofriendly consumer basket, sustainability would soon be achieved. This approach ultimately amounts to a commodification of ecological issues – i.e. you can *buy* your way to sustainability – and leaves the underlying capitalist system untouched.

On the other hand, there is a more fundamental approach based on what might be called an *austerity principle*. Here, ecologically sound behavior presupposes a considerable reduction of resource use.[15] By renouncing the long-accustomed conveniences of Western life, it seems, agents seek to take a stance against the capitalist growth mechanism. As outlined above, theorists such as Brand and Wissen insist that (marginal) individual sacrifices cannot hide that capitalism as a system necessarily depends on extensive consumption. Nevertheless, if their plea for counter-hegemonic struggles is taken seriously, alternative practices of everyday consumption might be one (of several) arena(s) where a fundamental switch in the production/consumption mode takes its course.

The rivalry between the above two approaches is obvious. However, field theory allows us to identify their common ground, and also calls attention to their positions and their prospect of being heard in the field. Whereas orthodox approaches (such as behavioral economics) possess a certain interpretational sovereignty – indicated, among others, by political programs and research funding – a fundamental critique of capitalism still appears to be the ecological field's marginal position, academically and non-academically.

The specific topology of the ecological field, the interrelation of the positions it offers, and its boundaries deserve much empirical attention. On a very small scale I sought to point out that the notion of the ecological field as an arena of negotiation and struggle clarifies how seemingly remote agents contribute to the same thing. The relational perspective that the field model applies allows one to investigate not only how ecological worldviews are shaped and perpetuated, but also how they are challenged. Concerning the crucial question of the socio-ecological *transformation*, we might conclude that such processes of change are (more) likely to happen when new actors enter the stage or field, or marginalized actors find ways to enforce new,

[15] Such efforts have been discussed under catchphrases such as *simple living*, *lifestyle of voluntary simplicity*, *intentional deceleration* (Rosa 2013), and *post-affluent society* (Etzioni 2004).

alternative narratives, change the rules of the game (e.g. allowing solutions outside capitalist commodification), and so on.

8 Summary and Concluding Remarks

The global ecological crisis inevitably yields socio-ecological change. In some cases, it might be possible to deliberately govern or design change in the long term; in others, change might consist of ad hoc reactions to rapidly transforming environments. As I tried to show, different theoretical approaches seek to address socio-ecological change and provide insights into the causes of the ecological crisis and remedies for it. The approaches that behavioral economics back and the approaches that criticism of capitalism informs make different – more or less sophisticated – assumptions about the social world and the mechanisms of social transformation. Whereas for the former, the overcoming of nonsustainable development, is only a technical problem of behavior change, the latter claims that individual behavior and its conversion merely represent the surface of fundamental structural flaws. Similarly, market-based strategies are inclined to presume fairly quick solutions for the scope of change, while capitalism's critics seem to be more aware of socio-political systems' perseverance and inertia.

Not only diverging theories follow from these different starting points, but also different policies. From a praxeological perspective, we observe that the political efforts to date (based on behavioral economics) display a certain unease with the complexities of social life, preferring fairly technocratic approaches to the ecological crisis. Nevertheless, whatever the pathway to sustainability will be paved with, from a praxeological position it is safe to assume that the present calls for the steering of transformation processes will not suffice in the long run. If we take the growing evidence that transforming unsustainable modes of living requires a *fundamental* reduction in resource use seriously, policies of environmental governance should address the *roots* of our resource usage patterns.

I sought to illustrate that these roots of nonsustainable lifestyles – or their various counterparts – can be conceived in terms of habitus, field, and practical sense. The cursory framework developed theoretically here, acknowledges that everyday practice is not performed in a vacuum, but is always part of a socio-cultural universe. Hence, from my perspective, a threefold enterprise is necessary to better comprehend the ecological crisis. *First*, a careful empirical examination of concrete (non) sustainable practices and their respective backgrounds will help us understand the perseverance of current nature-society relationships and identify sites to challenge these. *Second*, if we further conceive sustainability research as a venture in *transformation science*, we might be able to make better use of the insights into other (historical) change processes. While social experiences are unlikely to translate directly, there is, for instance, a striking analogy between the expectable socio-ecological change and the transformation processes Bourdieu observed in Algeria in the 1950s. In both situations, actors are (or will likely be) faced with a deep transformation of their lifeworld structures. *Third*, praxeological thinking calls for radical reflexivity. Following the notion of an ecological field, sustainability research does not only mean producing *scientific* knowledge (as opposed to a layperson's knowledge), it also means providing narratives and ideas that might actively contribute to socio-ecological transformation processes. Behavioral theories' crucial role in current sustainability policies, for instance, is a strong challenge of the view that social science is restricted to academic ivory towers. As I have argued, field theory enables us to see the common 'playground' of distinct actors. Thus, sustainability research also requires analyses of the ecological field's current state, of its actors, the dominant narratives, and the power relationships.

Acknowledgements I am greatly indebted to Dorothee Quade, who patiently provided helpful comments on earlier drafts.

References

Bourdieu, P. (1990a/1980). *The logic of practice*. Stanford: Stanford University Press.

Bourdieu, P. (1990b). *In other words. Essays towards a reflexive sociology*. Stanford: Stanford University Press.

Bourdieu, P. (1990c). The scholastic point of view. *Cultural Anthropology, 5*(4), 380–391.

Bourdieu, P. (1993). *Sociology in question*. London: Sage.

Bourdieu, P. (1995). *Outline of a theory of practice*. Cambridge: Cambridge University Press.

Bourdieu, P. (1998). *Practical reason. On the theory of action*. Cambridge: Polity Press.

Bourdieu, P., & Wacquant, L. J. D. (1992). The purpose of reflexive sociology (The Chicago workshop). In P. Bourdieu & L. J. D. Wacquant (Eds.), *An invitation to reflexive sociology* (pp. 61–215). Chicago: The University of Chicago Press.

Brand, U. (2012a). Green economy and green capitalism: Some theoretical considerations. *Journal für Entwicklungspolitik, 28*(3), 118–137.

Brand, U. (2012b). Green economy – The next oxymoron? No lessons learned from failures of implementing sustainable development. *GAIA, 21*(1), 28–32.

Brand, U., & Wissen, M. (2012). Global environmental politics and the imperial mode of living: Articulations of state-capital relations in the multiple crisis. *Globalizations, 9*(4), 547–560.

Carlowitz, H. C. von (2000). *Sylvicultura oeconomica*. Leipzig 1713. Reprint. Freiberg: TU Bergakademie Freiberg.

Dworkin, G. (2014). Paternalism. In Zalta, E. N. (Ed.). *The Stanford Encyclopedia of Philosophy* (Summer 2014 Edition). http://plato.stanford.edu/archives/sum2014/entries/paternalism/. Accessed 28 July 2014.

Etzioni, A. (2004). The post affluent society. *Review of Social Economics, 62*(3), 407–420.

Foster, J. B. (2000). *Marx's ecology. Materialism and nature*. New York: Monthly Review Press.

Friedman, T. A. (2007). The power of green. *New York Times Magazine, 15*(2007), 40–72.

Giddens, A. (1979). *Central problems in social theory. Action, structure and contradiction in social analysis*. Berkeley: University of California Press.

Giddens, A. (2012). *The constitution of society. Outline of the theory of structuration*. Cambridge: Polity Press.

Goleman, D. (2010). *Ecological intelligence. The hidden impacts of what we buy*. New York: Broadway Books.

Gram-Hanssen, K. (2011). Understanding change and continuity in residential energy consumption. *Journal of Consumer Culture, 11*(1), 61–78.

Grober, U. (2012). *Sustainability. A cultural history*. Cambridge: Green Books.
Haluza-DeLay, R. (2008). A theory of practice for social movements: Environmentalism and ecological habitus. *Mobilization: The International Quarterly, 13*(2), 205–218.
Henderson, D. R. (2014). Libertarian paternalism: Leviathan in sheep's clothing. *Society, 51*(3), 268–273.
Illouz, E. (2007). *Cold intimacies. The making of emotional capitalism*. Cambridge: Polity Press.
Kasper, D. V. S. (2009). Ecological habitus: Toward a better understanding of socioecological relations. *Organization & Environment, 22*(3), 311–326.
Lahire, B. (2011). *The plural actor*. Cambridge: Polity Press.
Leggewie, C. & Welzer, H. (2009). *Das Ende der Welt, wie wir sie kannten. Klima, Zukunft und die Chancen der Demokratie* [The end of the world as we know it. Climate, future, and democracy's prospects]. Frankfurt a. M.: S. Fischer.
Peters, G. (2011). The social as heaven and hell: Pierre Bourdieu's philosophical anthropology. *Journal for the Theory of Social Behaviour, 42*(1), 63–86.
Pykett, J. (2011). The new maternal state: The gendered politics of governing through behaviour change. *Antipode, 44*(1), 217–238.
Reckwitz, A. (2002). Toward a theory of social practices. A development in culturalist theorizing. *European Journal of Social Theory, 5*(2), 243–263.
Røpke, I. (2009). Theories of practice. New inspiration for ecological economic studies on consumption. *Ecological Economics, 68*(10), 2490–2497.
Rosa, H. (2013). *Social acceleration. A new theory of modernity*. New York: Columbia University Press.
Sahakian, M., & Wilhite, H. (2014). Making practice theory practicable: Towards more sustainable forms of consumption. *Journal of Consumer Culture, 14*(1), 25–44.
Schatzki, T. (1996). *Social practices. A Wittgensteinian approach to human activity*. Cambridge: Cambridge University Press.
Schatzki, T. (2001). Introduction: Practice theory. In T. Schatzki, K. Knorr Cetina, & E. von Savigny (Eds.), *The practice turn in contemporary theory* (pp. 1–14). London: Routledge.
Schatzki, T., Knorr Cetina, K., & von Savigny, E. (Eds.). (2001). *The practice turn in contemporary theory*. London: Routledge.
Scott, K., Bakker, C., & Quist, J. (2012). Designing change by living change. *Design Studies, 33*(3), 279–297.
Shove, E. (2010). Beyond the ABC: Climate change policy and theories of social change. *Environment and Planning A, 42*(6), 1273–1285.
Shove, E., & Spurling, N. (Eds.). (2013). *Sustainable practices. Social theory and climate change*. London: Routledge.
Shove, E., & Walker, G. (2010). Governing transitions in the sustainability of everyday life. *Research Policy, 39*(4), 471–476.
Shove, E., Pantzar, M., & Watson, M. (2012). *The dynamics of social practice. Everyday life and how it changes*. London: Sage.
Smith, M. (2001). *An ethics of place. Radical ecology, postmodernity, and social theory*. Albany: State University of New York Press.
Spaargaren, G. (2011). Theories of practices: Agency, technology, and culture. Exploring the relevance of practice theories for the governance of sustainable consumption practices in the new world-order. *Global Environmental Change – Human and Policy Dimensions, 21*(3), 813–822.
Steinberg, T. (2010). Can capitalism save the planet? On the origins of green liberalism. *Radical History Review, 107*, 7–24.
Sunstein, C. R., & Reisch, L. A. (2014). Automatically green: Behavioral economics and environmental protection. *Harvard Environmental Law Review, 38*(1), 127–158.
Thaler, R. H., & Sunstein, C. R. (2003). Libertarian paternalism. *The American Economic Review, 93*(2), 175–179.

Thaler, R. H., & Sunstein, C. R. (2008). *Nudge. Improving decisions about health, wealth, and happiness.* New Haven: Yale University Press.

Tienhaara, K. (2013). Varieties of green capitalism: economy and environment in the wake of the global financial crisis. *Environmental Politics, 23*(2), 187–204.

Wacquant, L. J. D. (1992). The structure and logic of Bourdieu's sociology. In P. Bourdieu & L. J. D. Wacquant (Eds.), *An invitation to reflexive sociology* (pp. 1–59). Chicago: The University of Chicago Press.

Wacquant, L. J. D. (2004). Following Pierre Bourdieu into the field. *Ethnography, 5*(4), 387–414.

WBGU (German Advisory Council on Global Change). (2011). *Flagship report: World in transition. A social contract for sustainability.* http://www.wbgu.de/fileadmin/templates/dateien/veroeffentlichungen/hauptgutachten/jg2011/wbgu_jg2011_en.pdf. Accessed 28 July 2014.

Wilhite, H. (2014). Sustainability as social practice. New perspectives on the theory and policies of reducing energy consumption. In S. Lockie, D. A. Sonnenfeld, & D. R. Fisher (Eds.), *Routledge international handbook of social and environmental change* (pp. 133–141). London: Routledge.

Wilkinson, T. M. (2013). Nudging and manipulation. *Political Studies, 61*(2), 341–355.

Wittmayer, C., Schulz, S., & Mittelstaedt, R. (1994). A cross-cultural look at the 'supposed to have it' phenomenon: The existence of a standard package based on occupation. *Advances in Consumer Research, 21*, 427–434.

Woehrle, L. M. (2010). Environmental/green cultural shifts: Dynamics of social change. *Sociology Compass, 4*(11), 936–946.

Part II
Knowledge Production and Action

From Knowledge Co-production to Transdisciplinary Research: Lessons from the Quest to Produce Socially Robust Knowledge

Juergen Weichselgartner and Bernhard Truffer

1 Introduction

Almost 50 years ago, Wood (1964) argued that scientists could be part of an "apolitical elite" triumphing in the political arena, since "effective policy-making required fast and swiftly what scientists believed they had to offer: objective shifting of the facts, balanced visions, thoughtful reflection and the mobilization of the best wisdom and highest competence" (Wood 1964, 64). Since then, interest in the validity of scientific claims has expanded to wider audiences, and public trust in experts' problem-solving competence has eroded. Policy and society have increasingly confronted science with uncomfortable questions regarding the soundness of its methods and the validity of its knowledge products.

While science has undergone many changes in its subjects, methods, and social functions over the centuries, today's science is characterized by complex internal dynamics, capabilities to absorb external goals, and generate reality-changing, or even reality-producing, knowledge. Increasingly, scientists have addressed (science-)*external* societal problems, studied them (science-)*internally*, and transferred the results to specific societal domains in order to contribute to (science-)*external* societal problem-solving. This thesis was presented in the so called "finalization of science" advocated by Böhme et al. (1973). Following Kuhn's theory of scientific paradigms (1962), but shifting the focus from the application of science's results to

J. Weichselgartner (✉)
Department of Architecture and Urban Studies, Politecnico di Milano, Milan, Italy

Department of Geography, Ludwig-Maximilians University Munich, Munich, Germany
e-mail: juergen.weichselgartner@gmail.com

B. Truffer
Swiss Federal Institute of Aquatic Science and Technology, Dübendorf, Switzerland

Faculty of Geosciences, University of Utrecht, Utrecht, The Netherlands
e-mail: bernhard.truffer@eawag.ch

B. Werlen (ed.), *Global Sustainability*, DOI 10.1007/978-3-319-16477-9_5

the application of its modes of operation, the finalization thesis identifies additional aspects in relation to the social contexts of science (Böhme et al. 1976, 1983).

Based on this broader view of science, the transdisciplinarity paradigm has evolved in environmental research due to a revised positioning of science in society. Under different labels, emerging concepts propagated a shift from academic, investigator-initiated, discipline-based research to a more context-driven, problem-focused, and multidisciplinary one. This different knowledge mode is produced in the context of application, thus improving research's applicability to societal problem-solving. Various scientists also argue that users of societal knowledge should have a say in the governance of science. Involving nonscientists in knowledge production process compels scientists to justify the societal benefits of their research to funders and the public.

The following section provides an overview of the different understandings and divergent styles of transdisciplinary research. Section 3 outlines transdisciplinarity's multifaceted dimensions and pathways. First, we present some of the essential conceptions and approaches, followed by a number of challenges in transdisciplinary research. In Sect. 4, some practical examples – past, recent, national, and international – are introduced to illustrate the emergence of approaches and programs in response to new intellectual and pragmatic developments. Section 5 concludes that transdisciplinarity still appears to be a domain of more or less different but complementary activity fields that transcend the standard competence domains of present-day academic institutions.

2 Rearrangements in the Science-Society Interface

During the past decades, a variety of approaches have emerged to understand and explain transformation processes in the science-society interface. Under different labels, these concepts have propagated a shift from academic, investigator-initiated, discipline-based research to more context-driven, problem-focused, and multidisciplinary research. These approaches provide the core motivational background for specific transdisciplinary initiatives. A brief sampling of the key accounts of changes in scientific practice includes: "the fifth branch" (Jasanoff 1990), "postnormal science" (Funtowicz and Ravetz 1993), "mode 2" (Gibbons et al. 1994), "postacademic science" (Ziman 1996), and "problem-oriented research" (Bechmann and Frederichs 1996). It should be noted that there are more alternative approaches than those mentioned in this section (e.g., Latour 1987: "science in action," Salter 1988: "mandated science," Etzkowitz and Leydesdorff 2000: "triple helix," and Rip 2004: "strategic research").[1] Furthermore, Weingart (1997, 2008), Etzkowitz and Leydesdorff (2000), Zierhofer and

[1] The order of appearance is chronological and does not reflect any judgment about their quality or importance.

Burger (2007), Hessels and van Lente (2008), and Hirsch Hadorn et al. (2008), for instance, provide valuable reflections on the new modes of knowledge production, particularly on "mode 2".

With modern society's expansion and complexity, another branch has accompanied the three traditional branches of government (executive, legislative, and judicial). Owing to its power to lead public opinion, the press, public interest groups, and independent administrative agencies usually refer to it as the fourth power. Likewise, the roles of experts as mediators of knowledge grew rapidly. As scientists expanded their role as consultants and advisers concerning policy and society, the scientific advisory process became so pervasive and influential that Jasanoff (1990) has called it "the fifth branch." By merging legal and institutional analysis with social studies of science, she challenges the conventional assumptions that science guarantees truth and that science experts' advice is unbiased, fail-safe, and thus indispensible to policy-making: How can scientists ensure the objectivity of science and "how can they maintain their authority as neutral experts, especially when challenged in the media or the courts?" (Jasanoff 1990, 9). Jasanoff articulates a social construction model that outlines the problems of objectivity and neutrality in science policy and explains how science has become increasingly vulnerable to the other actors in the network.

Another prescriptive approach – postnormal science, which originated from ecological economics – focuses on aspects of problem-solving often neglected in traditional accounts of scientific practice: uncertainty, value-loading, and the plurality of legitimate perspectives (Funtowicz and Ravetz 1991, 1993). In these sorts of issue-driven science, particularly with regard to complex systems, "typically facts are uncertain, values in dispute, stakes high, and decisions urgent" (Funtowicz and Ravetz 1993, 744). In such contexts, "normal science" in the sense of Kuhn (1962) is not an adequate knowledge production mode, since the traditional scientific mindset fosters expectations of regularity, simplicity, and certainty. Where uncertainties are crucial, value-loadings critical, and less information available than decision-makers in policy and practice desire, there can be no single privileged perspective for measurement, analysis, and evaluation. Thus, invoking "truth" as the goal of science is a distraction from real tasks. It is argued that quality is a more relevant and robust guiding principle – understood as a contextual property of scientific information. By unifying "facts" and "values" in a concept of problem-solving in these areas, and by replacing "truth" with "quality" as its core evaluative concept, postnormal scientific practice provides a framework for extended participation in decision-making, a path to the democratization of science, and a response to the current tendencies towards postmodernity.

Postnormal science shares a number of characteristics with another prominent concept: mode 2 knowledge production introduced by Gibbons et al. (1994). The underlying hypothesis is that "the old paradigm of scientific discovery (mode 1) – characterized by the hegemony of theoretical or, at any rate, experimental science; by an internally-driven taxonomy of disciplines; and by the autonomy of scientists and their host institutions, the universities – was being superseded by a new paradigm of knowledge production (mode 2), which was socially distributed,

Table 1 Attributes of knowledge production mode 1 and mode 2

Mode 1	Mode 2
Academic context	Context of application
Disciplinary	Transdisciplinary
Homogeneity	Heterogeneity
Autonomy	Reflexivity/social accountability
Traditional quality	Novel quality control

Hessels and van Lente (2008, 741)

application-oriented, trans-disciplinary, and subject to multiple accountabilities" (Nowotny et al. 2003, 179). Additionally, it was claimed that this new mode would lead to "socially more robust" knowledge, because it relied on a variety of knowledge sources represented by the different actors and their life contexts when compared to the rather narrow delimitations of monodisciplinary academic engagements with the world (Table 1). Despite criticism of this new knowledge production mode's empirical validity, conceptual strength, and political value (among others), the concept has become extremely visible and has influenced scientific practice. Seven years after mode 2's introduction, three of the authors published a second book (Nowotny et al. 2001), partly to answer the valid criticisms of the concept and partly to develop a broader thesis (see Nowotny et al. 2003; Hessels and van Lente 2008).

Starting with specialization and change in scientific careers, Ziman (1987, 1996, 2000) has incorporated elements from other diagnoses of knowledge production to describe and explain a set of developments in postacademic science, which has transformed the way science is organized, managed, and performed. According to the British physicist, science now fits neither the academic model – i.e. focuses on basic questions, supported by grants, driven by both curiosity and opportunity, disseminated widely through peer-reviewed publication – nor the industrial model – i.e. focuses on application, underwritten by real or anticipated profits, driven by a business agenda, closely held, and restricted. The emerging postacademic standard's primary characteristics are the facilitation of interdisciplinary inquiry, privatization of academic knowledge, openness to public scrutiny, and an increase in specialized knowledge production institutions.

Contrary to basic and applied research, Bechmann and Frederichs (1996) consider problem-oriented research, a new type of research between science, public policy, and society. While applied research is oriented towards the specification and application of available scientific knowledge for specified purposes, problem-oriented research deals with uncertainties related to prognosis, complexity, and contingency in order to legitimize decisions with no certain foundations but that typically cannot wait until basic science questions have been solved. Also, contrary to basic research, it is less interested in gaining new general scientific findings than in the utilization of knowledge for practical societal problems not structured according to academic disciplines. This implies not only inherent scientific, but also ethic and normative, uncertainties, discourses about interpretation and values, and the need for interdisciplinarity or transdisciplinarity.

3 Transdisciplinarity: A New Course of Action

3.1 Origin and Definitions

The call for a science system more explicitly focused on problem-oriented interdisciplinarity or transdisciplinarity can be traced back to the 1950s when the first statements about the urgency of socio-ecological problems were expressed in public (Miller et al. 2008). In the 1960s, the plea to democratize science and technology gained prominence, in particular after the publication of Carson's book "Silent Spring" in 1962. However, as a new means to organize scientific practice, transdisciplinarity was only formulated in the early 1990s. The emerging debate on global environmental problems and the need for sustainable development called for new forms of scientific activities that would provide better problem resolutions and would guide societal reform in the long run. Funtowicz and Ravetz (1991, 1992, 1993) were among the first to demand transdisciplinary knowledge production in environmental research. In a related field, namely technology assessment, scholars argued in favor of a nonlinear, pluralist, and reflexive approach to technology evaluation and innovation management. A prominent approach was developed and implemented in the Netherlands under the label "constructive technology assessment" (Schot and Rip 1997). These and other developments were synthesized by Gibbons et al. (1994) in their well-known call for new knowledge production modes, identifying transdisciplinary knowledge production as a potential new paradigm.

These fundamental writings share a number of starting points from which the necessity for a transdisciplinary approach to knowledge production was derived. First, they agree regarding their refutation of the "linear" knowledge production model, which postulates a very clear-cut role for academic research, namely the production of new, codified, and universally true knowledge (and technological principles). In the linear view, academia's best contribution to problem resolution lies in the adequate transfer (via communication, education, patenting, or publication) to other actors charged with implementing such knowledge in the form of products, procedures, regulations, entertainments, therapies, or problem solutions. On the other hand, a nonlinear understanding of knowledge production assumes that relevant knowledge can be produced by any kind of actor (academic or lay), who should be acknowledged for his or her specific perspective on a given problem. Different actors can therefore be at the origin of new ideas, and knowledge flows can go in all directions – from users to basic research, or from engineers to basic science, and so on (for an early critique of the linear model in innovation management, see Kline and Rosenberg 1986).

A second point that all the authors share is that the current science system is not well prepared to interact in heterogeneous actor networks of knowledge production. New capabilities would be required to interact with actors from different societal realms (economy, politics, media, or civil society). This relates not only to different "languages", but extends to criteria of relevance and success, the time structure and reliability of the research outputs, and the ways of handling value claims (Truffer 2007). As a result, transdisciplinarity would require academics to develop new sets of skills, such as public relations, marketing, negotiation techniques, project management, and political bargaining. Thirdly, due to the increasing degree of specialization and compartmentalization in academia, research projects that seek to address complex societal problems should synthesize evidence from more than one discipline. Therefore, interdisciplinary integration is a core task of transdisciplinary knowledge production.

Based on this broad requirement set, different specific proposals have been formulated over the past 20 years to define and finally institutionalize transdisciplinarity in academic contexts (e.g., Bunders et al. 2010; Jahn et al. 2012). Nowadays, transdisciplinarity is largely seen as a research approach or perspective (Jahn et al. 2012) and less as a theory (or even metatheory, for instance, as authors such as Max-Neef (2005) or Costanza et al. (1991) advocated it). In this view, transdisciplinarity still builds on academic disciplines and cannot ultimately supplant the conventional system of disciplines (Gibbons et al. 1994; Maasen et al. 2006).

Transdisciplinarity can therefore be defined as fulfilling many or all of the following characteristics (based on Jahn et al. 2012): address complex societal problems, encompass co-operation between disciplines and non-academic knowledge holders (particularly those who are relatively powerless, see Bunders et al. 2010), place emphasis on mutual learning between different societal actors (Scholz 2001), and enable the integration of different kinds of knowledge. Transdisciplinary research has mostly been applied to significant societal problems such as sustainability or public health problems. In more procedural terms, transdisciplinarity can be described as consisting of specific research tasks specifying how problems are identified, framed, and structured (Pohl and Hirsch Hadorn 2007). As a result, transdisciplinary knowledge production was defined as contributing particularly to the interrelated analysis of target knowledge (what type of goals should be defined?), system knowledge (how do societal, natural, and technical systems work?), or transformation knowledge (how may these systems be influenced?).

3.2 Challenges of Transdisciplinary Research in Practice

While the claim of the inevitability of a transdisciplinary reform of academia has been loud and recurrent, transdisciplinarity is still far from being a well-established and highly institutionalized activity domain (Pohl 2011); furthermore, there is no widely accepted definition (Jahn et al. 2012). Over the past 20 years, many industrialized countries have witnessed the generation, formulation, and implementation of an impressive number of publications, methodological research projects, funding structures, institutional strategy papers, and research programs related to transdisciplinarity. Nevertheless, the ensuing activities have often been short-lived, and only a few universities and research institutes can look back on a continuous history of successful transdisciplinary research activities. Some scholars explain this limited success as due to a fundamental mismatch of expectations between academic and other societal systems. They see the rise of transdisciplinary activities as a mere (mostly rhetoric) reaction to the current neoliberal pressure on the science system for more responsiveness, accountability, and control by other societal domains (Maasen and Lieven 2006).

At a more operational level, the lack of success at institutionalization has been attributed to transdisciplinarity still lacking widely accepted quality criteria and adequate institutional structures (Pohl 2011; Pohl et al. 2010; Bunders et al. 2010). In line with Sect. 2, transdisciplinarity has to fulfill formal excellence criteria in its specific disciplinary components. It also has to measure up to expectations and relevance criteria from other societal actors (extended peer review). This leads to the development of procedural approaches to assess research teams and processes' quality where substantial quality criteria are hard to come by. Pohl (2011) describes success in transdisciplinarity as a matter of generating novelty in the context of different lines of thought represented in a specific transdisciplinary project setup, whether academic or from other societal realms.

Clearly, it is difficult to combine understandings from multiple knowledge sources in order to produce more timely and context-appropriate solutions. Providing appropriate mechanisms to link solutions proposed by research to the articulated needs of policy-makers and practitioners is even more challenging. Studies on "assessing the assessment process" (e.g., Cash and Moser 2000; Cash and Clark 2001; Cash et al. 2003; Mitchell et al. 2006) conclude that especially saliency, credibility, and legitimacy are critical attributes of an assessment around which audiences make judgments, and which determine whether they will change their thoughts, decisions, and behavior. Research efforts are therefore more likely to be influential when they manage boundaries between knowledge and action in ways that simultaneously enhance the scientific credibility, political saliency, and practical relevance of the information they produce. According to Jasanoff (2003, 235), a major problem is "how to institutionalize polycentric, interactive, and multipartite processes of knowledge-making within institutions that have worked for decades at keeping expert knowledge away from the vagaries of populism and politics. The question confronting the governance of science is how to bring knowledgeable publics into the front-end of scientific and technological production – a place from which they have historically been strictly excluded."

Winowiecki et al. (2011) point to the key fact that, before any substantive collaborative research can take place, an understanding of the ontologies, epistemologies, methodologies, and the way that they serve as the "building blocks" of research, as well as tools to locate and overcome existing barriers, is needed. Given disciplinary specialization, the authors also argue that interdisciplinary projects can harness specialized knowledge, using core disciplines' rigorous depth and skills as an essential foundation to bridge them. By means of two multidisciplinary, problem-oriented, and policy-oriented social science research projects, Conrad (2002) outlines the limitations of interdisciplinary research concepts. He emphasizes that sufficient time should be provided to acquire and utilize a shared conceptual (theoretical) framework in multidisciplinary, problem-oriented research.

Emerging literature specifically points to barriers at the science-policy interface that inhibit efficient co-production of knowledge, hinder transfer into action, and aggravate the implementation of policy measures that are win-win situations for different societal actors. Examples of such barriers include divergent objectives, needs, scopes, and priorities; different institutional settings and standards; differing cultural values, understanding, and mistrust (Cash and Clark 2001; Kasperson and Berberian 2011; Lövbrand 2011; Naustdalslid 2011; Weichselgartner 2013). These factors have been grouped by Weichselgartner and Kasperson (2010, 273–274) into functional, structural, and social barriers. They contribute to and result in failures that typically occur when knowledge is transferred through the traditional pipeline mode in which scientists set the research agenda, do the research, and then transfer the results to potential users, assuming that they diffuse automatically through the policy and practice communities (Fig. 1). Dilling and Lemos (2011) indicate that, rather than being a straightforward outcome of the call for the production of usable science, an iterative process results from the actions of specific actors and organizations who "own" the task of building the conditions and mechanisms that help create it. According to Zierhofer and Burger (2007, 68), deliberate participation serves many objectives other than knowledge production and may subject stakeholder negotiations to scientific standards of rationality, thereby affecting the result of the process.

Jasanoff (2004, 2010) and Hulme (2010), among others, outline how the contemporary knowledge production mode of streamlining knowledge into one global signature hides far more than it discloses. Particularly geographical differentiation, cultural heterogeneity, and social plurality are relevant with regard to local practices and knowledge-making traditions. Produced in a specific science policy setting with particular institutional arrangements, global knowledge types yield global types of meaning-making and policy-making. Scientists have provided various criticisms of decontextualized, top-down planetary knowledge: It marginalizes place-based knowledge, detaches knowledge-making from meaning-making, restricts the framing and goals of scientific enquiry, offers a one-dimensional guide to policy, and may have considerably less purchase in problem-solving and policy-making than a multiplicity of diverse knowledge producers and tools (Weichselgartner and Marandino 2012).

Although it is only a limited sampling, these highlighted difficulties demonstrate that co-designing and co-producing knowledge face various challenges related to ontologies, epistemologies, methodologies, credibility, saliency, and relevancy. There is no golden bullet to achieve transdisciplinarity. Disentangling transdisciplinary research methodologically into specific types of research objectives and research instruments is essential in order to assess the diverse relationships between particular sorts of knowledge claims and research format types. Designing and operating transdisciplinary knowledge systems definitely require additional temporal, spatial, financial, and human resources.

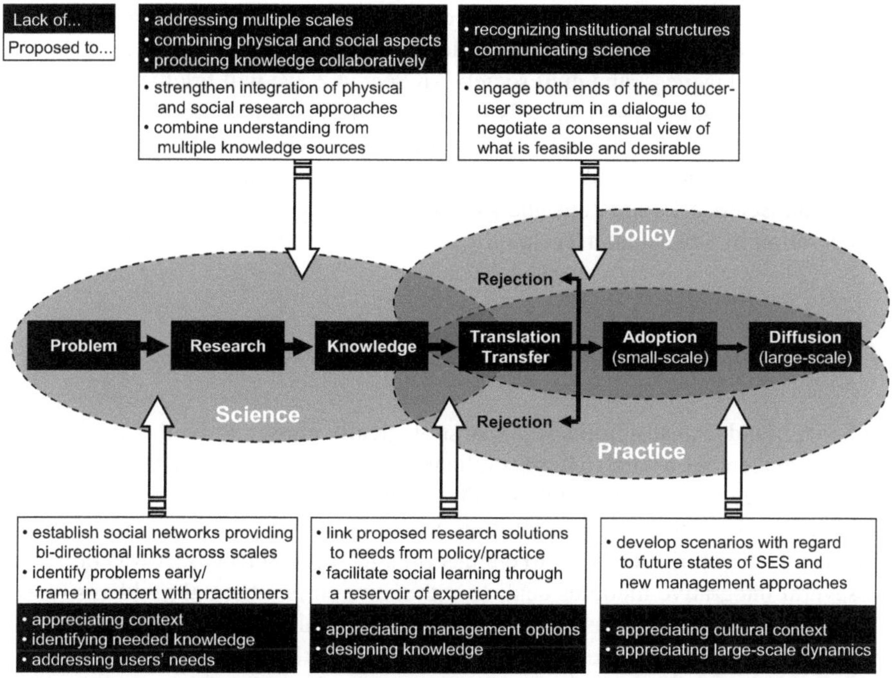

Fig. 1 Enhancing the effectiveness of research-based knowledge: Pitfalls and propositions. Pitfalls (in *black boxes*) typically occur when knowledge is designed in "academic isolation" and transferred through the traditional pipeline mode; proposed ways (in *white boxes*) to enhance effectiveness of research-based knowledge would result in more socially robust and context-sensitive knowledge (After Weichselgartner and Marandino (2012, 327)

4 Transdisciplinarity: A Guide for Science-Policy Interactions

Science-policy circles readily took up transdisciplinarity as a framework to organize research programs in different countries. However, it was not implemented as a precise concept, but as an umbrella term for a wide range of different activities (Pohl 2011). Discourses and initiatives were especially vivid in the German-speaking countries: Germany, Austria, and Switzerland (D-A-CH).[2] In Sect. 4.1, we briefly reconstruct the different interpretations and institutionalizations of transdisciplinarity in these countries. The concept has played a prominent role in science-policy circles in the D-A-CH countries for approximately 20 years. A closer look at the developments may therefore provide insights into specific opportunities and challenges that transdisciplinarity may involve. Likewise, two recent

[2] Germany, Austria and Switzerland are also known as the D-A-CH countries.

international-level activities exemplify current efforts to connect research activities, seek new resources for key questions, and produce solutions together with governments, business, and civil society. In Sect. 4.2, we will briefly introduce the International Council for Science's Earth System Visioning Process for the next decade of earth system research, and the Responses to Environmental and Societal Challenges for our Unstable Earth, a joint Frontiers of Science initiative of the European Science Foundation and the intergovernmental initiative for European Cooperation in Science and Technology.

4.1 National Initiatives: The Case of the D-A-CH Countries

In Switzerland, transdisciplinarity was strategically chosen to be the leading integrative concept of a large interdisciplinary research program on global environmental change that ran from 1992 to 2002. The program was called SPP-U (priority program environment), a multimillion Swiss Franc program covering research projects from natural sciences, to engineering, to social sciences. It was organized in several integrative modules dealing with major environmental problems (e.g., a larger interdisciplinary consortium dealing with global climate change) (see Cebon et al. 1998). The program leaders gave transdisciplinarity center stage as a means to integrate findings from different disciplines into their interaction with policy-makers and other societal actors. They believed that they could conjointly inform the emerging environmental policy debate in Switzerland, which had increased strongly after the Chernobyl disaster (Häberli et al. 2002; Balsiger 2005).

During this period, transdisciplinary practice and discourse expanded strongly. A large number of experiments, integrative concepts, and institutional strategies were developed that referred explicitly to transdisciplinarity as a new knowledge production mode. However, for the program leaders, it remained a challenge to bundle all these efforts into an overarching conceptual framework. Transdisciplinarity was for a long time perceived as a potentially promising but ill-structured list of very disparate activities. In response, the promoters and managers of SPP-U organized a large international conference in 2000 during which the Swiss transdisciplinarity award was first awarded. Exemplary projects were awarded to improve their academic credibility and visibility in the Swiss academic community. However, all these activities proved insufficient to establish transdisciplinarity as a "normal" competence domain in academia. Consequently, an expert platform, transdisciplinarity-net (later abbreviated as td-net), was established in 1999 (Hirsch Hadorn 2002) to store, further develop, and canonize the state-of-the-art in transdisciplinary research in Switzerland.

The td-net platform was first launched as an initiative of a Swiss association of concerned scientists, the Swiss Academic Society for Environmental Research and Ecology (SAGUF), but was transferred shortly after to the Swiss Academies of Sciences, where it received the status of an integration platform connecting the different academies (Hirsch Hadorn 2002). A small team of researchers located at the central office has subsequently conducted most of the research and coordinated the networking activities. They have published a series of methodology books and guidelines, have organized international conferences, have coordinated with activities emerging in other countries (mostly Austria and Germany), and have extended transdisciplinarity's application to new fields, such as public health and technology assessment. More recently, activities outside the original D-A-CH countries have gained importance, and td-net became a central player in the emerging global network of scholars in transdisciplinarity. td-net is supervised by a broadly based academic advisory board and is currently co-funded by the Swiss Academies of Sciences and a private foundation (Stiftung Mercator, Schweiz). It continues to issue the transdisciplinarity award on a biannual basis and has developed a broad basis of exemplary projects.

Through these activities, transdisciplinarity has become an established field of methodological research and practice: It has put transdisciplinarity on the agenda of major funding agencies, such as the Swiss National Science Foundation, has supported the introduction of core aspects of transdisciplinarity in the curricula of all major Swiss universities, and has enabled the exchange of experiences and methods in a wide variety of research contexts. In parallel, a number of university chairs with transdisciplinary foci were installed in the 1990s, and several national research institutes endorsed transdisciplinarity as a core element of their research strategies. Despite these very positive developments and the systematic expansion of the competence base in Switzerland, it is fair to say that transdisciplinarity is still largely perceived as a sideline activity of academic research. For many, it is just a modern, and somewhat convoluted, cover for universities' public relations and outreach obligations.

The development of transdisciplinary programs in Switzerland did not happen in isolation. Strong interactions with science policy actors in Germany and Austria also took place (Defila et al. 2006). Austria developed a similar priority program on landscape ecology between 1995 and 2003 in which transdisciplinarity served as a core integrative concept. In Germany, the sustainability flagship research program, called socio-ecological research, of the German Ministry of Education and Research (BMBF) was installed in 1999 and runs to this day with an annual budget of approximately EUR 10 million. More recently, transdisciplinarity has served as the strategic profile of an entire university (Leuphana-Lueneburg) and also gave rise to new inter-university consortia (Schneidewind and Augenstein 2012). One of the most important resources for discussing and presenting research on transdisciplinarity is the journal *GAIA*, which was launched in 1992 by Swiss and German universities (Winiwarter 2012). *GAIA* originally had a strong focus on German-speaking environmental research, but has increasingly established itself as an international journal and was recently ISI-listed.[3]

[3] The Institute of Science Index (ISI) offers bibliographic database services, e.g., maintenance of citation databases and citation indexing (for more information see, e.g., http://isi-thomsomreuters.net/).

4.2 International Initiatives: ICSU and RESCUE

At a global level, science has become more active in elaborating programs and activities to identify and address societal priorities. Among the many initiatives, the International Council for Science (ICSU) recently undertook two interesting exercises: a visioning process (ICSU 2010) and a foresight analysis (ICSU 2011). In light of the need for an overarching set of solution-focused and integrated research priorities, the ICSU and the International Social Science Council (ISSC) led a visioning process to rethink the focus and framework of earth system research. The consultative process resulted in five identified grand challenges for earth system science, a consensus list of the highest priorities, which would remove critical barriers to progress towards sustainable development (Reid et al. 2010). The challenges meet four criteria: (1) scientific importance, (2) the need for global coordination, (3) relevance for decision-makers, and (4) leverage. Several important research questions were identified for each challenge (Table 2) that have to be answered within the new 10-year initiative entitled Future Earth: Research for Global Sustainability.

Table 2 ICSU grand challenges and important research questions

Challenge 1: Forecasting – improve the usefulness of forecasts of future environmental conditions and their consequences for people 1.1 What significant environmental changes are likely to result from human actions? How would those changes affect human well-being, and how are people likely to respond? 1.2 What threats do global environmental changes pose for vulnerable communities and groups and what responses could be most effective in reducing harm to those communities?
Challenge 2: Observing – develop, enhance, and integrate the observation systems needed to manage global and regional environmental change 2.1 What do we need to observe in coupled social-environmental systems, and at what scales, in order to respond to, adapt to, and influence global change? 2.2 What are the characteristics of an adequate system for observing and communicating this information?
Challenge 3: Confining – determine how to anticipate, recognize, avoid and manage disruptive global environmental change 3.1 Which aspects of the coupled social-environmental system pose significant risks of positive feedback with harmful consequences? 3.2 How can we identify, analyze and track our proximity to thresholds and discontinuities in coupled social-environmental systems? When can thresholds not be determined? 3.3 What strategies for avoidance, adaptation and transformation are effective for coping with abrupt changes, including massive cascading environmental shocks? 3.4 How can improved scientific knowledge of the risks of global change and options for response most effectively catalyze and support appropriate actions by citizens and decision-makers?
Challenge 4: Responding – determine what institutional, economic and behavioral changes can enable effective steps toward global sustainability 4.1 What institutions and organizational structures are effective in balancing the trade-offs inherent in social-environmental systems at and across local, regional and global scales and how can they be achieved? 4.2 What changes in economic systems would contribute most to improving global sustainability, in the context of global environmental change, and how could they be achieved?

(continued)

Table 2 (continued)

4.3 What changes in behavior or lifestyle, if adopted by multiple societies, would contribute most to improving global sustainability, in the context of global environmental change, and how could they be achieved? 4.4 How can institutional arrangements prioritize and mobilize resources to alleviate poverty, address social injustice and meet development needs under rapidly changing and diverse local environmental conditions and growing pressures on the global environment? 4.5 How can the need to curb global environmental change be integrated with the demands of other inter-connected global policy challenges, particularly those related to poverty, conflict, justice and human security? 4.6 How can effective, legitimate, accountable and just, collective environmental solutions be mobilized at multiple scales? What is needed to catalyze the adoption of appropriate institutional, economic or behavioral changes?
Challenge 5: Innovating – encourage innovation (coupled with sound evaluation mechanisms) in developing technological, policy, and social responses to achieve global sustainability 5.1 What incentives are needed to strengthen systems for technology, policy and institutional innovation to respond to global environmental change and what good models exist? 5.2 How can pressing needs for innovation and evaluation be met in the following key sectors? (a) How can global energy security be provided entirely by sources that are renewable and that have neutral impacts on other aspects of global sustainability, and in what timeframe? (b) How can competing demands for scarce land and water be met over the next half century while dramatically reducing land-use greenhouse gas emissions, protecting biodiversity, and maintaining or enhancing other ecosystem services? (c) How can ecosystem services meet the needs for improving the lives of the world's poorest peoples and those of developing regions (such as safe drinking water and waste disposal, food security, and increased energy use) within a framework of global sustainability? (d) What changes in communication patterns are needed to increase feedback and learning processes to increase the capacity of citizens and officials, as well as to provide rapid and effective feedback to scientists regarding the applicability and reliability of broad findings and theoretical insights to what is observed in the field? (e) What are the potentials and risks of geo-engineering strategies to address climate change, and what local to global institutional arrangements would be needed to oversee them, if implemented?

After ICSU (2010, 10–16)

In addition to visioning, a foresight process – chaired by physicist Marks – was initiated to explore: (1) how science as a whole might develop over the next two decades in a changing economic, social, political, and environmental context, and (2) how international science collaboration can be supported to help science progress and benefit society. On the basis of predictable megatrends and uncertain key drivers of global science, a scenario approach was used to define four world-views differing in context (interest of states as global vs. national) and with different science positions (engaged vs. detached). The resulting four scenarios

Table 3 RESCUE theme foci and recommendations

Themes

1. Social-human: Contributions from social sciences and humanities in developing responses to challenges of the Anthropocene
2. Collaboration: Collaboration between the natural, social, and human sciences in global environmental change studies
3. Requirements: Requirements for research methodologies and data
4. Revolution: Education and capacity-building – towards a "revolution"
5. Interface: The interface between science and policy, communication, and outreach

Recommendations

1. Build an institutional framework for an open knowledge system
2. Re-organize research so disciplines share knowledge and practices, and, from the onset, work together with each other and with stakeholders
3. Initiate long-term integrated demonstration projects
4. Develop sustainability education and learning in an innovative, open knowledge system
5. Respond to the challenges and opportunities created by the internet for an open knowledge system ready for transitions towards sustainability
6. Create a dynamic, adaptive, and integrated information and decision-support system on global change issues

After ESF/COST (2011)

offer storylines about the potential challenges and consequences for international collaboration in science, which should stimulate creative thinking on future courses of action. Both ICSU endeavors provide an overarching structure for research and clearly illustrate that the relationship between science and society has a significant impact on the future of international science.

Another remarkable scientific foresight is Responses to Environmental and Societal Challenges for our Unstable Earth (RESCUE), a joint Frontiers of Science initiative of the European Science Foundation (ESF) and the intergovernmental initiative European Cooperation in Science and Technology (COST). According to RESCUE, the traditional disciplinary scientific approach cannot address the complex global change challenges but requires an integrated knowledge base, a new set of common practices, and integrated cooperation between the research community, policy-makers, society, and – ultimately – private individuals (ESF/COST 2011). Therefore, methods are needed that deal appropriately with uncertainty and unknowns, support the use of exploratory agent-based modeling, combine participatory and modeling approaches, explore the roles of human values and behavior, and stimulate change. RESCUE has therefore proposed a shared theoretical and operational framework for research and innovation across disciplines to achieve collaboration across disciplines. In addition, it has outlined a radically interdisciplinary and transdisciplinary research environment (RITE) model as a strategy for global environmental change research. This RESCUE research initiative's specific themes and recommendations are listed in Table 3.

5 Conclusion

Global societal and environmental changes and the discursive processing of these issues in the broader public have reshuffled the context of academia and research, particularly the conditions for research, the academic disciplines, and – as a result – their structures and methods. Despite these changes and the revised position of science in society, transdisciplinarity assembles heterogeneous sets of relationships between epistemic ends and epistemic means and is, therefore, still far from being a "historical shift" in the way knowledge production is organized. However, emerging conceptual thinking and approaches have illustrated that science is not only a producer of knowledge, but also an agent of social change.

Today, we can no longer assume science policy relationships in which science finds "truth", then speaks "truth" to "power", and, finally, a "truth-based policy" that follows "research-based knowledge products". We have to continuously revise our ways of thinking about the science-society nexus with regard to knowledge production and application. Both researchers and decision-makers in policy and practice need to address the role of different socio-economic settings in which behaviors are undertaken and the ways in which these relate to underlying social practices.

We strongly recommend that future research pay more attention to both the qualities necessary for an understanding of practical knowledge that enables knowledge realization, and the abilities that make the implementation of findings possible – which Stehr (2009, 264) calls the capacity to shape contrary to knowledge as the capacity to act. This includes mechanisms that facilitate the lay public's better understanding of scientific issues, and a better understanding of the lay public's concerns with regard to scientific knowledge and its social implications. Since a transdisciplinary way of co-designing and co-producing knowledge is capable of integrating different perspectives and is robust enough to maintain identity across knowledge producers and users, the current state of transdisciplinarity is certainly not final. The need for openness, flexibility, and attention to diversity and uncertainty in knowledge will certainly inspire further democratic ways to re-organize the science-society nexus.

References

Balsiger, P. W. (2005). *Transdisziplinarität* [Transdisciplinarity]. München: Wilhelm Fink Verlag.

Bechmann, G., & Frederichs, G. (1996). Problemorientierte Forschung: Zwischen Politik und Wissenschaft [Problem-oriented research: Between policy and science]. In G. Bechmann (Ed.), *Praxisfelder der Technikfolgenforschung: Konzepte, Methoden, Optionen* (pp. 11–37). Frankfurt: Campus.

Böhme, G., van den Daele, W., & Krohn, W. (1973). Die Finalisierung der Wissenschaft [Finalization of science]. *Zeitschrift für Soziologie, 2*(2), 128–144.

Böhme, G., van den Daele, W., & Krohn, W. (1976). Finalization of science. *Social Science Information, 15*(2–3), 307–330.

Böhme, G., van den Daele, W., Krohn, W., Hohlfeld, R., & Schäfer, W. (1983). *Finalization in science*. Dordrecht: Reidel.

Bunders, J. F. G., Broerse, J. E. W., Keil, F., Pohl, C., Scholz, R. W., & Zweekhorst, M. B. M. (2010). How can transdisciplinary research contribute to knowledge democracy? In R. J. in 't Veldt (Ed.), *Knowledge democracy* (pp. 125–152). Heidelberg: Springer.

Carson, R. (1962). *Silent spring*. Boston: Houghton Mifflin.

Cash, D. W., & Clark, W. C. (2001). *From science to policy: Assessing the assessment process* (John F. Kennedy School of Government Faculty Research Working Paper RWP01-045). Cambridge: Harvard University.

Cash, D. W., & Moser, S. C. (2000). Linking global and local scales: Designing dynamic assessment and management processes. *Global Environmental Change, 10*(2000), 109–120.

Cash, D. W., Clark, W. C., Alcock, F., Dickson, N. M., Eckley, N., Guston, D. H., Jäger, J., & Mitchell, R. B. (2003). Knowledge systems for sustainable development. *Proceeding of the National Academy of Sciences, 100*(14), 8086–8091.

Cebon, P., Dahinden, U., Davies, H., Imboden, D., & Jaeger, C. (1998). *Views from the Alps: Regional perspectives on climate change*. Boston: MIT Press.

Conrad, J. (2002). Limitations to interdisciplinarity in problem oriented social science research. *The Journal of Transdisciplinary Environmental Studies, 1*. http://www.journal-tes.dk/vol%201%20no%201/jobst%20conrad.html. Accessed 12 Aug 2012.

Costanza, R., Daly, H. E., & Bartholomew, J. A. (1991). Goals, agenda and policy recommendations for ecological economics. In R. Costanza (Ed.), *Ecological economics: The science and management of sustainability* (pp. 1–20). New York: Columbia University Press.

Defila, R., Di Giulio, A., & Scheuermann, M. (2006). *Forschungsverbundmanagement: Handbuch für die Gestaltung inter- und transdisziplinärer Projekte* [Management of research networks: Handbook for the design of inter- and transdisciplinary projects]. Zürich: vdf Hochschulverlag.

Dilling, L., & Lemos, M. C. (2011). Creating usable science: Opportunities and constraints for climate knowledge use and their implications for science policy. *Global Environmental Change, 21*, 680–689.

ESF/COST (2011). *Responses to environmental and societal challenges for our unstable earth (RESCUE). ESF forward look – ESF-COST 'Frontier of Science' joint initiative*. Strasbourg: European Science Foundation/European Cooperation in Science and Technology. http://www.esf.org/fileadmin/Public_documents/Publications/rescue.pdf. Accessed 14 Oct 2012.

Etzkowitz, H., & Leydesdorff, L. (2000). The dynamics of innovation: From national systems and "mode 2" to a triple helix of university-industry-government relations. *Research Policy, 29*(2), 109–123.

Funtowicz, S. O., & Ravetz, J. R. (1991). A new scientific methodology for global environmental issues. In R. Costanza (Ed.), *Ecological economics: The science and management of sustainability* (pp. 137–152). New York: Columbia University Press.

Funtowicz, S. O., & Ravetz, J. R. (1992). Three types of risk assessment and the emergence of post-normal science. In S. Krimsky & D. Golding (Eds.), *Social theories of risk* (pp. 251–274). London: Praeger.

Funtowicz, S. O., & Ravetz, J. R. (1993). Science for the post-normal age. *Futures, 25*(7), 739–755.

Gibbons, M., Limoges, C., Nowotny, H., Schwartzman, S., Scott, P., & Trow, M. (1994). *The new production of knowledge: The dynamics of science and research in contemporary societies*. London: Sage.

Häberli, R., Gessler, R., Grossenbacher-Mansuy, W., & Lehmann-Pollheimer, D. (2002). *Vision Lebensqualität. Nachhaltige Entwicklung: Ökologisch notwendig, wirtschaftlich klug, gesellschaftlich möglich* [Vision quality of life. Sustainable development: Ecologically necessary, economically wise, socially possible]. Zürich: vdf Hochschulverlag.

Hessels, L. K., & van Lente, H. (2008). Re-thinking new knowledge production: A literature review and a research agenda. *Research Policy, 37*, 740–760.

Hirsch Hadorn, G. (2002). Vom Sagufnet zum transdisciplinarity-net. *GAIA, 11*(3), 227–231.

Hirsch Hadorn, G., Hoffmann-Riem, H., Biber-Klemm, S., Grossenbacher-Mansuy, W., Joye, D., Pohl, C., Wiesmann, U., & Zemp, E. (Eds.). (2008). *Handbook of transdisciplinary research*. Dordrecht: Springer.

Hulme, M. (2010). Problems with making and governing global kinds of knowledge. *Global Environmental Change, 20*(4), 558–564.

ICSU (2010). *Earth system science for global sustainability: The grand challenges*. Paris: International Council for Science.

ICSU (2011). *ICSU foresight analysis report 1: International science in 2031 – Exploratory scenarios*. Paris: International Council for Science.

Jahn, T., Bergmann, M., & Keil, F. (2012). Transdisciplinarity: Between mainstreaming and marginalization. *Ecological Economics, 79*, 1–10.

Jasanoff, S. (1990). *The fifth branch: Science advisers as policy makers*. Cambridge: Harvard University Press.

Jasanoff, S. (2003). Technologies of humility: Citizen participation in governing science. *Minerva, 41*(3), 223–244.

Jasanoff, S. (2004). *States of knowledge: The co-production of science and social order*. London: Routledge.

Jasanoff, S. (2010). A new climate for society. *Theory, Culture & Society, 27*(2–3), 233–253.

Kasperson, R. E., & Berberian, M. (Eds.). (2011). *Integrating science and policy: Vulnerability and resilience in global environmental change*. London: Earthscan.

Kline, S. J., & Rosenberg, N. (1986). An overview of innovation. In R. Landau & N. Rosenberg (Eds.), *The positive sum strategy: Harnessing technology for economic growth* (pp. 275–305). Washington, DC: National Academy Press.

Kuhn, T. S. (1962). *The structure of scientific revolutions*. Chicago: Chicago University Press.

Latour, B. (1987). *Science in action: How to follow scientists and engineers through society*. Cambridge: Harvard University Press.

Lövbrand, E. (2011). Co-producing European climate science and policy: A cautionary note on the making of useful knowledge. *Science and Public Policy, 38*(3), 225–236.

Maasen, S., & Lieven, O. (2006). Transdisciplinarity: A new mode of governing science? *Science and Public Policy, 33*, 399–410.

Maasen, S., Lengwiler, M., & Guggenheim, M. (2006). Practices of transdisciplinary research: Close(r) encounters of science and society. *Science and Public Policy, 33*(6), 394–398.

Max-Neef, E. (2005). Foundations of transdisciplinarity. *Ecological Economics, 53*(1), 5–16.

Miller, T. R., Baird, T. D., Littlefield, C. M., Kofinas, G., Chapin, F. S., III, & Redman, C. L. (2008). Epistemological pluralism: Reorganizing interdisciplinary research. *Ecology and Society, 13*(2), 46.

Mitchell, R. B., Clark, W. C., Cash, D. W., & Dickson, N. M. (Eds.). (2006). *Global environmental assessments: Information and influence*. Cambridge: MIT Press.

Naustdalslid, J. (2011). Climate change: The challenge of translating scientific knowledge into action. *International Journal of Sustainable Development & World Ecology, 18*(3), 243–252.

Nowotny, H., Scott, P., & Gibbons, M. (2001). *Re-thinking science: Knowledge and the public in an age of uncertainty*. Cambridge: Polity Press.

Nowotny, H., Scott, P., & Gibbons, M. (2003). Introduction: 'Mode 2' revisited: The new production of knowledge. *Minerva, 41*(3), 179–194.

Pohl, C. (2011). What is progress in transdisciplinary research? *Futures, 43*(16), 618–626.

Pohl, C., & Hirsch Hadorn, G. (2007). *Principles for designing transdisciplinary research: Proposed by the Swiss Academies of Arts and Sciences*. München: Oekom.

Pohl, C., Perrig-Chiello, P., Butz, B., Hirsch Hadorn, G., Joye, D., Lawrence, R., Nentwich, M., Paulsen, T., Rossini, M., Truffer, B., Wastl-Walter, D., Wiesmann, U., & Zinsstag, J. (2010). *Questions to evaluate inter- and transdisciplinary research proposals* (Working Paper). Bern: td-net.

Reid, W. V., Chen, D., Goldfarb, L., Hackmann, H., Lee, Y. T., Mokhele, K., Ostrom, E., Raivio, K., Rockström, J., Schellnhuber, H. J., & Whyte, A. (2010). Earth system science for global sustainability: Grand challenges. *Science, 330*(6006), 916–917.

Rip, A. (2004). Strategic research, post-modern universities and research training. *Higher Education Policy, 17*, 153–166.

Salter, L. (1988). *Mandated science: Science and scientists in the making of standards*. Dordrecht: Kluwer.

Schneidewind, U., & Augenstein, K. (2012). Analyzing a transition to a sustainability-oriented science system in Germany. *Environmental Innovation and Societal Transitions, 3*, 16–28.

Scholz, R. W. (2001). The mutual learning sessions. In J. Thompson-Klein, W. Grossenbacher-Mansuy, R. Haberli, A. Bill, R. W. Scholz, & M. Welti (Eds.), *Joint problem solving among science, technology and society: An effective way of managing complexity* (pp. 117–129). Basel: Birkhäuser.

Schot, J. W., & Rip, A. (1997). The past and future of constructive technology assessment. *Technological Forecasting and Social Change, 54*(2–3), 251–268.

Stehr, N. (2009). What is socially relevant science? *Society, 3*, 262–266.

Truffer, B. (2007). Wissensintegration in transdisziplinären Projekten: Flexibles Rollenverständnis als Schlüsselkompetenz für das Schnittstellenmanagement [Knowledge integration in transdisciplinary research projects: The importance of reflexive interface management]. *GAIA, 16*(1), 41–45.

Weichselgartner, J. (2013). *Risiko – Wissen – Wandel: Strukturen und Diskurse problemorientierter Umweltforschung* [Risk – knowledge – change: Structures and discourses of problem-oriented environmental research]. München: Oekom.

Weichselgartner, J., & Kasperson, R. E. (2010). Barriers in the science-policy-practice interface: Toward a knowledge-action-system in global environmental change research. *Global Environmental Change, 20*(2), 266–277.

Weichselgartner, J., & Marandino, C. A. (2012). Priority knowledge for marine environments: Grand challenges at the society-science nexus. *Current Opinion in Environmental Sustainability, 4*(3), 323–330.

Weingart, P. (1997). From "finalization" to "mode 2": Old wine in new bottles? *Social Science Information, 36*(4), 591–613.

Weingart, P. (2008). How robust is "socially robust knowledge"? In M. Carrier, D. Howard, & J. Kourany (Eds.), *The challenge of the social and the pressure of practice: Science and values revisited* (pp. 131–145). Pittsburgh: University of Pittsburgh Press.

Winiwarter, V. (2012). 20 Jahre GAIA: Zurück und in die Zukunft. *GAIA, 21*, 1–4.

Winowiecki, L., Smukler, S., Shirley, K., Remans, R., Peltier, G., Lothes, E., et al. (2011). Tools for enhancing interdisciplinary communication. *Sustainability: Science, Practice, & Policy, 7* (1), 74–80.

Wood, R. (1964). Scientists and politics: The rise of an apolitical elite. In R. Gilpin & C. Wright (Eds.), *Scientists and national policy-making* (pp. 50–72). New York: Columbia University Press.

Zierhofer, W., & Burger, P. (2007). Disentangling transdisciplinarity: An analysis of knowledge integration in problem-oriented research. *Science Studies, 20*(1), 51–74.

Ziman, J. M. (1987). *Knowing everything about nothing: Specialization and change in scientific careers*. Cambridge: Cambridge University Press.

Ziman, J. M. (1996). "Postacademic science": Constructing knowledge with networks and norms. *Science Studies, 9*(1), 67–80.

Ziman, J. M. (2000). *Real science: What it is, and what it means*. Cambridge: Cambridge University Press.

Carving a Niche for the Social Sciences in Transdisciplinary Research on Climate Change Adaptation in Southern African Agriculture

Chipo Plaxedes Mubaya, Francis Themba Mugabe, and Sue Walker

1 Introduction and Background

Some of the initiated climate science-driven projects have mostly failed to address climate adaptation needs in the context of development needs (Hellmuth et al. 2007). Consequently, some of the products and services that the research community has developed have not been as useful as they could have been. It has become increasingly clear that an alternative approach to research is needed for development. For instance, natural scientists can identify and assess potential climate risks, can comment on the level of urgency with which they think actions should be taken, and can suggest appropriate technological solutions. However, it is up to the social sciences to describe, understand, and explain large-scale socio-economic transformations. Given the scale and urgency of the climate problem, the authors suggest that comprehensive and large-scale transformations are required if there is to be any chance of remaining within the 2 °C target agreed on in global climate negotiations. While the natural sciences can describe and highlight problems of environmental change, the social sciences are necessary to help develop solutions to these changes. It is therefore important to link the concept of

C.P. Mubaya (✉)
The African Climate Change Fellowship Programme (ACCFP) Secretariat, Institute of Resources Assessment (IRA), University of Dar es Salaam, Dar es Salaam, Tanzania

Current Affiliation: Directorate of Research and Resource Mobilisation, Chinhoyi University of Technology, Chinhoyi, Zimbabwe
e-mail: mubayacp@yahoo.com

F.T. Mugabe
Chinhoyi University of Technology (CUT), Chinhoyi, Zimbabwe

S. Walker
University of the Free State, Bloemfontein, South Africa

Current Affiliation: The University of Nottingham Malaysia Campus, Selangor, Malaysia

B. Werlen (ed.), *Global Sustainability*, DOI 10.1007/978-3-319-16477-9_6

sustainable development to climate change, since this provides an opportunity to explore long-term societal responses to global environmental change (Bizikova et al. 2008). This understanding has resulted in alternative efforts that advance an integrated approach, which incorporates climate science into multidisciplinary development planning and projects. This integrated approach has resulted in the natural and social sciences working together, including on local-level and national-level policy-making. This integration of the natural and social sciences translates into the multidisciplinary aspect to which we refer, rather than interdisciplinary and transdisciplinary aspects, which focus on making fundamental changes to disciplinary approaches to research, to worldviews and to bringing non-academics on board to identify local problems and design innovations together. These three terms are addressed in Sect. 1.1.

The close connection between society and nature necessitates an integrated transdisciplinary approach to environmental questions (Martens et al. 2009; Sorvari et al. 2009). In recent years, individual scientists and research groups, working at universities and research institutes seeking to understand global changes, including climate change, in the context of societal decision making, have made significant progress (Lowe and Lorenzoni 2006). These include reports by the Intergovernmental Panel on Climate Change (IPCC) and other ongoing global reporting activities such as Biodiversity Science (DIVERSITAS), the International Geosphere-Biosphere Programme (IGBP), the International Human Dimensions Programme (IHDP), the World Climate Research Programme (WCRP), and the International Polar Year (IPY). The Intergovernmental Panel on Climate Change (IPCC) (2007a) concludes that the "warming of the climate system is unequivocal" (IPCC 2007b, 2) and that this warming "is very likely due to the observed increase in anthropogenic GHG [greenhouse gas (CPM and FTM)] concentrations" (IPCC 2007b, 5). In addition, it was found that the projected consequences of climate change are being felt earlier and more intensely than previously thought. Therefore, timely and strong action is required to reduce further climate change (and the associated adaptation needs) and to avoid any further locking in of carbon-intensive infrastructures (IPCC 2007a). Consequently, it is important to understand societal dynamics, as well as the society-nature or human-environment interactions, and to recommend strategies for more effective environmental management and policy practices.

Most climate mitigation and adaptation efforts have been undertaken at the national and international policy levels. This is at odds with the general understanding that a participatory framework involving all stakeholders is crucial to the success of an integrated approach, since this will ensure that local needs are met (Hellmuth et al. 2007). It is critical to engage smallholder farmers, who are both the most climate-vulnerable group and the largest group of decision makers in the strategically important agricultural sector, to develop successful practices and policies. It is also important to note that adaptation to climate change is a challenge that involves not just individuals, but also complex and dynamic social interaction systems. Adaptation is inherently social in that it is based on attitudes and beliefs (Mubaya et al. 2012). In the past, research on climate change tended to focus on the biophysical components, but recent studies have begun to employ multidisciplinary, interdisciplinary, and transdisciplinary approaches, and to address social aspects. This article advances this position by outlining the role of social science research in a transdisciplinary project on climate change adaptation by addressing two key questions: (1) What role do the social sciences play in this transdisciplinary study? (2) What are the challenges and opportunities of social science research in such a context?

1.1 Models and Approaches to Crossdisciplinary Research

Various models in crossdisciplinary research have been termed interdisciplinary, transdisciplinary, or multidisciplinary. These approaches are not new and date back to Darwin's integrative theory of evolution, the European Enlightenment, or even further back to the Renaissance (Lyall et al. 2011). Scholars maintain that while there has indeed been a separation of the natural and social sciences, focusing on interdisciplinarity as 'new' opens it up to being dismissed as a trend (Thompson Klein 2004). The first use of the term interdisciplinarity is ascribed to the United States Social Science Research Council in the 1920s, although the Organisation for Economic Co-operation and Development (OECD) only formalized it in the early 1970s (Organisation for Economic Co-operation and Development and Centre for Educational Research and Innovation 1972). There is a general lack of consistency in the basic definitions of all three terms, which are also often used interchangeably in different contexts. Lyall and colleagues (2011) provide a framework for understanding the three terms, which we outline next.

Multidisciplinary research approaches a research problem from a range of disciplines' perspectives. However, researchers maintain that although their disciplinary positions may offer a few pieces of the 'jigsaw puzzle' minimal cross-fertilization of the disciplines occurs in the process. Moreover, there is little fundamental change, if at all, in the perception and understanding of knowledge-based outcomes. Lyall and colleagues (2011) further highlight that final reports on this process usually have separate sections for each discipline. There is little potential for optimal collaboration and changes to the researchers' worldviews, or in the functioning of academic institutions. On the other hand, interdisciplinary research has the potential to fill the gaps in multidisciplinary research in terms of enhancing fundamental changes to disciplinary worldviews, as well as institutional structures and functions (Lowe and Phillipson 2009; Lyall et al. 2011). The contributions of various disciplines are integrated to provide a holistic and potentially synergistic outcome. However, it is important to point out that such an integrated outcome is difficult to achieve. Transdisciplinary research transcends the traditional academic scope and includes non-academic stakeholders and various academic disciplines' knowledge production (Lyall et al. 2011; Nowotny et al. 2001). Based on this framework, the case study presented in this article has elements of the interdisciplinary and transdisciplinary approaches, although it mainly takes a transdisciplinary approach. This approach incorporates knowledge generation and exchange with non-academics when designing innovations and technologies that do not rely on a predetermined singular disciplinary foundation.

That is, the researchers conducted research within their disciplines, as well as in close collaboration with other disciplines.

2 Context of the Study

Social sciences and humanities research clearly has the potential to deliver the knowledge needed to achieve sustainability. A study on climate change adaptation in Zimbabwe and Zambia was based on this premise. The study involved multiple research institutions and was funded by the International Development Research Centre (IDRC) and the Department for International Development (DFID) through the Climate Change Adaptation in Africa (CCAA) program. Six institutions collaborated in this project: Midlands State University (MSU) in Zimbabwe, which acted as the lead institution; the International Crops Research Institute for the Semi-Arid Tropics (ICRISAT) in Zimbabwe; the International Centre for Tropical Agriculture (CIAT), also in Zimbabwe; Zambia Meteorology; the Zambia Agricultural Research Institute (ZARI); and the Commonwealth Scientific and Industrial Research Organisation (CSIRO), based in Australia.

The study was titled *Building adaptive capacity to deal with vulnerability due to climate change in Zimbabwe and Zambia* and was based on the understanding that Zambia and Zimbabwe are two of the countries worst affected by southern African droughts. Droughts and floods are also common and almost annual threats in Mozambique, Malawi, Southern Zambia, and Zimbabwe (Clay et al. 2003; Cooper et al. 2007; De Wit 2006). Furthermore, the study was based on the premise that improving the ability of institutions to train the 'future change agents,' who will subsequently help smallholder communities adapt their agricultural practices to the current climate variability, is the first step in building an adaptive capacity to deal with future climate changes. These 'future change agents' include agricultural and meteorological extension service providers and researchers at universities (Twomlow et al. 2008). The project sought to bring together experiences from national and international research and extension institutions working in Zimbabwe and Zambia, and to build on their existing skills, networks, and field activities with the aim of strengthening the regional capacities to adapt to climate change. Participatory on-farm research and climatic forecasting were integral project components in order to increase the smallholder farmers' competencies to deal with the current climatic variability and adapt to potential climatic change (Twomlow et al. 2008).

The objectives of the socio-economic component of the study were: (1) to establish smallholder farmers' existing perceptions of climate risk, which change agents and meteorological services can then use to better target their interventions; and (2) to determine how rural communities cope with existing climate variability and extremes, and if they have developed appropriate strategies for adapting to future climatic change. Objectives from other disciplines included:

- To develop, test, and disseminate climatic risk communication materials and interventions to prevent risk
- To use farmer participatory approaches (FPA) linked with simulation and climate forecasting to develop and evaluate scenarios for adaptation with farmers
- To build capacity and competency within Zambian and Zimbabwean institutions in the use of simulation and climatic forecasting tools.

3 Approach and Methods Employed for the Study

The project was based on interdisciplinary and transdisciplinary research that sought to build capacity among smallholders, agricultural, and meteorological extension service providers, and researchers at local universities. Discipline-specific studies focused on crop modeling, a socio-economic study, agro-meteorology, agronomy, agricultural economics, soil science, and hydrology. The study followed different methodological perspectives for data collection and analysis. Discipline-specific methodologies were used, which included agronomists' and soil scientists' applications of fertilizers in on-farm trials, agro-meteorologists' installations of weather stations on farmers' trial plots, and agricultural economists' econometric and statistical models. The discipline-specific methodologies employed in this study were all situated within the broader framework of the Integrated Agricultural Research for Development (IAR4D) through action research.

The IAR4D innovation system framework was proposed by the Forum for Agricultural Research in Africa (FARA) as a basis for the transformation of agricultural research in Sub-Saharan Africa (SSA). The IAR4D concept encourages multiple actors to promote the process of innovation in the agricultural system through effective interaction among the role-players and iterative learning for stakeholders (Hawkins et al. 2009). Among the advantages of this approach is that it effectively draws on the knowledge of the relevant actors at each stage and creates a network that encompasses technical, social, and institutional elements, ultimately generating innovation beyond mere research products (Hawkins et al. 2009).

The socio-economic study primarily employed the Participatory Rural Appraisal (PRA) approach, which provides a basis for dialogue, through which information is shared and researchers can learn from and with rural people. This method makes use of local knowledge and experience, with local people identifying their problems and needs, and encompasses Farmer Participatory Research (FPA). The researcher only plays a facilitative role (Bhandari 2003; Chambers 1997; Wall et al. 2006). Within the same stakeholder engagement framework, Farmer Participatory

Research (FPR) was conducted in the form of mother-baby on-farm trials,[1] seasonal climate forecasting (SCF), and community modeling. Two field trials were conducted in selected villages in the two countries' project districts. The soil samples for the on-farm trials were based on treatment options that the farmers had selected. This process was designed to identify problems and solutions together with the farmers. Econometric vulnerability assessments were also based on socio-economic data collected through a questionnaire survey, which was part of the project baseline study. A total of four districts were sampled in the two countries, two in southern Zambia and two in southwestern Zimbabwe. A total of 720 questionnaires were administered and collected for the survey, 360 in each country and in all the sampled communities.

4 What Role Did the Social Sciences Play in This Study?

Social scientists in this study set the scene, established a dialogue, and provided the context for the project, both in scientific research and in practical engagements with non-academic stakeholders such as farmers, agricultural extension workers, scientists, NGO practitioners, and other stakeholders. Social scientists were part of the project from the outset; that is, they were involved in the development of the proposal and in the familiarization of the project officials with the other stakeholders. Calvert and Martin (2009) assert that social scientists can either be contributors by entering the scene after natural scientific knowledge has been produced, or collaborators by transcending natural scientific research and interacting with it. In this case, social science was an integral component of the research-based development project from its inception.

Social scientists initiated the engagement with farmers by co-developing a participatory diagnosis with these farmers. This process involved helping the farmers understand and providing an analysis of their opportunities, constraints, livelihood strategies, and farming systems, among others. This process was important for all the stakeholders, including the natural scientists, who benefitted from the foundation that the social scientists laid: outlining the socio-economic background and providing an understanding of the issues of concern to local farmers. This also enabled researchers from various disciplines to make an informed choice and to design appropriate interventions in the context of the project implementation. For instance, the agronomic Agricultural Production Systems Simulator (APSIM)

[1] Mother-baby trials aim to improve the flow of information between farmers and researchers about technology performance and appropriateness under specific farm conditions (Snapp 1999). Specifically, this design is used to quantify farmer evaluation of technologies (Snapp 2002; Snapp et al. 2002). The trial design consists of two types of trials: mother and baby trials. "The 'mother' trial is replicated within-site to test a range of technologies and research hypotheses under researcher management ... The 'baby' trial comprises a number of satellite trials (each trial is one replicate of large plots under farmer management and farm resources" (Snapp 2002).

modeling was based on the farmers' crop choices, which the diagnoses had identified, and included the seasonal climate forecasting and subsequent on-farm trials. Research suggests that the technologies that multiple disciplines develop collaboratively are more often transformational than those that a single discipline has developed (Trewhella 2009). Therefore, solving practical problems requires building the capacity to draw on and synthesize information from a variety of sources.

In addition, the initial participatory diagnosis enabled agricultural economists to undertake a thorough vulnerability assessment based on the gathered socio-economic data. Furthermore, it also enabled farmers to engage in the collaborative scenario-building process. This outcome supports the long-standing assertion that there are strong links between elements that a specialized, monodisciplinary scientific analysis would separate. Going beyond the frontiers of particular disciplines is likely to provide a more complete picture (Barnett 2010).

In this regard, the project's social science component further enriched the broader study by explaining farmer experiences, providing an indication of the diversity of each of the sampled communities in Zimbabwe and Zambia, and by understanding the community social dynamics. Essentially, social science ensured that the information had meaning for other scientists and glued disciplines together regarding the implementation of the project. The IPCC (2007) has noted that communities' responses to climate change tend to be complex and diverse, based on needs, values, cultures, capacities, institutional forms, and environmental issues. Consequently, the social sciences are best suited to enhance understanding of these characteristics, especially given that there are relatively few studies on adaptation that are informed by evidence from socio-ecological systems and how they respond and adapt to change (Martens et al. 2009). Similarly, recent social science studies on farmer adaptation have drawn on a variety of disciplines in an effort to contribute to policy making (Barnett 2010).

5 What Were the Opportunities and Challenges for the Social Sciences and Other Disciplines in the Context of This Study?

Through awareness raising and scenario building, hence allowing adaptation planning, innovations strengthened the farmers. By providing training, the project contributed significantly to strengthening institutions in the agriculture and extension sector (AGRITEX), as well as the Ministry of Agriculture and Cooperatives (MACO) in Zimbabwe and Zambia. In addition, there is a high likelihood that the researchers developed team-building skills, as teams from both countries collected and analyzed data by using replication. The team members published a number of joint scientific papers. Trewhella (2009) recognizes team spirit, which breaks down the barriers of language and culture and creates knowledge that drives innovation, as a recipe for high team performance.

Social science findings emerged as crosscutting issues in the study. These issues are critical for policy makers. Key findings include:

- Indigenous forecast knowledge is critical where there is a paucity of (scientific) data for decision making on crop and environmental management (Mugabe et al. 2010)
- Engaging smallholder farmers in seasonal climate forecasting (SCF) helps with investment decisions and suggestions regarding crop management options. Responses to the SCF proved stable over two seasons and enabled the farmers to improve their management of the investment risk in relation to their fertilizer inputs (Mugabe et al. 2010)
- While farmers recognize climate change, they tend to assign other factors as contributing to the negative effects of climate change. In addition, significant responses to climate change and variability are primarily agriculture based rather than nonfarm based (Mubaya et al. 2012).
- The baby trials proved to be a good strategy for learning and for encouraging farmer involvement in the project activities.
- The study demonstrates that seasonal climate forecast is an ideal entry point to bring meteorological officials, researchers, and extension agents together to work with the farmers.

Experiences from the project indicate that there is an inherent opportunity for the stakeholders' personal growth. There were five doctoral and four Master's students on the project, and by the time the funding for the project ended, two doctoral theses and three Master's dissertations had been finalized. The students also had the opportunity to move between the collaborating institutions and even between the disciplines. This is consistent with suggestions by Trewhella (2009) and Lyall and colleagues (2011) that moving students across institutions when undertaking research could have great benefits regarding institutions' ability to train the next generation of interdisciplinary and transdisciplinary researchers and to overcome disciplinary boundaries. Clearly, there is potential for improvement in developing interdisciplinary approaches to understand climate change adaptation and to strengthen the individual disciplines concerned. The social sciences have an opportunity to broaden their horizon by understanding natural issues and concepts in relation to environmental change.

On the other hand, transdisciplinary research is hardly easy to implement. It is important to highlight the time-consuming nature of this kind of research, since multidisciplinary, interdisciplinary, and transdisciplinary teams are obliged to create a shared language and a framework for discovery and innovation. When the participatory research was conducted, the researchers and farmers needed much time to harmonize their ideas and methods – particularly for the on-farm trials and other participatory exercises – in order to implement the project. Nevertheless, this process proved to be invaluable before meeting the other stakeholders – particularly for the scientific team – since there was a need to reach a shared understanding of the terms and issues. Papaioannou and colleagues (2009) caution against concepts developed in one discipline being uncritically applied to another without a shared understanding, since this can lead to adverse consequences.

It is also important to understand that multidisciplinary, interdisciplinary, and transdisciplinary research is a challenge as it tends to be expensive to conduct. Some subprojects could not be fully completed in the limited funding period, as they required time series data. Students in these disciplines failed to finalize their research prior to the project completion. Social science data, on the other hand, was collected within the limited timeframe, and the students could complete their work. In addition, owing to the timeframe, some of the papers initiated during the project period may not have been finalized in time, given that the researchers may no longer be in constant touch. In this regard, it is important to note the disadvantages of such one-off studies, which may fail to make a lasting impact if they are not expanded.

The natural scientists' perceptions regarding the rigor of socio-economic studies tended to place other scientists – specifically those in the social sciences – in a precarious position. Natural scientists have an inherent perception that the social sciences' studies lack scientific rigor and are generally difficult to replicate. This is a long-standing perception that dominates the cross disciplinary research discourse. The natural sciences as a whole generally enjoy an a priori degree of confidence from the scientific community (Lyall et al. 2011). The project's social scientists had to operate within this framework, but this did not entirely interfere with their performance. Furthermore, the scientists from other disciplines tended to expect too much assistance from the social scientists. As a result, data sharing issues also challenged the research process, since the natural scientists expected to have access to all the data that the social scientists had collected, although some of the data was specific to researchers' theses and could not be shared. Simultaneously, this data was essential for natural scientists' modeling and understanding of socio-economic issues.

6 Conclusion

Experiences from this study show that social science research is essential to inform natural sciences regarding crossdisciplinary research. Social science research also provides an opportunity for client-driven research and development planning. While difficult in many respects, it is possible to fully integrate social science into crossdisciplinary research if attention is paid to the process. However, social scientists have much work to do in responding to and gaining respect from natural science perceptions. Overall, transdisciplinary research has the potential to draw on synergies that are better informed and significant for decision making.

Acknowledgements The authors thank the following institutions, which provided funding for this project: International Development Research Centre (IDRC), Department for International Development (DFID), Climate Change Adaptation in Africa (CCAA), and technical support from collaborators and researchers in the field work: researchers from Zambia Agricultural Research Institute (ZARI), Zambia Agro-Met, and researchers and students from Midlands State University (MSU), Zimbabwe, and ICRISAT Zimbabwe.

References

Barnett, J. (2010). Adapting to climate change: Three key challenges for research and policy – An editorial essay. *Wiley Interdisciplinary Reviews: Climate Change, 1*(3), 314–317.

Bhandari, B. B. (2003). *Participatory rural appraisal (PRA)*. Institute for Global Environmental Studies (IGES). Module 4: Patumwan, Bangkok, Thailand

Bizikova, L., Neale, T., & Burton, I. (2008). *Canadian communities' guidebook for adaptation to climate change. Including an approach to generate mitigation co-benefits in the context of sustainable development*. Vancouver: Environment Canada and University of British Columbia.

Calvert, J., & Martin, P. (2009). The role of social scientists in synthetic biology. *EMBO Reports, 10*(3), 201–204.

Chambers, R. (1997). *Whose reality counts: Putting the first last*. London: Intermediate Technology Publication.

Clay, E., Bohn, L., der Armas, E. B., Kabambe, S., & Tchale, H. (2003). *Malawi and Southern Africa: Climate variability and economic performance* (Working Paper Series, No. 7). Disaster Risk Management.

Cooper, P., Singh, P., Traore, P. C. S., Dimes, J., Rao, K. P. C., Gerard, B., Alumira, J., et al. (2007). *New tools, methods and approaches in natural resource management*. India: ICRISAT, Patancheru 502 324, Andhra Pradesh.

De Wit, M. (2006). *Climate change and African agriculture*. http://www.ceepa.co.za/docs/POLICY%2520NOTE%252011.pdf. Accessed 23 Mar 2009.

Hawkins, R., Heemskerk, W., Booth, R., Daane, J., Maatman, A., & Adekunle, A. A. (2009). *Integrated Agricultural Research for Development (IAR4D)* (A Concept Paper for the Forum for Agricultural Research in Africa (FARA) Sub-Saharan Africa Challenge Programme (SSA CP)). Accra: FARA.

Hellmuth, M. E., Moorhead, A., Thomson, M. C., & Williams, J. (Eds.) (2007). *Climate Risk Management in Africa: Learning from Practice*. International Research Institute for Climate and Society (IRI), New York: Columbia University.

Intergovernmental Panel on Climate Change (IPCC). (2007a). In T. C. W. Team, R. K. Pachauri, & A. Reisinger (Eds.), *Climate change 2007: Synthesis report. Contribution of Working Groups I, II and III to the fourth assessment report of the Intergovernmental Panel on Climate Change*. Geneva: IPCC.

Intergovernmental Panel on Climate Change (IPCC). (2007b). In T. C. W. Team, R. K. Pachauri, & A. Reisinger (Eds.), *Climate change 2007: Synthesis report: Summary for policymakers*. Geneva: IPCC.

Lowe, T. D., & Lorenzoni, I. (2006). Danger is all around: Eliciting expert perceptions for managing climate change through a mental models approach. *Global Environmental Change, 17*(1), 131–146.

Lowe, P., & Phillipson, J. (2009). Barriers to research collaboration across disciplines. *Environment and Planning A, 41*(5), 1171–1184.

Lyall, C., Bruce, A., Tait, J., & Meagher, L. (2011). *Interdisciplinary research journeys: Practical strategies for capturing creativity*. http://www.bloomsburyacademic.com/view/Interdisciplinary-Research-Journeys/book-ba-9781849661782.xml;jsessionid=273AEEA49E156DBEA7D730C1007DFB6A. Accessed 15 Sept 2012.

Martens, P., McEvoy, D., & Chang, C. (2009). The climate change challenge: Linking vulnerability, adaptation, and mitigation. *Current Opinion in Environmental Sustainability, 1*(1), 14–18.

Mubaya, C. P., Njuki, J., Mutsvangwa, E. P., Mugabe, F., & Nanja, D. H. (2012). Climate variability and change or multiple stressors? Farmer perceptions regarding threats to livelihoods in Zimbabwe and Zambia. *Journal of Environmental Management, 102*(2012), 9–17.

Mugabe, F. T., Mubaya, C. P., Nanja, D. H., Gondwe, P., Munodawafa, A., Mutswangwa, E., Chagonda, I., et al. (2010). Using indigenous knowledge systems for climate forecasting and adaptation in Southern Zambia and south western Zimbabwe. *Zimbabwe Journal of Technological Sciences, 1*(1), 19–30.
Nowotny, H., Scott, P., & Gibbons, M. (2001). *Re-thinking science: Knowledge and the public in an age of uncertainty*. Cambridge: Polity Press.
Organisation for Economic Co-operation and Development (OECD) & Centre for Educational Research and Innovation (CERI). (1972). *Interdisciplinarity: Problems of teaching and research in Universities*. Washington, DC: OECD.
Papaioannou, T., Wield, D., & Chataway, J. (2009). Knowledge ecologies and ecosystems? An empirically grounded reflection on recent developments in innovation systems theory. *Environment and Planning C, 27*(2), 319–339.
Snapp, S. (1999). Mother and baby trials: a novel trial design being tried out in Malawi. *Target Newsletter of the Southern African Soil Fertility Network, 17*(8), 411–431.
Snapp, S. (2002). Quantifying farmer evaluation of technologies: The mother and baby trial design. In M. R. Bellon & J. Reeves (Eds.), *Quantitative analysis of data from participatory methods in plant breeding* (pp. 9–17). Mexico: CIMMYT.
Snapp, S., Kanyama-Phiri, G., Kamanga, B., Gilbert, R., & Wellard, K. (2002). Farmer and researcher partnerships in Malawi: Developing soil fertility technologies for the near-term and far-term. *Experimental Agriculture, 38*(4), 411–431. doi:10.1017/S0014479702000443.
Sorvari, S., Tikka, P. M., Niemelä, J., Raivio, K., & Korhonen-Kurki, K. (2009). Breaking the boundaries: Multidisciplinary environmental research at the University of Helsinki. *Boreal Environmental Research, 14*(Supplement A), 1–4.
Thompson Klein, J. (2004). Prospects for transdisciplinarity. *Futures, 36*(4), 515–526.
Trewhella, J. (2009). *Multidisciplinary research – An essential driver for innovation*. http://globalhighered.wordpress.com/2009/06/26/multidisciplinary-research-an-essential-driver-for-innovation/. Accessed 15 Sept. 2012.
Twomlow, S., Mugabe, F. T., Mwale, M., Delve, R., Nanja, D., Carberry, P., & Howden, M. (2008). Building adaptive capacity to cope with increasing vulnerability due to climatic change in Africa – A new approach. *Elsevier-Physics and Chemistry of the Earth, 33*, 780–787.
Wall, E., Armstrong, M., & Manityakul, S. D. (2006). *Climate change and Canadian society: Social science research issues and opportunities*. C CIARN (Canadian-Climate Impact and Adaptation Research Network) Symposium Report. http://www.c-ciarn.ca/pdf/june_symposium.pdf. Accessed 20 Jul 2009.

Sustainability and Territory: An Approach to Shape Development from the Perspective of the Imaginary

Enrique Aliste

1 Introduction

Processes that lead to territorial and environmental transformations are complex. This implies the acceptance of research efforts with epistemological frames that are diverse and often non-convergent.

Some results and thoughts described here attempt to open new, relevant discussion on how territories are articulated and created in development discourses. We also examine how certain ideas or notions deeply affect the ways territories and identities are created, as well as how to understand social and environmental phenomena articulated within a given territory, particularly from the *social imaginaries* perspective. The imaginary is a concept used in sociology and philosophy (Castoriadis 1987; Habermas 1996; Taylor 2004; Anderson 1991) to refer to an intersubjectively shared notion. In this article, it refers to development: its meaning, aims, and objectives, as well as how it can best be achieved.

This article focuses on the development of Concepción, Chile, which is considered a prototype of a *growth pole* or *development pole* (*polo de crecimiento* o *polo de desarrollo*) (Perroux 1955). This case study will provide a reflexive analysis of the various meanings that *urban sustainability* and *sustainable development* concepts have today.

In this analysis, the initial approach is a geohistorical one. In the later stages of the analysis, this is complemented with elements from urban studies, as well as sociological and anthropological perspectives on cities, development, and discourse. From this, a new point of view and a new interpretation of urban

With the support of the National Fund for Scientific and Technological Research of Chile (FONDECYT) projects: 1090248 and 1120306.

E. Aliste (✉)
Department of Geography, University of Chile, Santiago, Chile
e-mail: ealiste@uchilefau.cl

B. Werlen (ed.), *Global Sustainability*, DOI 10.1007/978-3-319-16477-9_7

development is proposed; this will lead to a discussion of certain concepts previously considered universal. These may, however, yield unexpected outcomes.

2 Development Discourse and Its Implications for Territory

Development is one of the most interesting concepts from the second half of the twentieth century. During its evolution, it has become a type of new evangelizing, missionary process of increasing relevance, especially in so-called underdeveloped countries. The elements that deserve attention are the origin of this concept and the narratives it produces. A notable event in post-WWII history was US President Harry Truman's inaugural address in 1949, in which he demanded that "(. . .) *we must embark on a bold new program for making the benefits of our scientific advances and industrial progress available for the improvement and growth of underdeveloped areas*" (Rist 2007, 130; Truman 1949, n. p.).

In Chile, as in the rest of Latin America, international organizations have advanced strong ideas committed to the articulation and application of development strategies. The Cepalian[1] discussions, which have helped establish several important lines of thought on Latin American development, have played a crucial role.

This can be seen in diverse contributions to the debate, such as those by Faletto (2007), Prebisch (1983, 1984, 1985), Sunkel and Paz (1970), and Furtado (1991, 2008), among others. In this context, the significance that territory has in discourses on and notions of development needs to be reviewed, since a spatial perspective appears to be surprisingly absent from this general body of literature. Nevertheless, spatial notions are present in these discourses – if somewhat disguised. It is in our interest to visualize and interpret how these notions emerged. It is worth mentioning that, in most cases, this spatial point of view is a result of the exploration of comparative advantages, the presence of natural resources, or of a way of organizing regions by production purposes. In general, it relates to economics or to the functionality of a neoclassical economic model, which is strongly committed to providing structure, content, and technical coherence to development from its own perspective, although always in a nuanced way.

Tendencies affecting the Latin American school of development are quite different. Within a period of 50 years, the school underwent a gradual transformation from purely economic ideas to a more interdisciplinary notion of development (Di Filippo 2007), although, in spite of this interdisciplinary feature, spatial notions did not become an explicit subject. This is very relevant, since it implies the concept's systematic absence (Foucault 1971). This reinforces one of the theses proposed here: the existence of development without territory.

[1] This refers to the influence of CEPAL or the Comisión Económica para América Latina y el Caribe (United Nations Economic Commission for Latin America and the Caribbean, or ECLAC).

Indeed, notions of development produced at different historical moments have been implemented in Chile through several policies, plans, and strategies that, as an indirect consequence, have permanently and noticeably transformed the territory in a way which can only be described ex post.

Therefore, development as both discourse and practice has generated paradoxical consequences. In some cases, it has been introduced as the main argument and the solution to overcome social issues, in other cases, it has produced new problems that made the realization of its purpose impossible, made development an obstacle to its purpose; environmental degradation is a prime example here. It is said that, from its very origin, development is considered to have *"a real existence, as something solid, material* (...)", a claim that several disciplines have debated, "*but without questioning its ontological status*" (Escobar 2000, 19). This idea tends to be reinforced when one browses the literature produced in the Francophone (Wackermann 2005; Lacoste 1965) and Anglophone areas (Smith 2008).

The many different approaches of scholars such as Escobar (1996, 2000, 2008), Robbins (2004), Peet et al. (2011), among others, as well as other thoughts on the notion of space and its position and effects on society (Werlen 1993; Harvey 1996; Soja 1989; Di Méo 1998; Di Méo and Buléon 2005) are helpful when questioning development's ontological status. All these perspectives may also be used to evaluate the way development discourses have affected the formation of contemporary territories. This opens innovative opportunities to draft and redesign strategies for the environmental sustainability of territories from new, less orthodox perspectives.

Different development conceptions respond to different realities constructed over time.[2] Therefore, in each moment of history, one can recognize a proliferation of several different conceptions, which appear simultaneously with broader concepts that provide sense within a certain period of history. These perspectives invite us to approach the issue of development and the environmental evaluation of territory while considering the characteristics of knowledge related to this topic (mainly provided by natural sciences) and how the collective evaluation of these issues is established and evolves within society (Aliste 2011).

It is therefore crucial to also consider the complexity of contemporary societies where partial reality constructions (often generated by a multiplicity of value judgments) confront descriptions of society that may often be discordant – as is the case with several environmental problems (Arnold 2003).

Michel Foucault pioneered the study of reality construction through discourse practices and the way they become articulated and acted upon within different

[2] Throughout history, development has adopted multiple conceptions, ranging from the ideas of evolution and progress to theoretical conceptualizations, such as modernization theory, dependence theory, market-based development, self-centered development, territorial development, regional development, local development, endogenous development, human development, sustainable development, and eco-development (Escobar 1996; Sachs 1981; Gligo 2006).

dominions.[3] In discourse on development, some publications such as those by Bernard (1989), Coquery-Vidrovitch et al. (1988), Escobar (1996, 2000), Giraut and Antheaume (2005), Kitching (1982), Rist (2007), and Legouté (2001) focus on the circumstances that have historically enabled and structured a discourse on development under the influence of certain strong, central ideas that have led to the construction of economy-centered hegemonies.[4]

From this perspective, the impact of the predominant notion of development – which equates development with economic growth as measured by increases in GDP – on the global geopolitical configuration is twofold. First, countries are categorized, for example, as first world, second world, or third world; the underlying assumption is that the first world is the standard by which the second and third worlds are measured. Social, economic, and environmental variables serve to 'prove' the 'progress'/development, or lack thereof. That is, it is implied that second and third world countries lack something that the first world has and – most importantly – that this something is worth striving for. Such normative claims imply that development's goal should effectively be a global homogenization of population. In 1951, for instance, the United Nations (UN) published a document that a team of experts elaborated, concluding:

> There is a sense in which economic progress is impossible without painful adjustments. Ancestral philosophies must be eradicated; old institutions need to dissolve; the ties of caste, creed and race must be severed; and vast masses of people unable to follow the pace of progress will not satisfy their expectations for a comfortable life. Very few communities are willing to pay for the price of economic progress. (United Nations Department of Social and Economic Affairs 1951, cited in Escobar 1996, 20)

Second, this attempt to homogenize the global population is framed in economic terms. It also gives rise to a new social conceptualization: the categorized hierarchy of rich and poor countries generates a new socio-global order. As a result, this concept of development privileges particular symbolic systems (those of the first world), leading to their becoming hegemonic. That is, the political system that exercises power also configures certain territories and pushes them to resemble each other under a new socio-global model. Certain notions commonly used today in the field of urban studies, such as winner and loser territories (Merchand 2007), are originally based on this. Finally, taking this scenario into consideration, how should one understand the sustainable development concept, or the sustainability of development?

---

[3] Foucault (1971) proposes three main systems of exclusion that allow for understanding discourses: the forbidden word, the way of sharing (views about) insanity, and the will of truth. In this sense, every discourse responds to a way of organization around these systems of exclusion. Analytically, the discourse on development terms can also be viewed from this perspective.

[4] Gramsci's concept of hegemony focuses on the study of the cultural aspects of society as an element from which one could conduct a political action and as a way to create and reproduce it (Anderson 1976).

3 The City of Concepción (Chile): Icon and Challenge

Concepción[5] is a metropolitan area that underwent a significant urbanization process in the second half of the twentieth century. Concepción has been selected for this article because it is a prime example to illustrate the notion of growth poles (Perroux 1955; Armstrong 1973). It is also a city that, due to its economic structure, has suffered profound environmental modifications (Sanhueza and Azócar 2000) that bear witness to various phenomena associated with discourse and territorial transformation practices.

Founded in 1550 for military purposes, the city only consolidated its importance in the Chilean cities system around 1830 when it began to experience a process of expansion and economic diversification. However, Concepción only began to consolidate its industrial character after 1930, and especially since 1950, when the Huachipato steel plant was established in the city of Talcahuano. This may be the most important milestone in Concepción's urban history (Hernández 1983) (Fig. 1).

The Huachipato steel plant was commissioned in Talcahuano in 1950 after a national industrialization plan had been drafted. This effort was the materialization of the Chilean state's long-time desire: providing a self-sufficient domestic supply of steel. Until 1945 domestic steel production could satisfy only approximately 20 % of the national demand (Sánchez 1952; Echeñique and Rodríguez 1990). San Vicente Bay was chosen as the location owing to its multiple geographical advantages: its very good connection to a network of roads and railways, the potential for the port to be expanded, the coal mines in the cities of Lota and Coronel (both near Concepción), and the significant possibilities to exploit the hydroelectric resources of the Bio Bío River. These reasons eventually led to the strategic possibility to enhance Concepción as a steel production hub of nationwide relevance (Hernández 1983; Aliste et al. 2012a).

Within approximately 30 years, industrial activity was consolidated in the area. Three indicators verify this: changes in the quantity of industrial establishments, the number of employees in the industrial sector, and the production capacity. For example, there were 201 companies in 1942, a number that had increased by 32 % in 1957. Similarly, between 1957 and 1971, the number of companies in the area grew by 45 %, while between 1942 and 1971 industrial establishments in the Concepción Metropolitan Area (CMA) grew by 63 %, which clearly reflects the aforementioned strategy.[6] Compared to Santiago, and taking a nationwide perspective, what happened

[5] The city of Concepción refers to the metropolitan area that also includes Talcahuano. Nowadays, the municipalities of Concepción, Talcahuano, Hualpén, Penco, San Pedro de La Paz, and Chiguayante are considered part of this unit, since they define a large, single urban continuum. Other municipalities, such as Coronel, Lota, Tomé, Hualqui, and Santa Juana, are also closely linked to this urban sprawl.

[6] According to the industrial census by the National Institute for Statistics (INE) between 1942 and 1971.

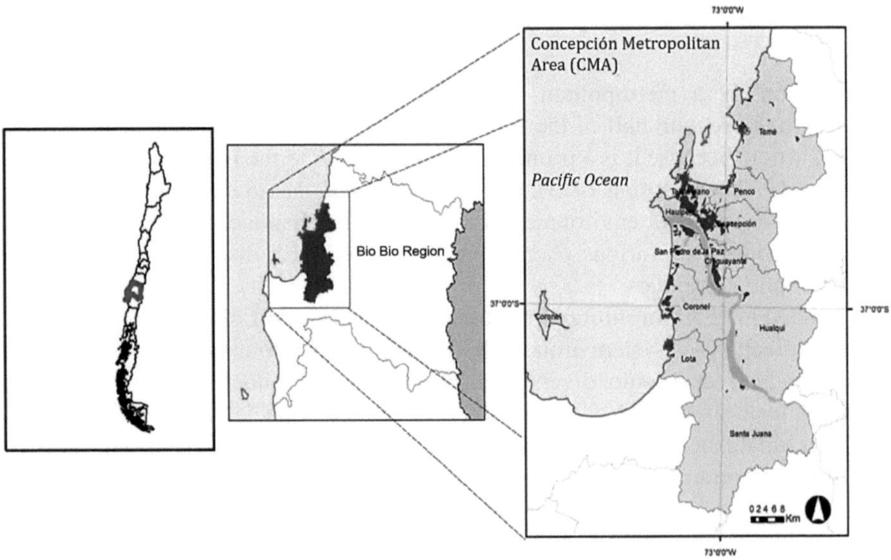

Fig. 1 Concepción Metropolitan Area (CMA), Bio Bío Region, Chile (Muñoz 2011)

in Concepción provides insight into the industrialization phenomenon, since the growth rates of the number of industrial establishments in the area are considerably higher than the national rates (Aliste and Almendras 2010).

Demographically, census statistics reveal that 1940–1952 was the period with the highest population growth. This can be explained by the area's new economic orientation towards a mainly industrial profile, which is in line with the mid-twentieth century developmentalist discourse (Aliste et al. 2012b). This growth process resulted in a high concentration of population in a territory that is not only quite small, but also geomorphologically complex, which resulted in a particular geography characterized by strong tensions between nature and society.

Several studies that the Centro EULA de Ciencias Ambientales (European-Latin America Environmental Science Center) conducted at the University of Concepción between 1992 and 2000 reveal the environmental condition of the Bio Bío basin during this period, particularly that of Concepción (e.g., Della Croce et al. 1992; De Fraja et al. 1993; Parra et al. 1999).

These studies indicate high air pollution and identify nearly 40 major industries: chemical, petrochemical, steel, metalworking, fishing, food, and services. There are also emissions from raw material and fuel storage, domestic activity (heating), bakeries, etc. One of the most sensitive issues concerning air quality is the presence of foul odors from activities in the fishing, petrochemical, and steel-metallurgy industries. In addition, water pollution has been detected in the Bio Bío River, which receives nearly 40 % of the wastewater of the Talcahuano area in its lower course. Also, most of the Bio Bío basin's wastewater is disposed of in the river, including industrial and oil refinery wastewater, which is carried in the sewage system. Various monitoring measures and analyses in the bays of Concepción and San Vicente revealed alarming pollution levels, particularly between 1996 and 1998. This coincided with the period when authorities began to take action in this regard. In spite of this, it was reported that the pollution levels were so high that reverting to previous levels was virtually impossible, and that the area was one of the most polluted places in the world (Valenzuela 2002).

But one of the most symbolic and important elements concerning environmental changes in the Concepción area relates to the effects of urban growth on the city's wetlands. Large wetland areas were lost between 1950 and 2010; in the case of the Rocuant-Andalién wetland (also known as Carriel), there was a surface loss of nearly 40 % over this period (Beltrán 2011). The loss of wetlands is associated with a significant deterioration in environmental quality, including a considerable reduction in biodiversity, ecosystem services, and landscape quality (Smith and Romero 2009).

The second focus element of this analysis is how discursive practices exert pressure on environmental issues. Therefore, in the discourse on growth and the consolidation of the development model, the occupation of marshy areas or swamps were the most evident sign of progress and the advance of civilization over barbarism as represented by untamed nature. For half a century, the path was clear and the city's expansion over such surfaces was a testimony of a march or advancement in the right direction: civilization succeeded in gaining space from nature as the absence of progress. Urbanization and industrialization were the most powerful signs that progress was increasing and moving in the politically correct path. What happened thereafter?

The emergence of environmental discourse is a clear indication of the emphasis placed on the 'civilizing' efforts in Chile since the 1990s. While this type of discourse first appeared in the 1970s, after the publication of *The limits to growth* (Meadows et al. 1972), it only became important in Chile in 1994 when the *Ley de Bases Generales del Medio Ambiente* (Law of General Bases for the Environment) was promulgated, which meant that environmental issues had acquired institutional, political, and social importance (Aliste 2010). From a short-term geohistorical perspective,[7] what comes next is the discordance between discourse and spatial practices. Development plans, regional development strategies, intermunicipal regulatory plans, and all general policy instruments that attempt to organize action on space will move in two dimensions that appear to be similar, but turn out to be very different: the imaginary and the tangible.

[7] While what is meant here is the concept of *geohistory* as proposed by French historian Fernand Braudel (1997), we use it with more subtlety and on a smaller time scale. This enables us to understand a process that, albeit brief (only half a century), has long-term effects on and very significant consequences for the spatial transformations of Latin American economies. These economies were significantly restructured after the shocks experienced in the nineteenth and twentieth centuries. These modifications led to the economy becoming strongly dependent on the exploitation of natural resources. This is why such processes' impacts on transformations and spatial dynamics are so important.

From here, beyond focusing on their effectiveness, the exploration of discourses leads to the questioning of development in the light of its spatial effects on the construction of territories. Likewise, it is possible to question and open a line of interpretation of the meaning and historicity of sustainability in emerging economies.

4 Spatial Practices and Modern Beliefs: Development, Sustainability, and Territory

The use of space is an indicator of how society conceives itself. The ways space is used and how these ways match (or not) political definitions regarding spatial planning, demonstrate the practical difficulties involved in development notions in the construction of public policies in Latin American contexts.

In the case of Concepción, a number of interesting examples illustrate this inconsistency. Nevertheless, instead of perceiving it as a difficulty or a problem, it is convenient to think of such inconsistency as an opportunity for reshaping the relationship between the natural system, the human-made objects/artifacts, the economic system, the social system, and the cultural system. This active symbiosis may be called the socio-environmental system (Fig. 2).

While the socio-environmental system is frequently defined as a complex phenomenon emerging from the environmental sciences (Musters et al. 1998; López-Ridaura et al. 2002), the definition proposed here seeks to reformulate the concept from a geohistorical perspective by considering[8]:

- Difficulties with the definition and description of the environment (from where is it observed? In what socio-historical context?).
- Biases from a biologist approach to environmental problematizations (Macnaghten and Urry 1995).
- Differences in the understanding of space, nature, and society (Di Méo 1991; Claval 2002).

This article will also analyze how dimensions associated with everyday life, human experiences, and living space may lead to a different view of how sustainability is understood in contexts with differing histories and spatial experiences, that are also strongly influenced by their basic geographical conditions, as is the case with Concepción.[9]

[8] In this case, we are inspired by Werlen's ideas in 'The subjective standpoint' included in *Society, actions and space* (Werlen 1993, 67–68).

[9] Historically, Concepción has been subject to natural hazards, including earthquakes, tsunamis, and floods. This established a condition that somehow defined and impregnated a particular way of living, constantly rebuilding and permanently under tension. During the twentieth and early twenty-first centuries, earthquakes in 1939, 1960, and 2010 led to catastrophic devastation, as did massive floods in 1899, 1951, 1965, and 2006.

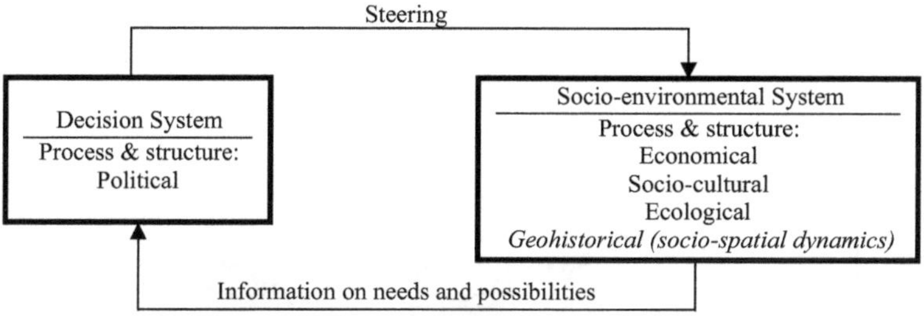

Fig. 2 Steering the socio-environmental system (own diagram, based on Musters et al. 1998, 244)

How do we understand the experience of living in a city intended for development if, in little more than 50 years, the results have been environmental problems? Years of planning and guidelines did not yield the expected results. The urban planning experience is a good example. Planning instruments created in 1963, 1980, and 2003 were unable to effectively steer the city's growth (Fig. 3).

One has to add the different forms of representation of welfare. Here, the production of discourses and images have shaped the articulation of new socio-environmental processes, such as the 'green view' of society and the elaboration of a new concept of welfare. The 'new welfare' is based on the importance of contact with nature, mass-represented by the color green.

The eruption of environmental problems in the area, which are tangible in everyday experience, created disappointment with modernity, particularly in Talcahuano. Environmental deterioration was an unanticipated adverse side effect of (industrial) work, progress, and economic growth.

As technical reports of that period indicated (Valenzuela 2002), air and water quality had reached critically low levels in the 1980s, to the extent that, in the early 1990s, a recovery program had to be created for Talcahuano (Programa de Recuperación Ambiental de Talcahuano, or PRAT) in order to assess the overall environmental conditions in the municipality and to design methods for gradual improvement (Aliste and Almendras 2010). When these plans to clean the air and water were implemented, a new appreciation of the environment began to emerge that was incompatible with the original purpose of urbanization, which was only designed along the lines of industrial production. Industry, which had attracted people to the area, was now viewed as the main cause of environmental problems. But the population is a problem too. Environmental analyses have indicated that the urbanized zones are in inappropriate locations. That is, urbanization has encroached on ecologically significant areas and/or areas subject to natural hazards; all of this has put pressure on the area's ecosystems (Smith and Romero 2009). Therefore, paradoxically, the pursuit of welfare and the actions taken to achieve it have caused problems that make welfare less accessible for the people living in the area. The promise of modernity dissolves as soon as new development discourse practices appear. Once the steps to achieve development appeared to move towards the correct path, ideological adjustment that sought sustainable development changed the course of things again. This caused significant frustration among important

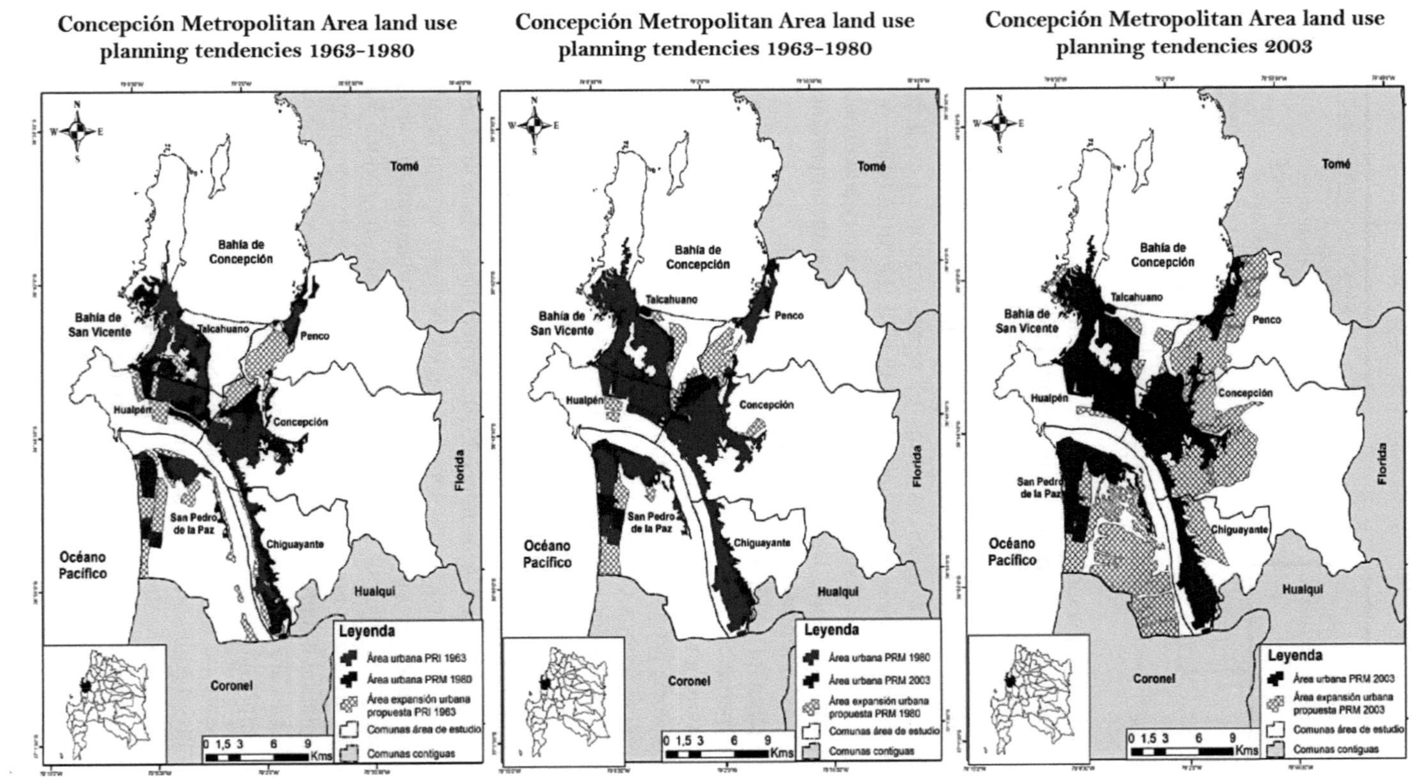

Fig. 3 Comparison between planning areas and actual growth areas in Greater Concepción, 1963–1980 (Muñoz 2011)

Fig. 4 Location and zoning of the logistics platform project (Bío Bío Plataforma Logistica 2004)

sectors of the population, especially those attached to developmentalist discourses (on the left and right of the political spectrum). Therefore, both the old (industrial) and the new paradigm of development (sustainable development) caused frustration. The new scenario is confusing. One of the first tangible results has been the tensions generated between the spatial practices and discourses of sustainable development, which were the result of contradictory actions that often led to inconsistent space usage decisions. Through these tensions, we can again explore the relevance of development imaginaries for the city's development over time. Here, discourse and action merge into what we can call factual development discourse.

A good example is the Plataforma Logística (Logistics Platform) project in Concepción. This large urban infrastructure project, which is intended for port activities, is located in an area that, according to previous studies, is of high ecological importance, which – notably – is used as an argument to justify the project. It is claimed that, with the implementation of this project, "it will be possible to add value to these areas" (Benoit and Pérez 2004, 3). On the other hand, this was one of the places that was hit hardest by the tsunami that affected Chile in February 2010 and flooded almost the entire proposed project area (Figs. 4 and 5).

As this figure shows, the project is located exactly in the areas most severely affected by the tsunami of February 2010. Furthermore, this land is precisely where territorial planning efforts were directed towards the preservation of ecological habitats and/or the restriction of land development for industrial or commercial use.[10] But beyond highlighting these contradictions between what is planned and what has been done, what matters here is the interpretation of the motives for these decisions and, therefore, what the sense is of environmental sustainability efforts.

How do we balance development and sustainability in emerging economies if the notion of development is in crisis?

[10] The regulatory plans and guidelines of the waterfront planning emphasize how significant these areas are for preservation and conservation purposes.

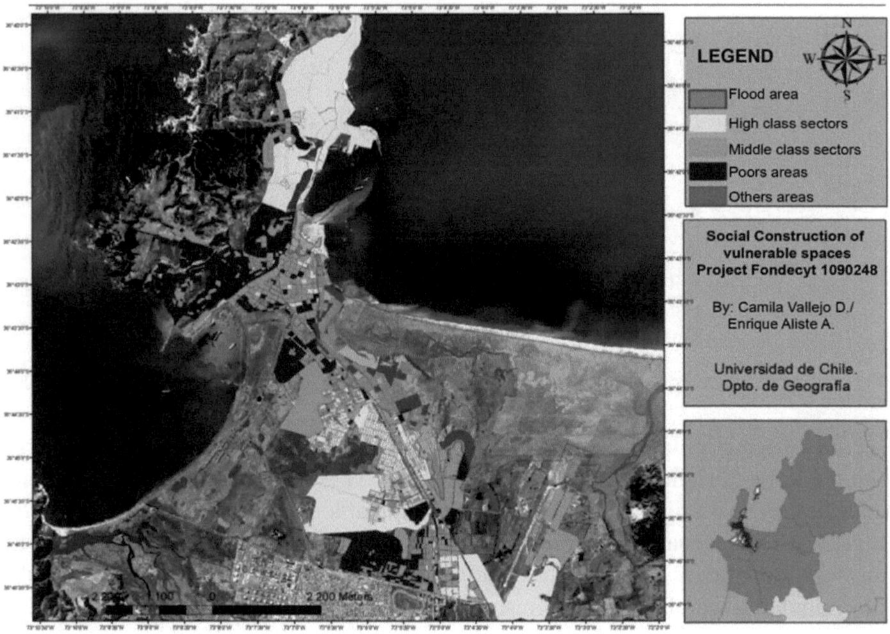

Fig. 5 Bay of Concepción, affected by the 2010 tsunami and urbanized areas around the wetlands of Rocuant-Andalién (Vallejo 2010)

5 Space, Science, and Modernity: Transdisciplinarity and New Geographies

After the case study described here, there may be a possibility to establish a transdisciplinary dialogue. From an analytical perspective, this notion will allow us to make progress regarding how environmental issues should be contextualized and how to propose sustainability strategies for cities in emerging economies.

In a transdisciplinary dialogue, many elements remain questions that need to be subjected to a thorough analysis and debate, because one of its basic aspects is the question: where do we observe from? This clarifies and describes the subject of our observations (Maturana 2009). Such an approach necessarily deals with the difficulty implied in the observations systems and with the validity of such observations.

In the previous section, I alluded to modern beliefs. Most decisions and actions that originated from developmentalism in the 1950s are oriented towards the implementation of measures that are strongly linked to rationality. Technological optimism and economic reasons for development drive government decisions and actions that are anchored in ideas and knowledge strongly committed to science and scientistic solutions to socio-environmental problems. Simultaneously, there are some ways of occupying and using space that, in spite of these rationalities, move in completely different directions. Given these facts, there is no doubt that the arrival of new ideas and concepts that require innovative frameworks from the public sphere – such as the idea of environmental sustainability – will not lack certain conflictive aspects, since meaning or interpretation may vary according to the interests involved and, particularly, according to each area's spatial experience.

This is easily observed in Concepción when it is explored from a geohistorical, transdisciplinary perspective; that is, by converging the analysis of geographic, sociological, anthropological, and ecological perspectives of the use and spatial trajectory of the city.

6 Conclusion

Given the above, one can conclude:

- In emerging economies, it is not viable to insist on a single perspective of the features that development models and ecological sustainability should meet, since development models depend on ideological adjustments and adequate timing. Further, such spaces may be very different, making it unviable to impose policies with one notion of sustainability or sustainable development onto very different socio-cultural contexts and geographical areas.
- A space perspective of the quest for environmental sustainability can help define appropriate sustainability criteria for a variety of socio-cultural contexts if it is sensitive to these contexts' historical and geographical specificities and situatednesses.
- Spatial knowledge is always the result of a dialogue that needs to pull together both social and spatial processes. Spatial forms are reflections of the social aspect, and the social aspect can be understood through spatial forms. If this is taken into consideration, the idea of environmental sustainability requires territorial approaches. In addition, geohistorical analysis provides an opportunity to reach these notions through transdisciplinary approaches.
- Such transdisciplinary approaches and analyses require observation criteria and broader interpretive frameworks that enable the processes that influence society when shaping territorial realities to be questioned, especially when these processes seek to pursue environmental sustainability. The perspectives provided by imaginaries of development and of sustainability are an opportunity to question how we observe phenomena and may help us focus on how we elaborate our dialogues and proposals regarding action in terms of territories and the implementation of sustainable development.

In short – based on the case presented in this article – it is argued that we need an active social geography. This social geography should seek to integrate time and space within a dynamic tailored for all territories and which considers the objective dimensions of classical spatial analysis while integrating them with interpretative observations. This social geography may help us analyze territories' hermeneutic levels and may help us understand the scope and possibilities of emerging discourses of postmodern geography in certain socio-cultural contexts. This opens a discussion about the dimension of sustainability and sustainable development in another perspective, perhaps outside the classical definition (such as the UNEP definition), to propose the need for other sources in different cultural and geohistorical contexts.

References

Aliste, E. (2010). Territorio y ciencias sociales: trayectorias espaciales y ambientales en debate. In E. Aliste & A. Urquiza (Eds.), *Medio ambiente y sociedad: conceptos, metodologías y experiencias desde las ciencias sociales y humanas* (pp. 55–76). Santiago de Chile: RIL Editores.

Aliste, E. (2011). Imaginarios del desarrollo en la dinámica del territorio del Gran Concepción, Chile: huellas de una transformación en la geografía social de la ciudad. *Revista Geográfica de América Central, 2*(47E), 1–14.

Aliste, E., & Almendras, A. (2010). Trayectoria territorial de la conurbación Concepción – Talcahuano: industria, asentamientos humanos y expresión espacial del desarrollo, 1950–2000. In L. Pérez & R. Hidalgo (Eds.), *Concepción Metropolitano (AMC). Planes, Procesos y Proyectos* (Serie Geolibros, pp. 123–149). Santiago de Chile: PUC-Centro EULA.

Aliste, E., Almendras, A., & Contreras, M. (2012a). La dinámica del territorio en la conurbación Concepción – Talcahuano: Huellas urbanas para una interpretación de las transformaciones ambientales durante la segunda mitad del siglo XX. *Revista de Geografía Norte Grande, 52*, 5–18.

Aliste, E., Contreras, M., & Sandoval, V. (2012b). Industrialización, desarrollo y ciudad: transformaciones socio-demográficas y espaciales en la geografía social del gran Concepción (1950–2010). *Revista INVI, 75*(27), 21–71.

Anderson, P. (1976). The antinomies of Antonio Gramsci. *New Left Review, 100*, 5–78.

Anderson, B. (1991). *Imagined communities: Reflections on the origin and spread of nationalism.* London: Verso.

Armstrong, W. (1973). Crítica de la teoría de polos de desarrollo. *EURE (Santiago), 3*(7), 113–123.

Arnold, M. (2003). Autoproducción de la amenaza ambiental en la sociedad contemporánea. *Revista MAD, 9*, 1–31.

Beltrán, M. (2011). *Transformaciones espaciales y valoración social del Humedal Rocuant-Andalién: una dinámica socio-ecológica de medio siglo.* Informe de práctica profesional (realizada en el marco del Proyecto FONDECYT 1090248). Santiago: FAU-Universidad de Chile. (unpublished)

Benoit, I., & Pérez, J. C. (2004). *Área de valor natural en el proyecto Plataforma Logística.* Informe técnico de la Subcomisión Infraestructura, Territorio y Medio Ambiente. Concepción (Chile).

Bernard, P. (1989). *Histoire du développement économique.* Paris: Éditions Marketing – École Polythécnique.

Bío Bío Plataforma Logística. (2004). *Bío Bío Plataforma Logística Chile.*http://www.plataformalogistica.cl/pages-proyecto-masterplan.html. Accessed 24 June 2013.

Braudel, F. (1997). *Les ambitions de l'histoire.* Paris: Éditions de Fallois.

Castoriadis, C. (1987). *The imaginary institution of society.* Cambridge, MA: MIT Press.

Claval, P. (2002). El enfoque cultural y las concepciones geográficas del espacio. *Boletin de la AGE, 34*, 21–39.

Coquery-Vidrovitch, C., Hemery, D., & Piel, J. (1988). *Pour une histoire du développement. États, sociétés, développement*. Paris: L'Harmattan.

De Fraja, E., Munari, S., Vismara, R., Zampaglione, D., Blanco, H., López, A., Malpei, F., & Ragazzi, M. (1993). *Estudio de prefactibilidad. Saneamiento de la cuenca hidrográfica del río Bio Bío y del área costera adyacente*. Concepción: Centro EULA, Universidad de Concepción.

Della Croce, N., Parra, O., Stuardo, J., Arrizaga, A., Ahumada, R., Chong, J., & Oyarzún, C. (1992). *El río Bio Bío y el mar adyacente como unidad ambiental*. Concepción: Centro EULA, Universidad de Concepción.

Di Filippo, A. (2007). La Escuela Latinoamericana del Desarrollo: tensiones epistemológicas de un movimiento fundacional. *Cinta de Moebio, 29*, 124–154.

Di Méo, G. (1991). *L'homme, la société, l'espace*. Paris: Anthropos.

Di Méo, G. (1998). *Géographie sociale et territoires*. Paris: Nathan Université.

Di Méo, G., & Buléon, P. (2005). *L'espace social. Lecture géographique des sociétés*. Paris: Armand Colin.

Echeñique, A., & Rodríguez, C. (1990). *Historia de la Compañía de Aceros del Pacífico S.A. Huachipato: consolidación del proceso siderúrgico chileno 1905–1950*. Santiago: CAP S.A. de Inversiones.

Escobar, A. (1996). *La invención del tercer mundo. Construcción y deconstrucción del desarrollo*. Barcelona: Grupo Editorial Norma.

Escobar, A. (2000). El lugar de la naturaleza y la naturaleza del lugar: ¿globalización o postdesarrollo? In A. Viola (Ed.), *Antropología del Desarrollo* (pp. 113–143). Barcelona: Paidós.

Escobar, A. (2008). *Territories of difference: Place, movements, life, redes*. Durham: Duke University Press.

Faletto, E. (2007). *Dimensiones sociales, políticas y culturales del desarrollo. Antología, selección e introducción de Manuel Antonio Garretón*. Santiago de Chile: Catalonia – FLACSO.

Foucault, M. (1971). *L'ordre du discours*. Paris: Gallimard.

Furtado, C. (1991). *Os ares do mundo*. Rio de Janeiro: Paz e Terra.

Furtado, C. (2008). *Criatividade e dependência na civilização industrial*. São Paulo: Companhia das Letras.

Giraut, F., & Antheaume, B. (2005). Au nom du développement, une (re)fabrication des territoires. In F. Giraut & B. Antheaume (Eds.), *Le territoires est mort. Vive les territoires!* (pp. 9–36). Paris: IRD Éditions.

Gligo, N. (2006). *Estilos de desarrollo y medio ambiente en América Latina, un cuarto de siglo después. Serie medio ambiente y desarrollo*. Santiago de Chile: Naciones Unidas, CEPAL.

Habermas, J. (1996). *Between facts and norms: Contributions to a discourse theory of law and democracy. Studies in contemporary german social yhought*. Hoboken: Wiley.

Harvey, D. (1996). *Justice, nature & the geography of difference*. Cambridge, MA: Blackwell Publishing.

Hernández, H. (1983). El Gran Concepción: desarrollo histórico y estructura urbana. Primera parte. Génesis y fundación. De las fundaciones militares a la conurbación industrial. *Informaciones Geográficas, 30*, 47–70.

Kitching, G. (1982). *Development and underdevelopment in historical perspective. Populism, nationalism and industrialization*. London: Methuen.

Lacoste, Y. (1965). *Géographie du sous-développement*. Paris: Presses Universitaires de France.

Legouté, J. (2001). Définir le développement: historique et dimensions d'un concept plurivoque. *Cahier de recherche* (Économie Politique Internationale) 1(1)

López-Ridaura, S., Masera, O., & Astier, M. (2002). Evaluating the sustainability of complex socio-environmental systems: The MESMIS framework. *Ecological Indicators, 2*(1–2), 135–148.

Macnaghten, P., & Urry, J. (1995). Towards a sociology of nature. *Sociology, 29*(2), 203–220.

Maturana, H. (2009). *La realidad: ¿objetiva o construida? Fundamentos biológicos de la realidad* (Vol. I). Barcelona: Anthropos.
Meadows, D., Meadows, D., Randers, J., & Behrens, W. (1972). *The limits to growth*. New York: Universe Books.
Merchand, M. A. (2007). Convergencia entre teorías que explican por qué hay territorios ganadores y otros perdedores. *Análisis Económico, 4*(12), 195–222.
Muñoz, F. (2011). *Instrumentos de planificación territorial del Área Metropolitana de Concepción (AMC) y su relación con los modelos de desarrollo (1960–2010)*. Memoria para optar al título profesional de Geógrafo (realizada en el marco del Proyecto FONDECYT 1090248). Santiago: FAU – Universidad de Chile. (unpublished)
Musters, C. J. M., de Graaf, H. J., & ter Keurs, W. J. (1998). Defining socio-environmental systems for sustainable development. *Ecological Economics, 26*(3), 243–258.
Parra, O., Acuña, A., & Olea, J. (1999). *Síntesis ambiental prospectiva del territorio de la Región del Bio Bío: Sistemas Naturales Claves*. SERPLAC Región del Bio Bío. Serie Estudios Prospectivos. Documento N° 4.
Peet, R., Robbins, P., & Watts, M. (2011). *Global political ecology*. New York: Routledge.
Perroux, F. (1955). Notes sur la notion de pole de croissance. *Economie Appliquée (Paris), 8*, 307–320.
Prebisch, R. (1983). Intereses de los países desarrollados y el desarrollo de América Latina. *Capítulos del SELA, 1*, 16–22.
Prebisch, R. (1984). Cinco etapas en mi pensamiento sobre el desarrollo. In G. Meier & D. Seers (Eds.) *Pioneros del desarrollo* (pp. 178–193). BIRF, Madrid: Tecnos.
Prebisch, R. (1985). Sobre la dependencia y el desarrollo. In Muñoz, H. (Ed.) *Crisis y desarrollo alternativo en Latinoamérica* (pp. 11–16). Academia de Humanismo Cristiano. Centro de Estudios de la Realidad Contemporánea; Instituto de Cooperación Iberoamericana. Santiago: Editorial Aconcagua/CERC/ICI.
República de Chile. Dirección de Estadísticas (INE). (1957). *III Censo de Manufacturas*. Santiago de Chile: Dirección General de Estadísticas.
República de Chile. Dirección General de Estadísticas (INE). (1942). *Industrias, 1942*. Santiago de Chile: Dirección General de Estadísticas.
República de Chile. Instituto Nacional de Estadísticas (INE). (1971). *IV Censo Nacional de Manufacturas*. Santiago de Chile: Instituto Nacional de Estadísticas (INE).
Rist, G. (2007). *Le développement: histoire d'une croyance occidentale*. Paris: Presses de la Fondation Nationale des Sciences Politiques (Sciences Po).
Robbins, P. (2004). Fixed categories in a portable landscape. The causes and consequences of land cover categorization. In K. Zimmerer & T. Bassett (Eds.), *Political ecology. An integrative approach to geography and environment-development studies* (pp. 181–2000). New York: The Guilford Press.
Sachs, I. (1981). Ecodesarrollo: concepto, aplicación, beneficios y riesgos. *Agricultura y Sociedad, 18*, 9–32.
Sánchez, C. (1952). *Evolución histórica de la industria siderúrgica chilena e ibero-americana*. Santiago de Chile: Nascimento.
Sanhueza, R., & Azócar, G. (2000). Transformaciones ambientales provocadas por los cambios económicos de la segunda mitad del siglo XIX; provincia de Concepción. *Revista Geográfica de Chile Terra Australis, 45*, 181–194.
Smith, N. (2008). *Uneven development. Nature, capital and the production of space*. Georgia: The University of Georgia Press.
Smith, P., & Romero, H. (2009). Efectos del crecimiento urbano del Área Metropolitana de Concepción sobre los humedales de Rocuant-Andalien, Los Batros y Lenga. *Revista de Geografía Norte Grande, 43*, 81–93. http://www.scielo.cl/scielo.php?pid=S0718-34022009000200005&script=sci_arttext.
Soja, E. (1989). *Postmodern geographies. The reassertion of space in critical social theory*. London: Verso.

Sunkel, O., & Paz, P. (1970). *El Subdesarrollo Latinoamericano y la Teoría del Desarrollo*. México: Siglo XXI.
Taylor, C. (2004). *Modern social imaginaries*. Durham: Duke University Press.
Truman, H. S. (1949). *Truman's inaugural Address*. Speech held at the swearing-in of the 33rd President of the United States of America, Capitol, Washington, DC. http://www.trumanlibrary.org/whistlestop/50yr_archive/inagural20jan1949.htm. Accessed 10 Dec 2012.
United Nations Department of Social and Economic Affairs. (1951). *Measures for the economic development of underdeveloped countries: Report by a group of experts appointed by the Secretary-General of the United Nations*. New York: United Nations.
Valenzuela, J. (2002). *Estudio de caso GWP: Recuperación Ambiental Integral de la Comuna de Talcahuano*. Global Water Partnership (GWP), Integrated Water Resources Management.
Vallejo, C. (2010). *Construcción social de territories vulnerables*. Informe de práctica profesional (realizada en el marco del Proyecto FONDECYT 1090248). Santiago: FAU-Universidad de Chile. (unpublished)
Wackermann, G. (2005). *Géographie du développement*. Paris: Éditions Ellipses.
Werlen, B. (1993). *Society, action and space. An alternative human geography*. New York: Routledge.

Part III
Society Nature Interfaces

Terrestrial Ecosystem Dynamics in Senegal's Agro-silvopastoral Center-East Region in the Second Half of the Twentieth Century

Aliou Diouf and Matthew G. Hatvany

1 Introduction

Terrestrial ecosystems are vital for humankind. They provide the essential life support systems we need to live: food, clean water, clean air, etc. However, in Africa, these terrestrial ecosystems have experienced unprecedented negative transformation rates over the past 60 years. Approximately one-fourth of the earth's terrestrial surface has been transformed into cultivated land systems. More wooded land was converted into cropland in the 30 years after 1950 than in the 150 years between 1700 and 1850. The pace and magnitude of these changes were such that ecosystems' sustainability and humankind's survival were in question. This prompted the UN to order a global study – called the Millennium Ecosystem Assessment (MEA) – on ecosystems' conditions, services, and their role in human welfare in 2005. The MEA concludes that the changes ecosystems are undergoing have globally resulted in a deterioration of ecosystems conditions and services since the beginning of the second half of the twentieth century. These transformations have resulted in a loss of two-thirds of ecosystems services in the last half of the twentieth century alone (MEA 2005).

To understand these current environmental changes, the issues of the scale of the study and the drivers of change are among the more significant questions addressed. Regarding scale, the importance of small-scale studies is recognized as a major way to better understand ecosystems dynamics. These studies are especially apt, given that the daily interactions between humans and ecosystems are often local in scale. Concerning drivers, globalization is now perceived as the main cause of ecosystems

A. Diouf (✉)
Enda Energie-Environnement-Développement, 54 rue Carnot, BP: 3370 Dakar, Senegal
e-mail: aliou_diouf@hotmail.com

M.G. Hatvany
Département de géographie, Université Laval, Pavillon Abitibi-Price, Québec, Québec G1V 0A6, Canada

B. Werlen (ed.), *Global Sustainability*, DOI 10.1007/978-3-319-16477-9_8

dynamics (Lambin et al. 2001). Furthermore, this capitalism-related process explains local environmental changes (Blaikie and Brookfield 1987). Indeed, globalization is closely related to the expansion of capitalism. By interconnecting the circulation of goods, capital, information, and people, as well as by fostering the sharing of knowledge, lifestyles, and institutions, globalization has engendered social, economic, demographic, political, and cultural changes. These changes have led to environmental transformations throughout the world (Global Land Project 2005, 10). Thus, linking colonial rule with the expansion of capitalist economic systems in the late nineteenth century has induced transformative changes in lifestyles and terrestrial ecosystems across Africa (Heilig 1993; Bassett 2006).

This historical pattern has been observed and studied in Senegal. Gonzalez (2001) used a vegetation inventory together with socio-economic surveys, and noted a decline in floristic richness and density in north-western Senegal. He found that plant species richness declined by 33 % between 1945 and 1993, while the density of trees with a height equal to or greater than 3 m decreased by 23 % between 1954 and 1989. Using a similar approach, Diouf (2001) showed that the vegetation density in southern Ferlo (Thieul), Senegal's main pastoral area, declined significantly between 1970 and 2000. In the south, Mbaye (1998, 236) analyzed land cover, showing that woody cover had declined significantly in both the southern and the northern parts of the southern Sudanese ecosystem. Wood and colleagues (2004) analyzed land use and land cover changes, revealing that natural land had been converted into agricultural land at a rate of 27.77 %, corresponding to an annual rate of 4 ha between 1973 and 1999.

Regarding the Sudan center region of Senegal, which corresponds to our study area, few studies have focused on the state of terrestrial ecosystems, even though this region includes some of the richest land in Senegal, which is recognized as a privileged area for agro-silvopastoralism.[1] Agriculture, livestock, and lumbering to make charcoal have taken place in this region for many decades. In addition, since the 1970s, the region has seen the establishment of large flows of pastoralists, loggers, and especially farmers. The convergence in this region of these migration flows and of three land use forms, often with conflicting interests, but using and competing for the same natural resources, necessarily leads to important environmental transformations. Tappan and colleagues (2004) conducted a study on the state of natural resources in this region. They noted a decrease in woody cover from

[1] Agro-sylvopastoralism is a production system that combines agricultural, forestry, and pastoral forms of production. It does not consist of a juxtaposition of three production systems, but is instead based on maximizing the synergies between these various production regimes to take advantage of the complementarity between trees, crops, and animals (Translation Bureau 2009). In this article, the term "agro-silvopastoral" is associated with an area and not a unit of socio-economic production (household or concession). Thus, agro- sylvopastoralism refers to the synergies of agriculture, forestry, and pastoralism within a geographical area. An "agro-silvopastoral region" is an area where agriculture, forestry, and pastoralism are the main livelihood.

14 %–20 % to 6 %–10 % in the western part of the agro-silvopastoral center-east region of Senegal, and to 1.9 % in the eastern part between 1965 and 1994.

Despite these insightful findings, local-level terrestrial ecosystems dynamics have not been properly investigated in Senegal, particularly in Senegal's agro-silvopastoral region. The woody cover that Tappan and colleagues (2004) examined is inappropriate for measuring the evolution and state of the vegetation. This environmental indicator is too poor in information content to help us understand the dynamics and state of the ecosystem. Tappan and co-workers (2004) provide no information on the dynamics and state of other ecosystem components, such as the soil and surface water, nor on the natural vegetation diversity. Natural vegetation should not be addressed as a homogeneous ecosystem component. The dynamics of this region's soil, surface water, and different vegetation units in the twentieth century is not yet well known.

How did terrestrial ecosystems evolve in the twentieth century? We assume that, in Senegal's agro-silvopastoral center-east region, terrestrial ecosystems were degraded in the second half of the twentieth century due to the region's link to the international capitalist market. This article assesses the dynamics of Senegal's agro-silvopastoral center-east region's ecosystem components, including its vegetation, soils, and surface water in the last half of the twentieth century in.

2 Methods and Tools

2.1 Region, Sub-region, and Study Sectors

Senegal's agro-silvopastoral center-east region covers 13,000 km^2 and is characterized by gravelly ferruginous tropical soils (Michel 1973, 671) and sizable dense savannah, which gives it a forest area character. The human perspective shows that in the agro-silvopastoral center-east region of Senegal, farming, pastoralism, and logging are the main land uses.

Given the size of the study area, a spatial sampling was undertaken on two levels to choose a less extended area for the study. First, three homogeneous sub-areas were identified: an agricultural sub-area dominated by peanut cultivation in the west, a pastoral one in the north-east, and a logging sub-area in the south-east. After this first level of sampling, study sectors had to be chosen where data would be collected: one for each sub-region. To identify these study sectors, a village center had to be chosen for its economic and demographic dominance of the surrounding villages (Fig. 1).

Thus, after triangulating data from the local government services, local authorities, and people, *Koumbidia Soce* was the chosen village center in the agricultural sub-region, *Wouro Seeno* in the pastoral sub-region, and *Fadiyacounda* in the logging sub-region. Once the village centers had been identified, three study sectors were defined around each village center. Each study sector had to be sufficiently

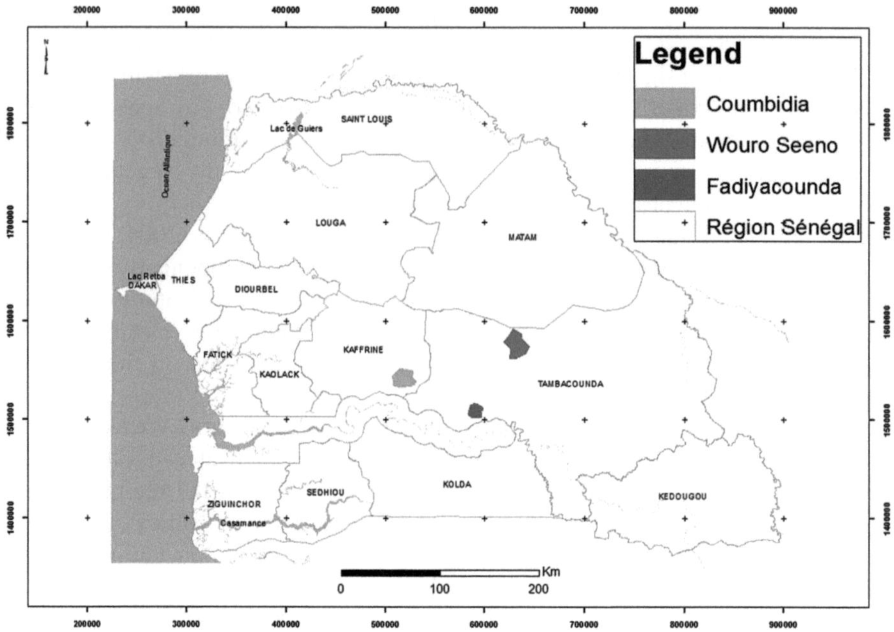

Fig. 1 Study sectors (Stancioff et al. 1986, 496)

extensive (at least a radius of 10 km) to encompass the sub-area's main socio-economic and ecological characteristics.

2.2 Sources, Data Collection, and Data Analysis

Aerial photographs from 1954 (AOF Mission 1954a, b, c) and 1982 (Organisation pour la Mise en Valeur du Fleuve Gambie 1982a, b, c), satellite images from 1999, topographic and thematic maps, as well as bibliographic and socio-economic data were used to analyze the dynamics of the land systems in Senegal's agro-silvopastoral center-east region. Both sets of aerial photographs were taken in the visible range of the electromagnetic spectrum, but the 1954 set had a scale of 1:40,000, while that of 1982 was 1:25,000. The satellite images used were those of scene 204-50 of November 13, 1999 of the *Koumbidia Soce* sector and those of scenes 203-50 and 203-51 of October 21, 1999 of the *Wouro Seeno* and *Fadiyacounda* sectors. The thematic maps used were all from 1986.

Data was collected using a global positioning system (GPS) device and a biophysical inventory form. The GPS was used to record the geographic coordinates of human settlements and biophysical inventory stations. These stations were chosen on the basis of satellite images' spectral signatures, which show several thematic unities. Each thematic class corresponds to a soil and vegetation type, or surface water, or pond unity. In each thematic class, at least two inventory stations were selected for a representative coverage of the unity. The biophysical inventory form was used to note observations on vegetation density, vertical stratification, air recovery by tree crowns, soil surface and pond characteristics (Aubreville 1957; Programme de Gestion Durable et participative des Energies traditionnelles et de substitution 1997; Mbaye 1998). These ecological parameters were measured in respect of a square of 10,000 m^2. Thereafter, the vegetation physiognomy, soil surface state, and pond sizes were evaluated and indicators derived.

Data analysis was undertaken by means of photo interpretation, satellite image processing, assemblage, georeferencing, interpreting, and vectorizing the spatial thematic unities identified on the photographs. The vectorizing of the thematic unities was done by combining stereoscopy and photo interpretation on computer screens. The satellite image processing entailed creating false color composites, a linear calibration set with a saturation threshold of 5, initial automatic classification, field verification, the selection of areas, or polygon training, supervised classification by means of the *maximum likelihood* algorithm, image classification assessment by means of a confusion matrix, and filtering of the classified image. Mapping comprised topological structuring, associating attributes, and calculating the surfaces of the vectorized spatial thematic unities from the aerial photographs. The final phase of the mapping was to assign colors, build the UTM grid coordinates, establish the legend, specify the scale and projection system, as well as determine the source of the data, the date this was collected, and adding the name of the map author.

2.3 Indicator and Approaches

The environmental indicator approach is closely associated with environmental and development issues. To measure the state of the environment and its relationship with economic development, the OECD developed environmental indicators in 1989 (Organisation for Economic Co-operation and Development 2003, 6). The indicators in this study were derived from ecosystems' main components – vegetation, soil, and surface water. The indicator construction was based on criteria found in the field and on the satellite and aerial images. These criteria included the soil surface state, pond sizes (for surface water), and vegetation density on several levels. The same criteria were used for the aerial and satellite images, based on their spectral signature, shape, location, texture, and structure.

Four environmental indicators were determined for vegetation: the proportion of very dense vegetation surface with an estimated density of 80–100 %, the proportion of dense vegetation surface with a density of 60–80 %, the proportion of moderately dense vegetation surface with a density of 40–60 %, and the proportion of sparse vegetation surface with a density of 1–20 %. The soil indicator was the proportion of bare soil surface. The water surface indicator was the proportion of pond surface. Other spatial indicators were the proportion of Koungheul urban agglomeration surface and the proportion of agricultural area surface, which includes crop fields, young fallows, average fallows, logging areas, and villages' surface areas.

Once the environmental indicators were defined to assess the state of the land cover components, two indicator categories were identified. One category measured the qualitative and quantitative content of each land cover component and provided information on its (im)balanced state. The qualitative content was obtained through spectral signature interpretation or through an evaluation of the pond or soil surface state. The quantitative content was determined through spectral signature interpretation and a field estimation of the vegetation density. This category included *environmental balance indicators* and *environmental imbalance indicators*. In this article, *balance* does not exclude the existing temporary disturbances in the land cover components. A balanced state is not a static or stationary state; it can vary from a norm (in the sense of a normal value) within certain limits. A balanced environmental indicator mainly implies the absence of sustainable and radical perturbations that lead to changes in structures and functioning. On the other hand, *imbalance* refers to the change and introduction of new features and new identity that can result in rupture of the system (Lapointe 1993). In this article, the *environmental balance indicators* comprise the following indicators (Table 1):

The proportion of dense vegetation surface corresponds to physiognomic units called *dense gallery forest* and *dense wooded savannah*, which constitute the proportion of moderately dense vegetation units. The proportion of sparse vegetation units includes the following physiognomic units: *sparse shrub savannah* and *sparse gallery forest*.

Another category includes natural and human land cover classes and aims to apprehend the degree of humanization of each of the three studied sectors. The term natural land cover class is specifically applied to terrestrial components whose existence is in no way linked to human intervention and whose nature has not been destroyed. In contrast, any terrestrial component whose original existence is due to humans, or which humans have radically transformed, is termed a human land cover class (Table 2).

The *degree of humanization* refers to the ratio between the human land cover classes area and the total sector area. The state of the terrestrial components is assessed by their degree of humanization and their environmental content or quality, which refer to the vegetation unit density level, to the pond sizes in respect of water surface, and to the surface state in respect of soil.

A diachronic analysis highlighted the land cover dynamics in the second half of the twentieth century by analyzing images from 1954, 1982, and 1999.

Table 1 Environmental balance indicators and corresponding land cover units

Balance indicator	Corresponding land cover units
Proportion of very dense vegetation surface	Very dense gallery forest
	Very dense wooded savannah
Proportion of dense vegetation surface	Dense gallery forest
	Dense wooded savannah
Proportion of pond surface	Pond surface
Imbalance indicator	**Corresponding land cover units**
Proportion of moderately dense vegetation surface	Moderately dense wooded savannah
	Moderately dense shrub savannah
Proportion of sparse vegetation surface	Sparse gallery-forest
	Sparse shrub savannah
Proportion of bare soil surface	Bare soil
Proportion of agricultural area	Agricultural area
Proportion of urban agglomeration surface	Urban area

Table 2 Natural and human land cover classes

	Very dense vegetation surface
Natural land cover classes	Dense vegetation surface
	Pond surface
	Moderately dense vegetation surface
	Sparse vegetation surface
Human land cover classes	Agricultural area surface
	Koungheul urban agglomeration surface
	Bare soil surface

3 Results

3.1 *State of Terrestrial Ecosystems in 1954*

3.1.1 Degree of Humanization

Aerial photograph analysis showed moderate humanization in 1954 in *Koumbidia Soce*, a sector where rain-fed agriculture was predominant. Human indicators – the Koungheul urban agglomeration surface and agricultural area – covered 18.23 % of the sector area. In contrast, natural indicators, such as the ponds and various vegetation units, comprised 81.77 % of the sector area. In *Fadiyacounda*, aerial photograph analysis showed that the degree of humanization in 1954 was relatively low. The human land cover class, corresponding to the agricultural area, comprised 17.28 % of the sector. Natural land cover classes – the pond surface, dense vegetation units, and very dense vegetation units – comprised 82.72 % of the sector area (Fig. 2).

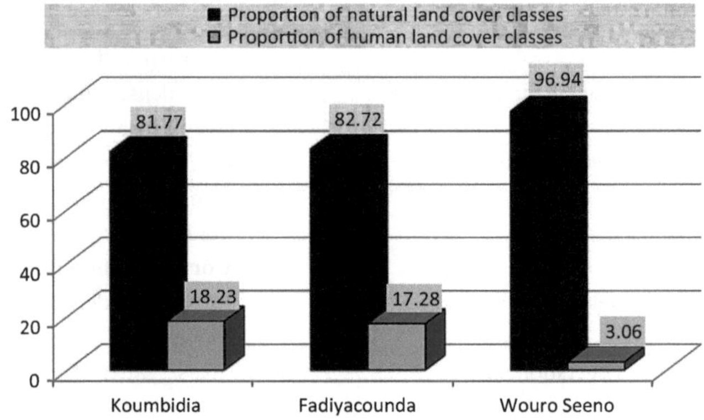

Fig. 2 Proportion of surface of human and natural land cover classes in 1954 (Own diagram based on aerial photographs (Koumbidia Soce, Fadiyacounda, Wouro Seeno) by AOF Mission (1954a, b, c))

In *Wouro Seeno*, aerial photograph analysis showed a very low degree of humanization in 1954. The agricultural area, which was the only human indicator, comprised a mere 3.06 % of the sector.

3.1.2 Environmental State

The environmental balance indicators covered only 50.28 % of the *Koumbidia Soce* sector, while the environmental imbalance indicators (including moderately dense vegetation, agricultural areas, and the Koungheul urban agglomeration surface) comprised 49.72 %, an area almost as large as that of the environmental balance indicators. Environmental imbalance was most prevalent in the vegetation and in the humanized areas. The environmental degradation occurred through the decreasing wood density and the conversion of wooded areas into agricultural areas. In *Fadiyacounda*, there was only one environmental imbalance indicator – the agricultural area –, which comprised 17.28 % of the sector. The natural indicators found in *Koumbidia Soce* in 1954 – the very dense vegetation unit, dense vegetation unit, and pond surface – were balance indicators. In the *Wouro Seeno* sector, vegetation comprised the main land cover unit, more than 95 % of the sector area. The vegetation unit was composed of two classes – dense wooded savannah and very dense wooded savannah – characterized by a truly balanced state (Fig. 3).

3.1.3 Spatial Distribution of Indicators

It is important to underline that in *Koumbidia Soce*, the environmental balance indicators were located on low hardpan plateaus. They were far from humanized areas, occupying slopes with a vegetation imbalance.

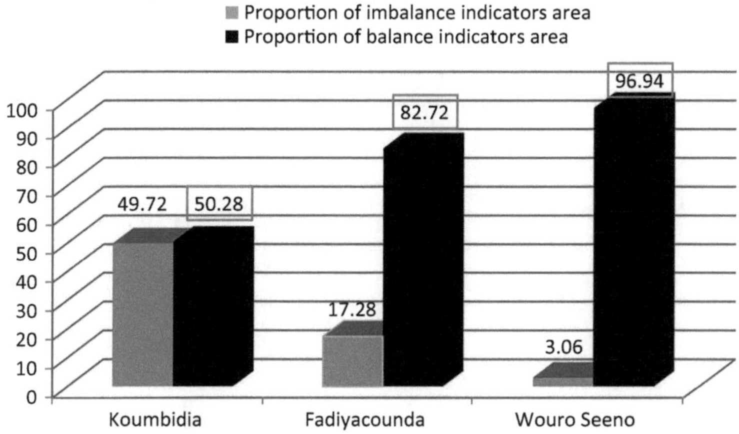

Fig. 3 Proportion of area of environmental balance and imbalance indicators in 1954 (Own diagram based on aerial photographs (Koumbidia, Fadiyacounda, Wouro Seeno) by AOF Mission (1954a, b, c))

In *Fadiyacounda*, all the balance indicators were confined to low gravel or hardpan plateaus, to the *Sandoungou* alluvial valley, and other small valleys. Humanized areas, such as villages and fields, were located on slopes adjacent to the alluvial valleys. Agricultural areas, characterized by irregular geometric shapes, were distributed symmetrically on both sides along farming tracks. The predominance of the balance indicators indicated an environmentally balanced state. In *Wouro Seeno*, the spatial distribution was similar to that of *Fadiyacounda*. Most of the balance indicators were situated on the low hardpan or gravel plateaus. The agricultural area also lay on the sides of valleys. It showed irregular geometric shapes and was striking due to its spatial distribution into islets around villages. The preeminence of natural and balance indicators in the land cover showed that the *Wouro Seeno* sector was in a state of environmental balance (Fig. 4).

3.2 *State of Terrestrial Ecosystems in 1982*

3.2.1 Degree of Humanization

In 1982, the *Koumbidia Soce* sector had a high humanization level. The humanized area covered about half of the entire sector: 49.15 % (vs. 18.23 % in 1954). It comprised the Koungheul urban agglomeration surface, bare soil surface, and agricultural area surface; the latter was also the main cause of the humanization. The *Fadiyacounda* sector showed a low degree of humanization. The human land cover classes, comprised of the agricultural area surface and bare soil surface, covered 25.77 % of the sector. Most of the humanized area (85.06 %) was

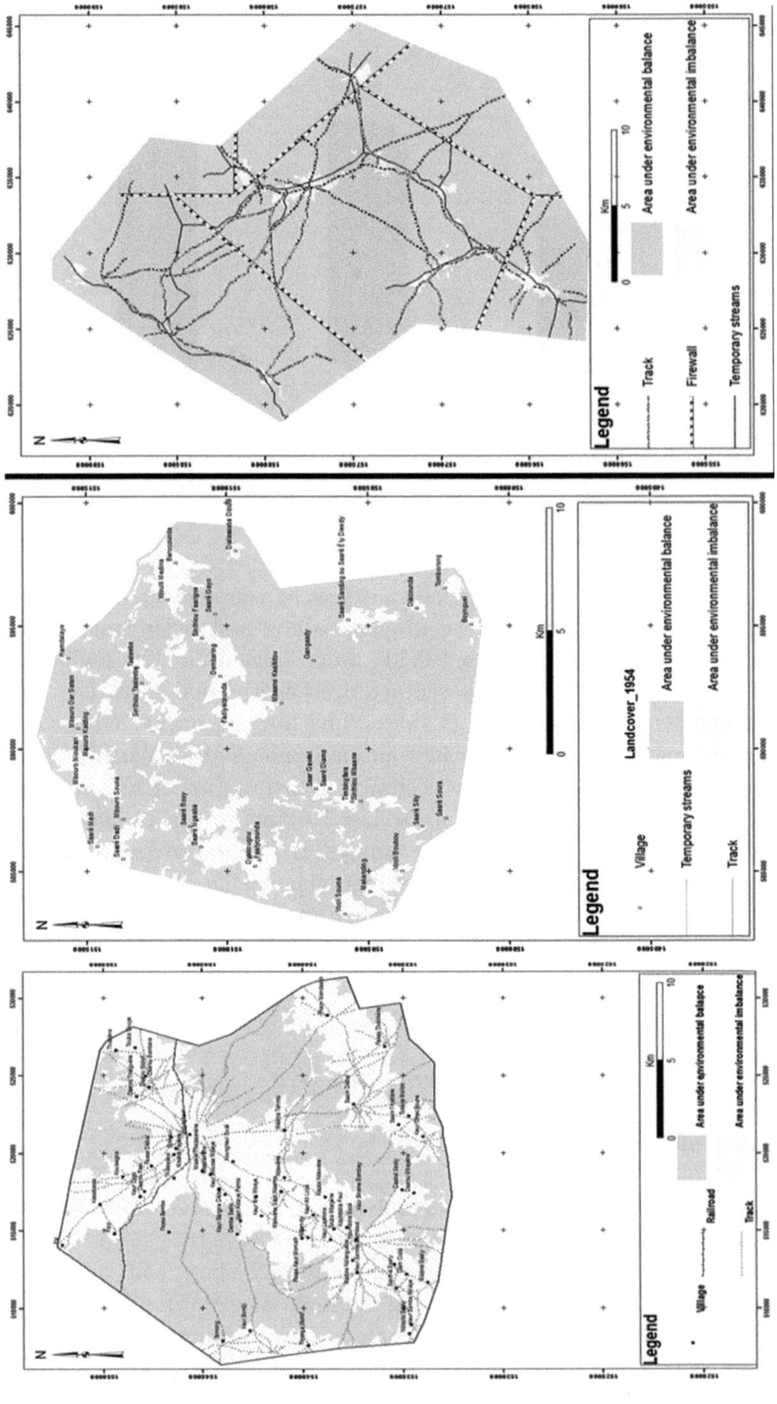

Fig. 4 Land cover map of the three study areas in 1954 (Own maps based on aerial photographs (Koumbidia, Fadiyacounda, Wouro Seeno) by AOF Mission (1954a, b, c))

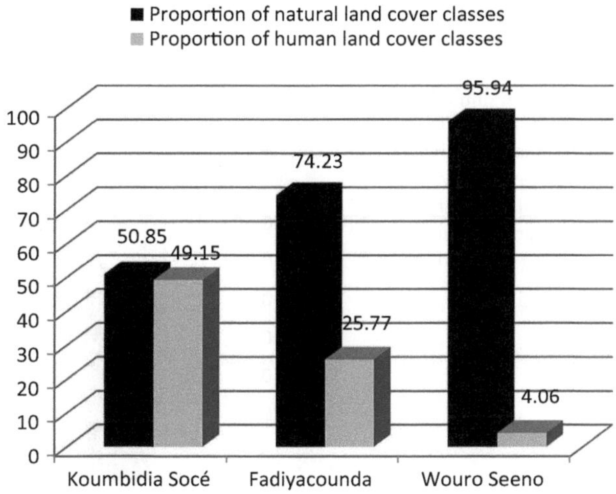

Fig. 5 Proportion of the surface of human and natural land cover classes in 1982 (Own diagram based on aerial photographs (Koumbidia, Fadiyacounda, Wouro Seeno) by Organisation pour la Mise en Valeur du Fleuve Gambie (1982a, b, c))

comprised of agricultural area surface. Natural land cover classes were still predominant at the time (Fig. 5).

A low degree of humanization characterized *Wouro Seeno*, the pastoral sector, since human activities had only radically transformed 4.06 % of the sector surface.

3.2.2 Environmental State

The predominance of environmental imbalance indicators characterized *Koumbidia Soce:* 52.79 % of the sector. The agricultural area surface, which comprised 90.90 % of the total area covered by imbalance indicators, dominated the indicators. Other imbalance indicators were: the urban agglomeration, bare soil surface, and sparse vegetation surface (respectively 0.67 %, 0.48 %, and 3.64 % of the sector). Besides the pond surface, the only balance indicator was the dense vegetation surface, which covered 47.16 % of the *Koumbidia Soce* sector. In the *Fadiyacounda* sector, the environmental balance indicators covered 74.31 % of the sector. The environmental imbalance was related to the agricultural area (21.89 %) and bare soil, which covered 3.80 % of the entire sector. In *Wouro Seeno*, the environmental imbalance indicators, which the agricultural area represented, covered only 4.06 %. The balance indicators – very dense vegetation, dense vegetation surface, and ponds – covered 95.94 % of the sector. The photo interpretation showed that a balanced environmental state characterized the *Wouro Seeno* sector in 1982 (Fig. 6).

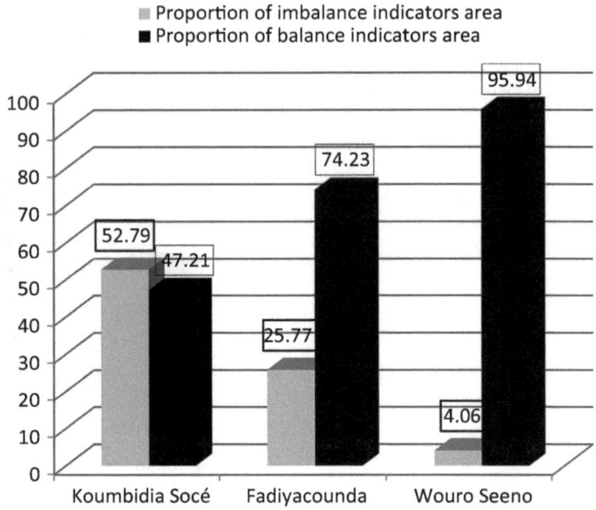

Fig. 6 Proportion of area of environmental balance and imbalance indicators in 1982 (Own diagram based on aerial photographs (Koumbidia, Fadiyacounda, Wouro Seeno) by Organisation pour la Mise en Valeur du Fleuve Gambie (1982a, b, c))

3.2.3 Spatial Analysis

In terms of spatial analysis, the agricultural area surface was distinguished by its spatial development. Besides sparse vegetation, there were no longer vegetation islets in the area of the slopes. Furthermore, imbalance indicators characterized the valley slopes. On the other hand, away from villages and field areas, on the low hardpan plateaus, balance indicators were identified. Given the larger area that the imbalance indicators covered, the *Koumbidia Soce* sector experienced environmental degradation in 1982. In *Fadiyacounda*, all the balanced vegetation indicators were found on hardpan plateaus, on slopes, and in alluvial valleys. However, they were absent in the areas surrounding villages. Bare soil surfaces were often located in the upper slopes of valleys. Ponds were confined to the middle of plateaus, while the agricultural area comprised islets around villages on low slopes, both thus characterized by spatial discontinuity. In *Wouro Seeno*, all the environmental balance indicators related to vegetation and surface water were confined to geomorphologic units: plateaus, slopes, and valleys. However, ponds and agricultural area were located on plateaus and on valley slopes around human settlements (Fig. 7).

3.3 State of Terrestrial Ecosystems in 1999

3.3.1 Degree of Humanization

In 1999, satellite image analysis showed that humanized indicators covered more than half of the *Koumbidia Soce* sector (up to 54.20 %). The *agricultural area class*

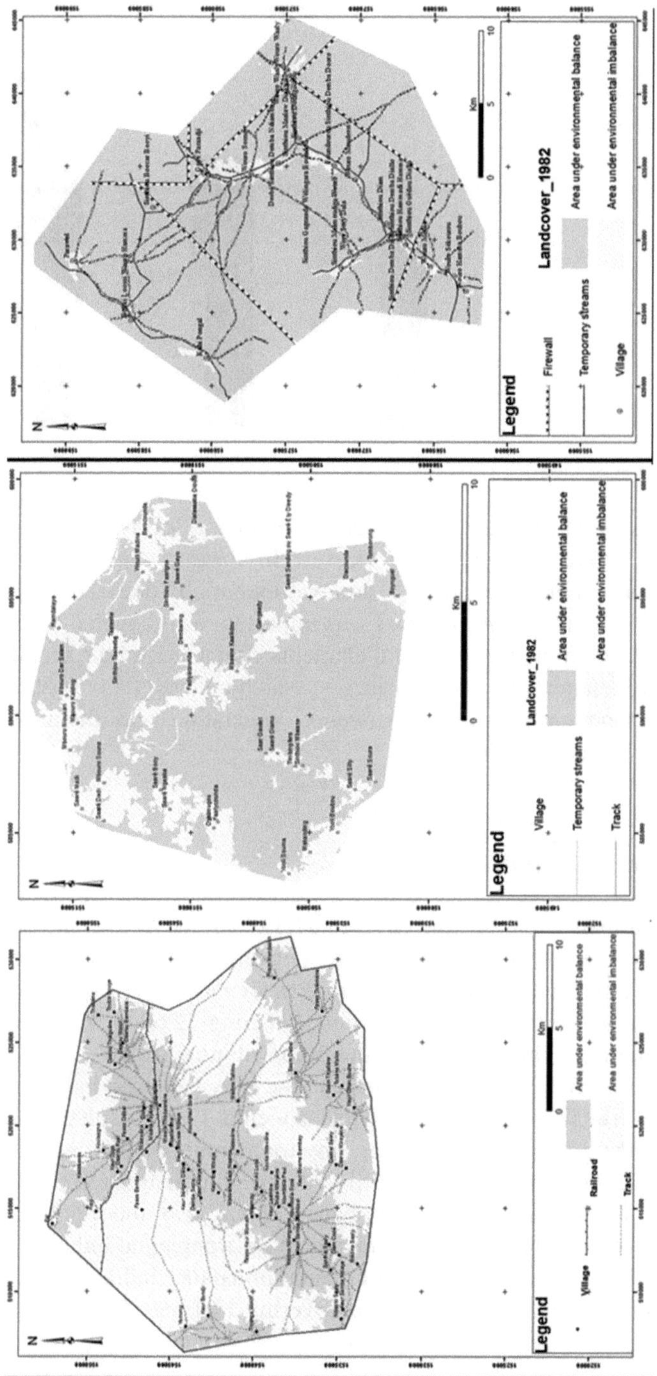

Fig. 7 Land cover map of the three study areas in 1982 (Own maps based on aerial photographs (Koumbidia, Fadiyacounda, Wouro Seeno) by Organisation pour la Mise en Valeur du Fleuve Gambie (1982a, b, c))

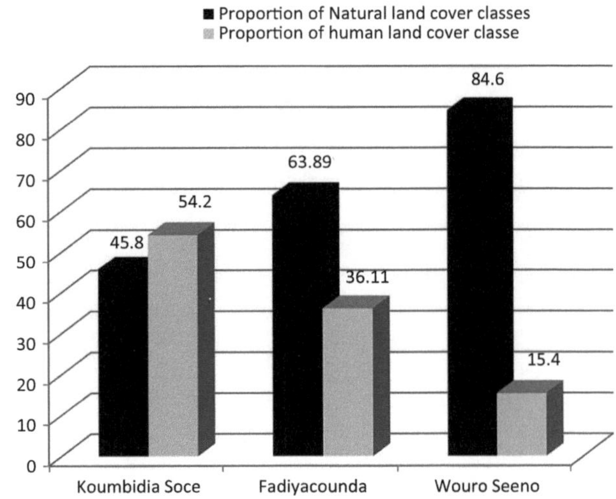

Fig. 8 Proportion of surface of human and natural land cover classes in 1982 (Own diagram based on Global Land Cover Facility (1999a, b, c))

(52.22 % of the sector) was the main humanization factor, followed by the *bare soil* and *Koungheul urban agglomeration* classes. In *Fadiyacounda*, the degree of humanization was evaluated at 36.11 % and was essentially due to the *agricultural area class*. However, the *bare soil class* was important as it covered 6.05 % of the entire sector. A predominance of natural elements characterized the sector. Natural components – vegetation and water surface –, which covered 84.6 % of the sector, dominated the land cover of *Wouro Seeno*. Vegetation was the main spatial component in this study area (Fig. 8).

3.3.2 Environmental State

Environmental imbalance indicators covered 87.33 % of the *Koumbidia Soce* sector study area. In this sector, the agricultural area covered the largest part (52.22 %). The vegetation imbalance indicators *proportion of moderately dense vegetation unit* and *proportion of sparse vegetation unit*, covering 32.67 % of the study area, were, besides the agricultural area, the main contributors to the state of environmental degradation. Besides the pond surface, the only balanced indicator was the proportion of *dense vegetation units*, which covered 12.34 % of the *Koumbidia Soce* sector. In the *Fadiyacounda* sector, environmental balance indicators comprised 61.07 % of the sector. The proportion of agricultural area (30.06 %) was the main indicator. The surface occupied by bare soil was also important, comprising 6.05 % of the sector. In *Wouro Seeno*, imbalance environmental indicators covered 99.98 % of the sector. Most of the environmental imbalance indicators were related to vegetation units, with *agricultural area* covering 14.16 % (Fig. 9).

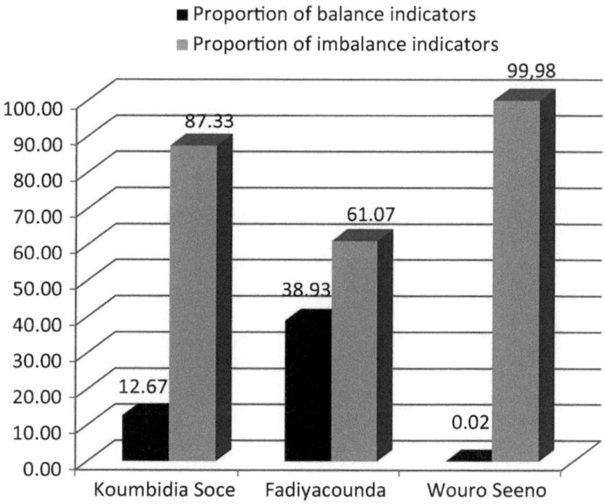

Fig. 9 Proportion of area of environmental balance and imbalance indicators in 1999 (Own diagram based on Global Land Cover Facility (1999a, b, c))

3.3.3 Spatial Analysis

In terms of spatial analysis, the agricultural area in *Koumbidia Soce* occupied almost all the slopes and valleys; it had also spread beyond the slopes and occupied the low gravel plateaus, reducing the vegetation cover to a set of spatially scattered relics on the plateaus. These vegetation cover fragments were the *Maka Yopp* and *Koungheul* protected areas. Bare soils were identified in the agricultural area and on the upper slopes. Most ponds were found in the sector's main valley. The few relics of vegetation were concentrated on hardpan and graveled plateaus.

In *Fadiyacounda*, all the vegetation indicators – which were balanced – were found on hardpan plateaus, on slopes and in alluvial valleys. However, they were absent in the areas surrounding villages. Bare soil surfaces were often located on the upper slopes of valleys. Ponds were confined to the middle of plateaus, while the agricultural area comprised islets around villages on low slopes. Therefore, spatial discontinuity characterized the vegetation indicators. In *Wouro Seeno*, on the plateaus, slopes, and valleys, imbalance indicators covered almost the entire sector (Fig. 10).

4 The 'Death of the Bush' in the Twentieth Century

This article studied the dynamics of terrestrial ecosystems in the Senegalese agro-silvopastoral center-east region during the second half of the twentieth century based on the level of humanization and the environmental content or quality.

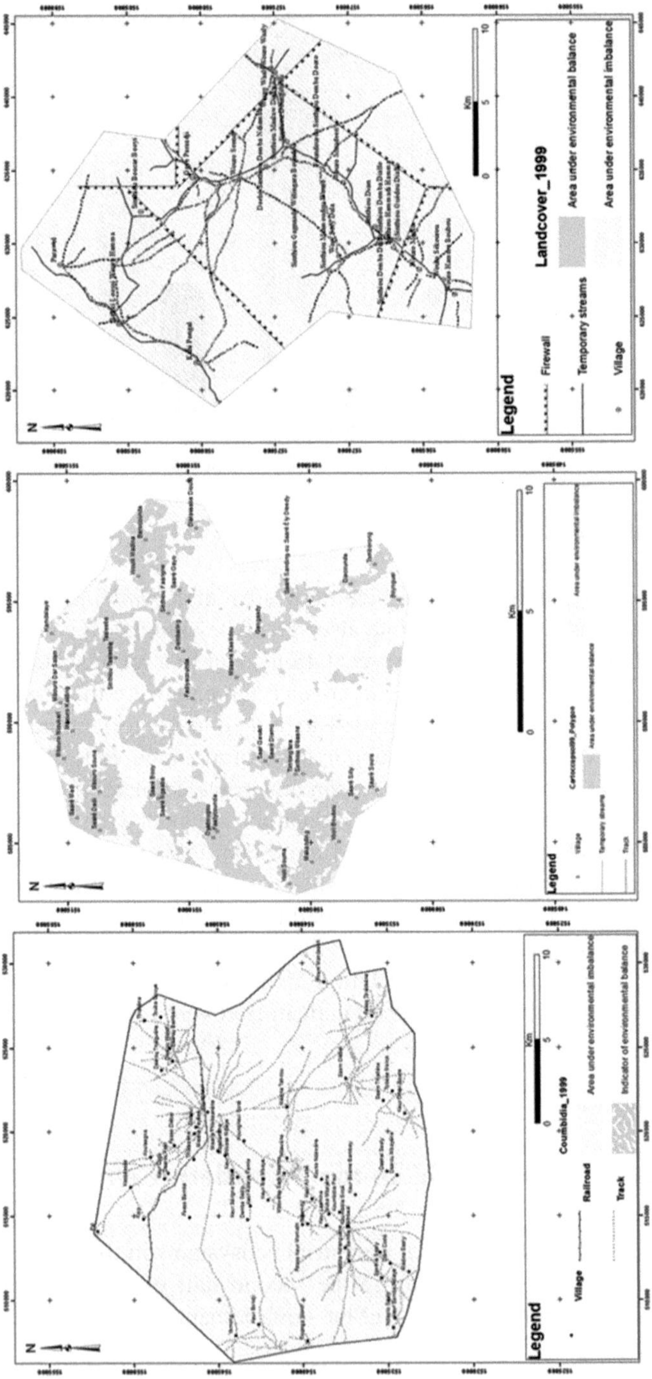

Fig. 10 Land cover map of the three study areas in 1999 (Own maps based on Global Land Cover Facility (1999a, b, c))

4.1 An Increasing Process of Humanization

The results of the study showed a low degree of humanization in 1954. Humans had only transformed 13.86 % of the region. At this time, a predominance of natural ecosystems characterized Senegal's center-east region. This confirms the findings of Pélissier (1966), who described the region as an area left to the hegemony of nature. However, it should be noted that the degree of humanization was not evenly distributed in the study area. In *Wouro Seeno*, the pastoral area, the natural features were prominent. Several factors contributed to this characteristic. The low population density in the area compared to that of *Koumbidia Soce* and *Fadiyacounda*. Further, pastoralism is a low-footprint activity, which also explains the low degree of humanization in the pastoral domain.

By 1982, the degree of humanization in the study area had doubled, compared to 1954. From 13.86 % humanization in 1954, human activities in the region increased to 26.33 %. However, in 1982, a preponderance of natural ecosystems still marked the region. It is important to emphasize the specific case of the area where rain-fed agriculture was the main activity – *Koumbidia Soce*. Human activities had profoundly transformed almost half (54.2 %) of this sector.

In 1999, the degree of humanization in all three sectors had become more important, covering 36.24 %. While the overall expansion rate of the area under permanent human footprint was 10 % between 1982 and 1999 (as opposed to 12.47 % between 1954 and 1982). It is important to note the average annual expansion rate was higher between 1982 and 1999 than between 1954 and 1982. From 1954 to 1982, the degree of humanization increased by 0.44 % per year, while it grew by 0.55 % per year between 1982 and 1999.

4.2 Increasing Environmental Imbalance

In terms of environmental content or quality, an analysis of the indicators of the environmental balance and imbalance in the three study areas in 1954 showed an overall state of dominant environmental balance. The proportion of the area in ecological balance in the region was estimated at 76.65 %. It should be noted that the *Koumbidia Soce* sector, which is dominated by rain-fed agriculture, showed a more advanced level of environmental imbalance – almost half (49.72 %). In terms of environmental quality, Senegal's agro-silvopastoral center-east was generally characterized by environmental balance in 1982. Nearly three-quarters (72.46 %) of the region was in balance. In contrast, the *Koumbidia Soce* sector was in a state of imbalance with 52.79 % of the surface area in a state of environmental degradation. In 1999, the study area was in a state of environmental imbalance – 70.69 %.

Human activities' gradual conversion of the ecosystems marked the study area. However, the most remarkable finding is the high degradation dynamics that characterized the study area's terrestrial ecosystems between 1954 and 1999. This degradation trend mainly started in the 1980s. The extent of the environmental imbalance reflected the importance of the annual expansion rate in the area between 1982 and 1999. During this period, the area of environmental imbalance increased by 2.40 % per year in contrast to the 0.14 % per year between 1954 and 1982. This gradual degradation of the ecosystems that characterized the region between 1954 and 1999 can be largely attributed to the following socio-economic factors.

4.3 The Proliferation of Socio-economic Units

Socio-economic units are frameworks of ownership, as well as the management of natural resources, social reproduction, and economic production. Until the twentieth century, the socio-economic units in the region comprised several households, with the oldest person heading dozens of people. It is at this level that social reproduction, economic production, and natural resource management occurred, which, from the mid-twentieth century became much more dislocated, leading to smaller units. Households replaced concessions in the social and economic organization and production. This resulted in the fragmentation of land into many plots. Individual appetites for land were especially high in a context in which householders replaced concession leaders as the payers of taxes. Given that extensive groundnut agriculture is the source of cash income to pay tax, a race for land developed that contributed to the increasing pressure on and degradation of the natural resources.

4.4 Representation of the Environment: From the Spiritual to the Economic

In the traditional Mandinka, Fula, and Wolof ethnic groups in the study area, ritual incantations preceded any action in nature, even the removal of bark, to obtain the approval of the spirits, which were regarded as the true governors of nature. This spiritual relationship with nature weakened considerably from the second half of the twentieth century onward. Ritual practices preceding business operations in nature have now all but disappeared. Nature is seen as a cornucopia of economic goods that people can procure profitably through their skills and intelligence. Land, formerly infused with spiritual significance, has become materially debased due to the capitalist logic of continued production growth.

4.5 From a Meet Local Needs Agriculture to a Meet International Needs Agriculture

Until the 1950s, agriculture's role was to satisfy the local economic needs, but this changed abruptly. The demand, norms, and values of the indigenous society had determined production, consumption, and distribution. During the second half of the twentieth century, the logic of local food security and social solidarity, which were initially strong indigenous concerns, were no longer the major determinants of agricultural production. The international market, the logic of wealth accumulation, and the continuous search for profits – the major characteristics of the capitalist economic system– drove production practices and the commercialization of agricultural goods. Agriculture became commercial and capitalist, sparking a growing demand for and increased pressure on agricultural land. Economics also impacted national policies and development institutions.

4.6 Land Management Institutions and Policies

The national development institutions, which the colonial and national independent authorities established, and natural resource management policies, which they implemented, were characterized by their interventionist and central nature. Until 1960, natural resource management had been a local concern. Communities had been responsible for natural resource management strategies. But with the establishment of the colonial and independent administrations, local communities were stripped of their powers to develop and implement their natural resource management visions. The indigenous logic of natural resource management was abandoned in favor of the government-inspired capitalist logic (Contreras-Hermosilla 2000). The colonial and independent governments' national policy development and natural resource management focused on increasing agricultural production, especially industrial crops. Therefore, many public development organizations were established in the study area. The main task of these public structures was to constantly increase agricultural production. Indigenous needs were no longer the central concern of these governmental structures. Their major interest was to meet the international market demand. Although the government interventionist approach was revised in the 1980s to include more local empowerment, the orientation of the policies and institutions remained productivist, and caused natural resource degradation. These different policies resulted in a change in the socio-economic units, especially in the production methods and techniques of central-east Senegal.

4.7 New Exploitation Methods and Techniques Undermining Ecological Regeneration

The farming, pastoral, and forestry methods and techniques underwent a significant change. The methods and techniques promoting the regeneration of natural resources in the study region changed to more environmentally aggressive techniques of production.

- **Abandonment of the long fallow**: The long fallow, a major feature of local agriculture, has almost disappeared. Once a widespread practice in the agro-silvopastoral region, only 1.52 % of farmers practice it today.
- **More mechanization**: A total of 98.04 % of farmers practice deep and frequent tillage and animal traction. Previously mechanization was essentially manual with animal traction introduced during the 1960s, which more than 98 % of farmers in the region practice today. This animal traction allowed increasing of plowing frequency and depth.
- **Pruning of fodder trees for increasing herds**: The progress made in veterinary medicine has helped to substantially increase the number of herds in the region and in the silvopastoral area adjacent to the study area. As a result, livestock growth outstrips fodder resources. To cope with this, pastoralists have developed new techniques for pruning tree branches to provide fodder for their animals. This pastoral practice has become common placing significant pressure on trees and, in several cases, leading to their disappearance.
- **Logging to meet urban charcoal demands**: The growing energy demand of urban centers, which developed together with the national economy's integration into the capitalist market, increased the demand for charcoal. Thus, the logging and sale of wood have replaced the traditional activity of a levy on bark, leaves, roots, and fruit of trees, leading to logging and a reduction in plant density.
- **Population Growth**: Senegal's agro-silvopastoral center-east region is part of this dynamic. Since 1900, the region has seen a constant population growth, which has accelerated since the 1950s. From less than two inhabitants per square kilometer, the density increased to 2.6 inhabitants per square kilometer in 1951, 2.7 inhabitants in 1958, and reached 10 inhabitants per square kilometer in 2008. From 1958 to 2008, the population density increased by 0.2 % per year. This population growth is linked to the land colonization policy in the region that Senegal's government initiated during the 1970s. This constant growth exerts a strong pressure on the natural resources.

5 Conclusion

Globally, the establishment of colonial capitalist modes of production in the twentieth century, and especially since 1950, has resulted in a socio-ecological rupture with market-oriented modes of production replacing subsistence activities, resulting in environmental degradation in Senegal's agro-silvopastoral center-east region. The results indicate that external factors over the past 60 years have increasingly driven the local agricultural systems and practices. This is consistent with the findings of Agrawal and Yadama (1997) and Bassett (2006). Despite the increasing market production, agricultural systems continue to exploit the original land and are still extensive instead of intensive. The results show a major trend towards environmental degradation in the region. In *Koumbidia Soce*, environmental imbalance is associated with increased farming areas. In *Fadiyacounda* and *Wouro Seeno*, environmental imbalance is associated with vegetational changes. Anthropogenic factors are directly driving environmental degradation in *Koumbidia Soce*. In *Fadiyacounda* and *Wouro Seeno*, although climatic factors regulate the ecosystems, vegetation density reduction is associated with natural resource management comprising logging, pruning, and bush fire practices; thus, human activities are a primary cause of the vegetation degradation. Local and international anthropogenic factors are driving environmental degradation in the region.

The acceleration of environmental degradation since the 1950s is a result of extensive methods of production linked to a productivist logic resulting from the integration of the Senegal's agro-silvopastoral center-east region into international economic systems and markets. It has been a change in lifestyles, rather than demographic growth that has caused the degradation of terrestrial ecosystems (Keys and McConnell 2005). Previously, investments in agriculture intensification were very low and limited to the introduction of rudimentary mechanization. With demand for agricultural products (groundnuts) growing in the international market, extensive farming practices are maintained and even increased, because without investment in and intensification of these practices to meet the growing demand, agricultural areas are spreading rapidly. This expansion of cultivated areas is a response to the increasing demand for agricultural products in the international market. However, the environmental impact of the colonial economic system has been gradual, as it did not start off by changing the environment in the region but eventually resulted in socio-economic change that has affected the environment. With the integration of the study area into the circuit of capitalist economy, previously existing worldviews, ambitions, attitudes, habits, social and economic practices have been disrupted.

References

Agrawal, A., & Yadama, G. N. (1997). How do local institutions mediate market and population pressures on resources? Forest Panchayats in Kumaon, India. *Development and Change, 28*(3), 435–465.

AOF Mission. (1954a). Vue aérienne du secteur d'etude de Koumbidia, 1/40 000, ND – 28 – X, numbers 057 to 063, 045 to 051, 460 to 466, 140 to 150, and 190 to 198.

AOF Mission. (1954b). Vue aérienne du secteur d'etude de Fadiyacounda, 1/40 000, ND – 28 – X, numbers 014 to 019, 043 to 048, and 054 to 059.

AOF Mission. (1954c). Vue aérienne du secteur d'etude de Wouro Seeno, 1/40 000, ND – 28 – XVII, numbers 1853-III-NE, 1853-II-NW, 1853-I-SW, and 1853-IV-SE.

Aubreville, A. (1957). Accord à Yangambi sur la nomenclature des types africains de vegetation. *Revue Bois et Forêts des Tropiques, 51*(Janvier-Février 1957), 23–27. http://bft.cirad.fr/cd/BFT_051_23-27.pdf. Accessed 25 Mar 2014.

Bassett, T. J. (2006). *The peasant cotton revolution in West Africa: Côte d'Ivoire, 1880–1995*. Cambridge: Cambridge University Press.

Blaikie, P. M., & Brookfield, H. C. (1987). *Land degradation and society*. London: Methuen.

Contreras-Hermosilla, A. (2000). *The underlying causes of forest decline* (Occasional Paper No. 30). Center for International Forestry Research. http://www.cifor.org/publications/pdf_files/OccPapers/OP-030.pdf. Accessed 26 Mar 2014.

Diouf, J. C. (2001). *Dynamique du peuplement ligneux dans l'aire du forage de Thieul (Ferlo-sud, Sénégal).* Dakar: Université Cheikh Anta Diop de Dakar, faculté des sciences et techniques.

Global Land Cover Facility (GLCF). (1999a). Landsat ETM+ vue satellitaire, scene 203-50 du 21 octobre 1999. www.glcf.umiacs.umd.edu/. Accessed 25 Jan 2003.

Global Land Cover Facility (GLCF). (1999b). Landsat ETM+ vue satellitaire, scene 203-51 du 21 octobre 1999. www.glcf.umiacs.umd.edu/. Accessed 25 Jan 2003.

Global Land Cover Facility (GLCF). (1999c). Landsat ETM+ vue satellitaire, scene 204-50 du 13 novembre 1999. www.glcf.umiacs.umd.edu/. Accessed 25 Jan 2003.

Global Land Project. (2005). *Science plan and implementation strategy* (IGBP Report No. 53/IHDP Report No. 19, 64p). Stockholm: IGBP Secretariat. http://www.globallandproject.org/arquivos/report_53.pdf. Accessed 15 Apr 2013.

Gonzalez, P. (2001). Desertification and a shift of forest species in the West African Sahel. *Climate Research, 17*(2), 217–228. http://www.int-res.com/abstracts/cr/v17/n2/. Accessed 2 April 2012.

Heilig, G. K. (1993). *Neglected dimensions of global land-use change: Reflections and data* (IIASA Working Paper WP-93-073). Laxenburg: International Institute for Applied Systems Analysis. http://www.iiasa.ac.at/publication/more_WP-93-073.php. Accessed 2 May 2012.

Keys, E., & McConnell, W. J. (2005). Global change and the intensification of agriculture in the tropics. *Global Environmental Change, 15*(4), 320–337. doi:10.1016/j.gloenvcha.2005.04.004.

Lambin, E. F., Turner, B. L., Geist, H. J., Agbola, S. B., Angelsen, A., Bruce, J. W., Coomes, O. T., et al. (2001). The causes of land-use and land-cover change: Moving beyond the myths. *Global Environmental Change, 11*(4), 261–269. doi:10.1016/S0959-3780(01)00007-3.

Lapointe, J. (1993). L'approche systemique et la technologie de l'education. *Les fondements de la technologie éducative Educatechnologiques, 1*(1), 54–87. http://www.sites.fse.ulaval.ca/reveduc/html/vol1/vol1_no1.html. Accessed 26 Nov 2006.

Mbaye, M. (1998). *Gestion actuelle des pâturages naturels forestiers soudaniens en Casamance: conséquences sur l'alimentation du bétail et la productivité de l'élevage* (Thèse de doctorat de 3ème cycle, FLSH, département de géographie). Dakar: Université Cheikh Anta Diop de Dakar.

Michel, P. (1973). *Les bassins des fleuves Sénégal et Gambie: étude géomorphologique* (Vol. 1. Mémoires ORSTOM no. 63). Paris: Office de la Recherche Scientifique et Technique Outre-Mer (ORSTOM). http://www.documentation.ird.fr/hor/fdi:06464. Accessed 30 Jan 2012.

Millennium Ecosystem Assessment. (2005). *Ecosystems and human well-being: Biodiversity synthesis.* Washington, DC: World Resources Institute. http://www.maweb.org/documents/document.354.aspx.pdf. Accessed 27 Jan 2012.

Organisation for Economic Co-operation and Development (OECD). (2003). *OECD environmental indicators: Development, measurement and use* (Reference Paper). Paris: OECD. http://www.oecd.org/environment/indicators-modelling-outlooks/24993546.pdf. Accessed 10 June 2011.

Organisation pour la Mise en Valeur du Fleuve Gambie (OMVG). (1982a). Vue aerienne du secteur d'etude de Koumbidia, 1/25 000, Mission n° 1652 IV NE, 1652 IV NW, 1652 IV SE, 1652 IV SW, 1652 I NW, 1653 III SE, 1653 III SW, 1653 II SW.

Organisation pour la Mise en Valeur du Fleuve Gambie (OMVG). (1982b). Vue aerienne du secteur d'etude de Fadiyacounda, 1/25 000, Mission. n° 1752-II-NE, 1752-II-NW, 1752-II-SE, 1752-II-SW.

Organisation pour la Mise en Valeur du Fleuve Gambie (OMVG). (1982c). Vue aerienne du secteur d'etude de Wouro Seeno, 1/25 000, Mission, n° 1652 IV NE, 1652 IV NW, 1652 IV SE, 1652 IV SW, 1652 I NW, 1653 III SE, 1653 III SW, 1653 II SW.

Pélissier, P. (1966). *Les paysans du Sénégal. Les civilisations agraires du Cayor à la Casamance.* Thèse de Doctorat d'État ès Lettres. Fabrègue. Saint-Yrieix (Haute-Vienne). http://data.over-blog-kiwi.com/0/47/01/11/201306/ob_d9ab16_pelissier-les-paysans-du-senegal.pdf. Accessed 12 Feb 2012.

Programme de Gestion Durable et participative des Energies traditionnelles et de substitution (PROGEDE). (1997). *Mise en place d'un Système d'Information Ecologique, Forestier et Pastoral (SIEF)*. Rapport de fin de mission, Tambacounda: PROGEDE.

Stancioff, A., Staljanssens, M., & Tappan, G. (1986). *Cartographie et télédétection des ressources de la république du Sénégal: étude de la géologie, de l'hydrologie, des sols, de la végétation et des potentiels d'utilisation des sols*. Dakar: Direction de l'Aménagement du Territoire, United States Agency for International Development, Remote Sensing Institute. http://library.wur.nl/isric/fulltext/ISRIC_16108.pdf. Accessed 13 Jun 2013.

Tappan, G. G., Sall, M., Wood, E. C., & Cushing, M. (2004). Ecoregions and land cover trends in Senegal. *Journal of Arid Environments. Special Issue: Land Cover, Biomass and Soil Carbon Trends in Senegal: Management Options and Climate Sensitivity, 59*(3), 427–462. doi:10.1016/j.jaridenv.2004.03.018.

Translation Bureau. (2009). *Termium plus: The government of Canada's terminology and linguistic data bank*. Agrosylvopastoralisme. http://www.btb.termiumplus.gc.ca/tpv2alpha/alpha-eng.html?lang=eng&i=1&index=frw&srchtxt=agrosylvopastoralisme. Accessed 31 Mar 2014.

Wood, E. C., Tappan, G. G., & Hadj, A. (2004). Understanding the drivers of agricultural land use change in south-central Senegal. *Journal of Arid Environments. Special Issue: Land Cover, Biomass and Soil Carbon Trends in Senegal: Management Options and Climate Sensitivity, 59* (3), 565–582. doi:10.1016/j.jaridenv.2004.03.022.

Integrated Global Change Research in West Africa: Flood Vulnerability Studies

Ibidun Adelekan

1 Introduction

One of the biggest challenges in human history is the current threat of human-induced climate change with potentially catastrophic consequences (Intergovernmental Panel on Climate Change (IPCC) 2007; United Nations Development Programme (UNDP) 2010). Different parts of the world are experiencing major droughts, floods, extreme temperatures (heat waves and cold spells), strong winds, and frequent and severe weather-induced fires. Coastal storms and storm surges are also increasing in frequency and intensity. Climate scientists have described this pattern of extreme weather and natural hazard events as a key characteristic of climate change (e.g., Scheraga and Grambsch 1998). These climate hazards and extreme events have severe consequences, especially for societies in developing nations where the populations are more vulnerable and have a comparatively low adaptive capacity. As a result, climate change has been described as the "defining human development issue of our generation" (UNDP 2007, 1), because of its capacity to reverse major achievements in human development across a range of sectors if it is not managed well (UNDP 2007). In the developing societies of West Africa, climate hazards such as floods can literally wipe out decades of development work within moments (Dow 2005).

Humans, as an integral part of the global climate system, play a dual role in the transformation of the earth's climate (Crutzen and Stoermer 2000; Steffen et al. 2007) and in the provision of solutions to the ensuing outcomes. Steffen and colleagues (2007, 619) note that interdisciplinary studies on human-environment systems have helped raise awareness of the anthropogenic influence on the earth system. In risk management, there is a growing acknowledgement of the need to understand the bases of human actions since these are not directed by knowledge

I. Adelekan (✉)
Department of Geography, University of Ibadan, Ibadan, Nigeria
e-mail: ibiadelekan@yahoo.com

B. Werlen (ed.), *Global Sustainability*, DOI 10.1007/978-3-319-16477-9_9

alone, but are also influenced by perceptions, attitudes, and social conditions. This explains why, although natural scientists have described global warming as perhaps the most significant environmental issue confronting the world in the twenty first century and as deserving of urgent action, social scientists have also noted that people respond to hazards on the basis of their perception of risks. Technically, the risk of any hazardous event is a product of the probability or likelihood of a hazard event and its consequences (Carter et al. 2007, 139–140; Lavell et al. 2012, 43). This can be expressed quantitatively as:

$$\begin{aligned}\text{Technical Risk} = &[\text{the likelihood of a hazard}] \\ &\times [\text{the adverse consequences of the hazard}]\end{aligned}$$

Nonetheless, an individual or society's assessment of risk affects other factors, including cultural, social and cognitive processes, affect the judgments of risk by an individual or society which results in perceived risk (Lavell et al. 2012, 45). The social conditions include basic needs, cultural diversity, tradition, the state of technology, access to resources, the balance of power, institutional restrictions, ethics, and social networks that influence the attention and sense of urgency the public perceive regarding climate risks. Unless these complex interrelationships are studied, a significant contribution towards the solution and prevention of environmental problems will elude us. For instance, Cutter (2003) notes that the social science community seeks to explain why people settle in risky places, thereby increasing their vulnerability to environmental threats. She also points out the need to understand the "contextual factors [that] decrease individual responsibility for the consequences of stupid locational decisions" (Cutter 2003, 8). Globally, flooding due to climate change is currently one of the main threats to human security, health, and livelihoods (IPCC 2007). Extreme rainfall events have the potential to increase the probability of flood occurrence resulting from an intensification of the global water cycle expected under climate change (Milly et al. 2002). Climate change is therefore worsening flooding situations in areas already prone to flooding (Wheeler 2007; Satterthwaite 2008). Other factors that contribute to flood incidence are unplanned rapid urbanization, alteration in land cover, and poor watershed management, particularly in flood plains of rivers. A combination of these factors and other local factors is redefining flood events across the world (Harvatt et al. 2011; Mitchell 2003; Few et al. 2004). The human-environment relationship has thus become complex, especially in urban areas, in which more than half of the world's population now live (United Nations Human Settlements Programme (UN-Habitat) 2007, 169). Though researchers are learning more about floods, and hydrological science and technology are advancing, the number of casualties from floods has remained constant at best, but is probably increasing (Drobot et al. 2007; Jha et al. 2012). Brilly and Polic (2005) argue that this is due mainly to the gaps in our understanding of flood management's social dimensions (see also Tapsell et al. 2002; Harvatt et al. 2011). On the one hand, the knowledge and behavior of flood-affected societies are not at such a level that they will lessen the impacts of flooding. On the other hand, decision-makers often lack an understanding of the social vulnerability, including the public's perception of flood risk and their behavior in response to flooding. This invariably limits the success of flood management plans. Given the multifarious dimensions of the flooding issue (hydrological, meteorological, sociological, economic, political, engineering), the best efforts to manage flood risks should be integrative and should involve all stakeholders (Few et al. 2004). In this respect, the importance of a long-term perspective on flood risk management, with strategies grounded in an understanding of the exposure to the hazard, hazard characteristics, and vulnerability patterns, as well as the relationship between the different stakeholders and their respective flood risk perception, has been highlighted (Brown and Damery 2002).

In this context, it is noted that approaches to flood risk management in more developed societies have undergone a shift in the past two decades; such shifts involve the integration of nonstructural approaches into flood mitigation, for which an understanding of flood risk's social dimensions is an important aspect (e.g., Correia et al. 1998; Few et al. 2004; Messner and Meyer 2006; Merz et al. 2010). Despite the importance of social factors in environmental and climate risk management, very few studies of climate hazards and risks in the African context have taken the social dimension into account when assessing the vulnerability of the social systems under consideration. In particular, the important link between assessments of exposure to flooding and social vulnerability issues in the context of rapid urbanization has hardly been explored with regard to flood hazards in West Africa. This article examines the underlying socio-cultural factors in risk perception and risk-producing behaviors of populations in flood-affected districts in Ibadan and the implications of these for flood risk management in the city. Specifically, the article seeks to understand how residents of flood-affected districts estimate flood risk and how risk cognition translates into tangible or empirically observable actions.

2 Conceptual Framing: Social Vulnerability and Risk Perception

The concept of vulnerability, as applied to social systems, could be interpreted as "a set of conditions and processes resulting from physical, social, economic and environmental factors, which increase the susceptibility of a community to the impact of hazards" (United Nations International Strategy for Disaster Reduction (UN/ISDR) 2003, 21). Cutter and colleagues (2009, 2) define vulnerability as "the susceptibility of a given population, system, or place to harm from exposure to the hazard and directly affects the ability to prepare for, respond to, and recover from hazards and disasters." In particular, Wisner and colleagues (2004, 11) define vulnerability as "the characteristics of a person or group and their situation that influence their capacity to anticipate, cope with, resist and recover from the impact of a natural hazard." The value of the concept of vulnerability lies in it enhancing understanding of the social, climatic, geographic, political, and economic processes that make a socio-ecological system susceptible to particular stressors (Wisner et al. 2004). Hence, the need has been underscored to appreciate the set of socio-economic conditions that are identifiable in relation to a particular hazard risk which play a predictive role in vulnerability and thus assist with risk reduction (Cannon 2008). The IPCC's fourth Assessment Report defines vulnerability in relation to climate change as "the degree to which a system is susceptible to, or unable to cope with, adverse effects of climate change, including climate variability and extremes. Vulnerability is a function of the character, magnitude, and rate of climate change and variation to which a system is exposed, its sensitivity, and its adaptive capacity" (IPCC 2007, 883). Smit and Pilifosova (2003) consider vulnerability a function of the exposure of a system or community to climate change effects and its adaptive capacity to deal with such exposure. In this context, exposure describes a community's property relative to climatic conditions (Ford and Smit 2004, 393), while adaptive capacity refers to a community's ability to address, plan for, and adapt to exposure (Smit and Pilifosova 2003). Besides the factor of exposure, there may also be non-climatic factors (e.g., political, economic, technological, demographic, social, and environmental) which determine the sensitivity of a system or community to climate change. While the interaction of environmental and social forces determines exposure and sensitivity, the various social, cultural, political, and economic forces impacting on a social system or community shape its adaptive capacity (Smit and Wandel 2006). The greater a system's exposure and/or sensitivity to a climate stimulus or a hazard relative to its adaptive capacity, the more vulnerable the system is (Fig. 1).

Adger (1999) explained a social group's vulnerability and resilience in terms of its environmental risk perception and social factors. What the public perceive as a risk, why they perceive it this way, and how they will subsequently behave are central in determining an individual or social group's vulnerability; these questions are therefore crucial for planners, decision-makers, and policy-makers attempting to address global change issues. Risk perception relates to the perceived probability of being exposed to climate change impacts and to the appraisal of how harmful these impacts would be on things and actor values (perceived severity) relative to the appraisal of how harmful and urgent other problems or challenges in life are (Grothmann and Pratt 2003). Risk perception has also been defined as the cognition of or belief in the seriousness of the threat of an environmental extreme and the subjective probability of experiencing a damaging environmental extreme (Mileti 1980). Similar to other definitions, Bubeck et al. (2012a) describe perceived risk as the combination of the perceived probability and perceived consequences of an event (e.g., a flood) or activity. Miceli and colleagues (2008, 165) define flood risk perception more explicitly as "the subjective evaluation of the likelihood of a future event occurring and of personal and material damage deriving from it." They also emphasize that risk perception should be seen "as a complex process which encompasses both cognitive and affective aspects" (Miceli et al. 2008, 165). Risk perception therefore plays an important role in shaping environmental management

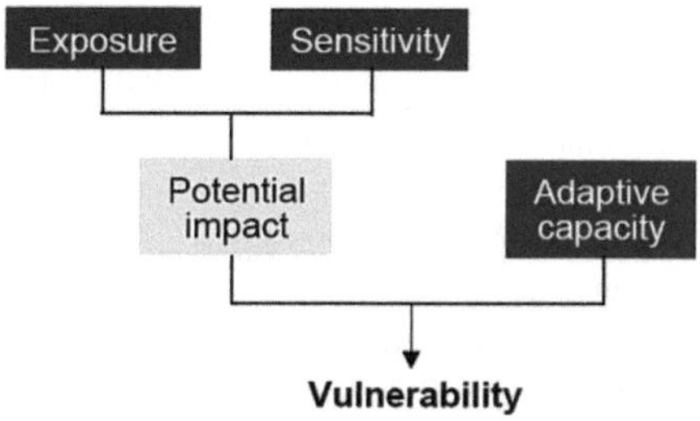

Fig. 1 Vulnerability and its components (Allen Consulting Group 2005, ix)

and disaster response systems, because the extent to which individuals understand the causes and consequences of environmental phenomena, including climate risks, and the extent to which these are perceived as a threat to their wellbeing, may correspond to their personal lifestyle decisions (Brody et al. 2008). An important decision that has the potential to expose an individual to risk is the choice of where to live or undertake economic activity. Social scientists note that societies' responses to hazards depend on their perceptions of the risks associated with a particular hazard. Environmental risk perception is strongly influenced by demographic, attitudinal, socio-cultural, and contextual variables (Brody et al. 2008; O'Connor et al. 2002; Bord et al. 1998; Peters and Slovic 1996; Kempton 1993) and is therefore both place-specific and group-specific.

3 West African Urbanization and Flood Hazards

Many African cities' vulnerability to disasters compared to those on other continents, especially Latin America and Asia, has only been recognized in recent years. This is partly due to Africa until recently being considered largely rural and vulnerability assessments and risk reduction therefore focusing mainly on these areas (Dodman et al. 2012). Nevertheless, a number of common characteristics and underlying social, economic, and cultural factors have led to the accumulation of risks in African cities (Wisner and Pelling 2012). In African societies, climate risks and natural disasters are mostly not the result of particular incidents in the natural environment that prove catastrophic for the local population, but the result of significant human transformations of the natural environment. This is compounded by poor environmental management, worsening socio-economic conditions, and inadequate public policy choices. The nature of West African urbanization and the consequences of urban growth under conditions of global environmental change are important considerations in the context of the increasing frequency and severity of extreme events, including severe floods, in the twenty first century. With the exception of Mauritania and Mali, West African countries have urbanized at a very rapid rate over the past 50 years (Agence Française de Développement (AFD) 2009). While the region's population increased 3.8-fold (at an average demographic growth rate of 2.7 %) during the period 1950–2000, the urbanization rate (i.e., the

Table 1 West Africa urbanization characteristics (1950–2020)

Year	1950	1960	1970	1980	1990	2000	2010	2020
Agglomerations >10,000 inhabitants	125	220	366	538	745	992	1,224	1,431
Urban inhabitants (millions)	4	10	19	31	48	78	97	123
Urbanization (%)	7.5	13	19	23	27	31	33	34

Source: own table compiled from data in AFD (2009)

percentage of the population living in urban areas) increased 4.2-fold over the same period (from 7.5 % in 1950 to 31.5 % in 2000) (AFD 2009, 18). The combination of these two factors has led to an almost 20-fold increase in the total urban population from 1950 to 2000 (AFD 2009, 18). Porto Novo, Ouagadougou, Accra, Niamey, Lagos, and Lomé are all confronted with over-urbanization, with populations growing faster than the local economies (UN-Habitat 2010, 100) (Table 1).

Changes have been observed in West Africa's hydrology that are not necessarily related to rainfall amounts. In the past half century, land use and land cover across the sub-region have changed under the combined impact of climatic effects and human activities (Ruelland et al. 2008). Land use tends to be more important than rainfall for runoff characteristics, since marked rain deficits have coincided with increased water runoff (e.g., Mahé et al. 2003; Séguis et al. 2002; Boulain et al. 2006; Mahé and Paturel 2009). Descroix and colleagues (2009) showed that the runoff in Sahelian basins has been increasing for three decades, despite a 20–25 % decrease in observed rainfall during the 1968–1997 period. Urbanization, land use, land cover changes, and weather extremes have contributed to increasing flood volumes, especially in West African cities. Since 2007, flood events in West Africa have been unprecedented in frequency and severity, and have affected unprecedentedly large areas. The Dartmouth Flood Observatory (DFO) ranked the flood event in West Africa during the 2007 wet season as one of the three most devastating global flood events in that year (Conway et al. 2009; Tschakert et al. 2010). Since 2007, exceptionally heavy flood-producing seasonal rains have been experienced in a number of West African countries (e.g., Burkina Faso, Chad, Nigeria, Benin, and Ghana), with severe social and economic consequences for the affected societies. Prolonged rainfall periods have also made it difficult for humanitarian aid to reach the flood-affected areas. Several rivers, including the Pendjari, Niger, Volta, and Senegal rivers have overflowed their banks, causing widespread damage to infrastructure, including buildings, bridges, roads, and extensive farmland areas. Some locations in the semi-arid Sahel region, which suffered severe and widespread droughts in the 1970s and mid-1980s after very low rainfalls, have also been at the receiving end of severe floods, with limited or no capacity to respond. In several West African cities, anthropogenic influences are compounding the climatic changes, leading to aggravated flood hazard outcomes (Adelekan 2010, 2011; Diagne 2007; Karley 2009; Mbow et al. 2008). In Senegal, for example, Mbow and colleagues (2008) observe that human factors and drivers are the single biggest cause of floods in Dakar, Senegal, while Diagne (2007) notes that although Saint

Louis in Senegal is located in a region with high flood risk due to weather extremes, the city's vulnerability to floods mainly results from human practices. Furthermore, Mbow and colleagues (2008) observe that the increased annual rainfall totals in Senegal in recent years – hailed as the return of good rainfall years – are nevertheless not comparable to the much higher annual totals recorded during the 1950–1969 period when there was no incidence of flooding. Therefore, urban development and urban sprawl onto marginal lands and relatively high-risk areas with no overall development plans explain the frequency of urban floods to a large extent and are placing an increasing proportion of these countries' economies and populations at risk (Adelekan 2010; Dodman et al. 2012; IPCC 2007).

4 Case Study: Flood Vulnerability in Ibadan, Nigeria

Ibadan, Oyo State's capital, is the second most populous city after Lagos in southwest Nigeria, and is acknowledged as the largest indigenous urban center in sub-Saharan Africa. Ibadan's population grew rapidly from 70,000 in 1856, to 2.55 million in 2006. However, urban growth has not been adequately managed, and development control in the city is not effective. A major result of this is human settlements in risky/marginal locations in the city, including the flood plains of the rivers Ogunpa, Kudeti, Ogbere, and Ona, and the hillside locations of Oke-Are, Oke-Aremo, Sapati, and Mokola. Compounding the flood plain development is the large-scale deforestation within the city limits. Vegetation cover in the city, including original forest reserves located mainly in the hilly areas in and around the city, has virtually disappeared as a result of urban development (Oluseyi 2006). This pattern of urban development has increased ecosystems, physical infrastructure, and large populations of city dwellers' exposure to flood hazards.

Ibadan, where the first flood event was recorded in 1902, is familiar with floods. Since then, noteworthy flood events were recorded in 1960, 1963, 1978, and 1980 (Areola and Akintola 1980; Ayoade 1979; Ayoade and Akintola 1980; Oguntala and Oguntoyinbo 1982). These earlier floods resulted from the overflowing of the Ogunpa River, and the impacts were largely confined to areas on its flood plain. The Ogunpa flood event of 1980 resulted in more than 250 deaths. On August 26, 2011, however, the city experienced yet another flood disaster. The immediate cause was 7 h of torrential rainfall, which devastated most parts of the city and its environs. The outcome was the loss of more than 100 lives and extensive damage to the physical infrastructure, buildings, property, farms, and enterprises across the city. The total rainfall was 187.5 mm, with 140.63 mm (75 %) falling in 1 h (between 7 pm and 8 pm) (Fig. 2). The rainfall event was acknowledged as the highest recorded in the past 30 years. The magnitude and extent of the flood characteristics are described in Table 2.

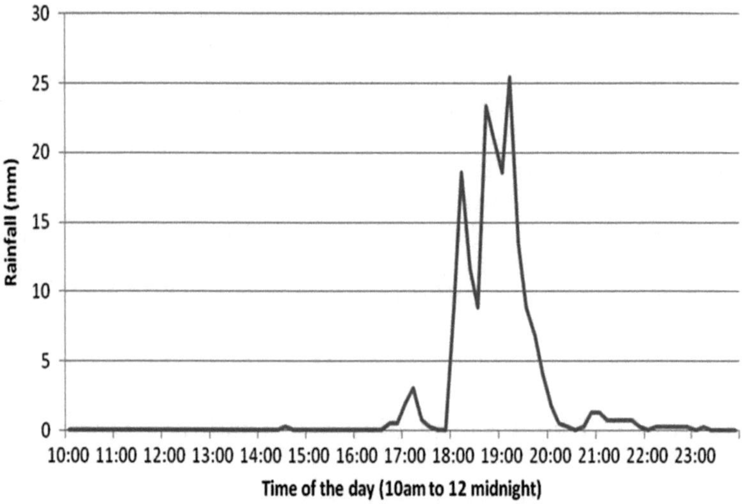

Fig. 2 Hyetograph, Ibadan climate station, of the rainfall event that caused the flood of August 26, 2011 (own diagram using data supplied by the International Institute of Tropical Agriculture (IITA))

Table 2 Flood description: Ibadan, August 26, 2011

Characteristic	Description
Cause	Torrential rain: 187.5 mm
Duration	14 days
Severity class[a]	1.5
Magnitude[b]	5.6
Affected region	20,013.55 km^2
Displaced persons	3,000
Deaths	20

Own table using data from Brakenridge (2013)
Notes:[a]Severity Class (NB: flood severity is reported in three classes: 1, 1.5, and 2.):
Class 1: large flood events: significant damage to structures or agriculture; fatalities; and/or one to two decades-long reported interval since the last similar event
Class 1.5: very large events: recurrence interval greater than 20 years but less than 100 years, and/or a local recurrence interval of at 10–20 years
Class 2: Extreme events: estimated recurrence interval greater than 100 years
[b]Flood magnitude = LOG (duration × severity × affected area)

4.1 Data and Methods

Data was obtained from both primary and secondary sources. Trained field assistants undertook the primary data gathering by means of a specifically designed questionnaire administered to a random selection of 300 households[1] in 11 of the city's flood-affected areas (Fig. 3). The study locations were selected on the basis of newspaper reports of areas severely affected by the flood, as well as a reconnaissance survey of all the city's flood-affected areas to validate these reports. A semi-structured questionnaire was designed to elicit information on the respondents' household experience of the August 2011 floods, their personal perceptions of flood risk, past experiences of flood events, and indicators of vulnerability. Specifically, the schedule of questions addressed the following:

1. Socio-demographic and household characteristics: sex, age, education, occupation, residential status, number of people in household, etc.
2. Exposure indicators: distance of building from river/stream, drainage structure, domestic water supply source, prior flood experience.
3. Sensitivity indicators: type of building structure, duration of stay in flood-affected areas, perception of flood risk and preparedness for possible flood occurrences in the future.

In addition, in-depth interviews were conducted with certain respondents; these provided further insight into area-specific situations in the different communities. The field survey was carried out in September 2011 – at a time when most residents had returned to their homes, with only a few buildings still uninhabitable. Secondary data from publications on the flood event were sourced from daily newspapers, the Oyo State Emergency Management Agency, and the Oyo State task force's report on the flood event. The records of the Dartmouth Flood Observatory also provided information on the flood event's characteristics.

4.2 *Results and Discussion*

4.2.1 Respondents' Demographic, Socio-economic, and Housing Characteristics

The demographic and socio-economic characteristics of the 300 sampled respondents in the 11 flood-affected districts surveyed are described in Table 3. Note that the selection of the sample was not guided by the aim to provide a statistically representative coverage. Instead, the extent of the flooded area and the flood impacts informed the number of households sampled in each area. Respondents of different age categories were interviewed; 53 % of them were male and 47 % female. Only 14 % did not have any formal education, while others' education levels varied between the primary and tertiary level. The respondents were engaged in different occupational activities; approximately one-third were traders (34.3 %),

[1] Note that one respondent per household was interviewed so that 300 households in this study are equivalent to 300 respondents.

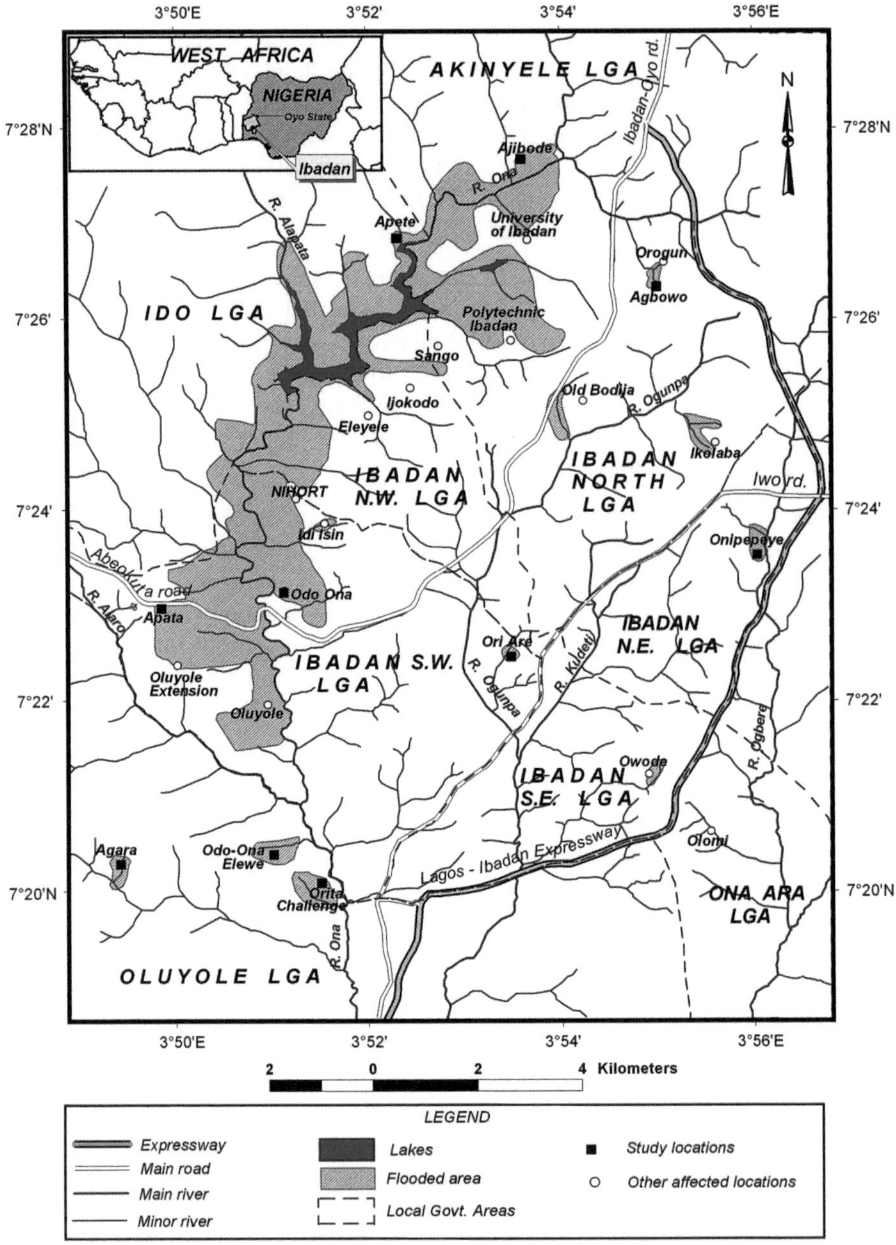

Fig. 3 Flooded areas of Ibadan in August 2011 and study locations (Source: own map adapted from United Nations Platform for Space-based Information for Disaster Management and Emergency Response (UN-SPIDER) (2011): Flooded areas extracted from TerraSAR image (August 31, 2011) and overlaid on SPOT 5 image (August 1, 2011) © German Aerospace Center (DLR) (2011). Map produced by NASRDA with support from the International Charter Space and Major Disaster, September 2011). *Note: LGA* Local Government Area

Table 3 Respondents' characteristics and household description

Variable	Results
Age (years)	16–25: 59 (19.7 %); 26–35: 81 (27 %); 36–45: 68 (22.7 %); 46–55: 42 (14 %); > 55: 50 (16.6 %)
Gender	Male: 159 (53 %); Female: 141 (47 %)
Level of education	No formal education: 41 (13.7 %); Primary: 49 (16.3 %); Secondary: 97 (32.3 %); Tertiary: 110 (36.7 %); Other (e.g., Koranic education): 3 (1.0 %)
Occupation	Civil servant: 42 (14 %); Craftspeople: 48 (16 %); Trader: 103 (34.3 %); Farmer: 12 (4 %); Others (incl. the unemployed, students, house workers, retired persons/the elderly): 82 (27.4 %)
Residential status of respondent	Landlord: 126 (43 %); Tenant: 150 (50 %); Family house: 18 (6 %); Squatting: 6 (2 %)
Distance of building from river/stream	<100 m: 167 (55.7 %); 100–500 m: 87 (29 %); 500 to 1,000 m: 28 (9.3 %); > 1,000 m: 18 (6 %)
Type of building	One story (concrete): 246 (82 %); Two or more stories (concrete): 30 (10 %); One story (wooden): 13 (4.3 %); One story (mud): 11 (3.7 %)
Number of households residing in building	1–3: 124 (41.3 %); 4–6: 120 (40 %); 7–9: 38 (12.7 %); 10 and more: 18 (6 %)
Length of stay in flood-affected area (years)[a]	<2: 41 (13.7 %); 2–5: 86 (28.7 %); 6–10: 85 (28.3 %); 11–20: 14 (13.3 %); > 20: 48 (16.0 %)
Residence[b] affected by flood?	Yes: 192 (64 %); No: 108 (36 %)
Ever experienced any flood event in the past?	Yes: 57 (19 %); No: 243 (81 %)
Past flooding experience in present location	Yes: 43 (14.3 %); No: 257 (85.6 %)
Knowledge of Ogunpa flood 1980	Yes: 81 (27 %); No: 219 (73 %)
Knowledge of flood warning	None: 259 (86.4 %); Yes: 41 (13.6 %)

[a]The question inquired into the total length of stay in the affected area, and people might have moved within that area during this time
[b]'Residence' denotes the respondents' building including the associated infrastructure such as fences around the section on which the building is situated

while others were craftspeople, civil servants, farmers, and professionals in different fields. There were about as many homeowners and occupants of family residences as tenants residing in the sampled flood-affected districts. A negligible percentage (2 %) of respondents lived in squatter settlements. The majority (82 %) of buildings that the respondents occupied were one-story (bungalows), while another 10 % had two or more floors. Concrete was the main building construction material (92 %), with wood (4.3 %) and mud (3.7 %) the other building materials.

4.2.2 Exposure

The State regulations guiding the minimum riparian setbacks on either side of watercourses ranges from 15 to 46 m, depending on the size of the river or stream (Oyo State 2005). These regulations notwithstanding, 26,533 buildings in the city were identified within the statutory setbacks for rivers and streams (Oyo State 2011). More than half the households (56 %) sampled were located less than 100 m from a river or stream, and another 29 % between 100 and 500 m. Only 6 % of buildings were located more than 500 m from a river, but were still affected by the flood event. The respondents provided different reasons for residing in the flood-affected areas. This included proximity to workplace or educational institution (23.3 %), the property was handed down in the family and the current occupants did not have an influence on the location of the family property (13 %), the availability of land for purchase (13 %), and affordable rent (11.3 %). Other reasons given included the suitable environment (19 %) and low land cost (1.7 %).

The respondents' tenure in the flood-affected districts was relatively short, since the majority (42.4 %) had been residing in the flooded districts for less than 5 years. Only 16 % of the respondents, most of whom were landlords, had been living in the flood-affected districts for more than 20 years. This category of respondents observed that the rivers that overflowed their banks in the August 2011 flood were small streams that were easily crossed by foot at the time they acquired the land for building several years before. The majority (85.6 %) had no previous experience of floods in the affected districts, and only 57 (19 %) indicated that they had flood experience prior to August 2011. Of those with previous flood experience, 43 respondents (75 %) experienced the flooding in their present residential location. Only 27 % of respondents indicated any experience of the 1980 Ogunpa flood disaster – the last big flood event to affect the region.

The respondents' proximity to a river was directly related to damage or harm. While 76 % of the respondents living less than 100 m from a stream or river indicated that their residence was affected by the flood event, only 53 % of those living 100–500 m away reported that their residence was affected. Similarly, the costs of damage and loss had been higher for the residents living closer to the rivers. Of those respondents who lived less than 100 m from a river, 66 % recorded damage of varying costs. The floodwater depth in the residential buildings was not uniform across the flood-affected areas. While 68 % of respondents indicated that the floodwater depth inside their houses had not exceeded 2 ft, 9.3 % and 2.7 % of them noted depths of 2.1–4 ft, and 4.1–6 ft, respectively. However, there were a small number of cases where the depth exceeded 6 ft.

The majority of the respondents (74 %) resided in areas with inadequate drainage infrastructure. This further aggravated the flood impact's severity, because in most cases the storm water drainage had been ineffective, resulting in flooding. More than a third of the respondents (35 %) indicated that there was no drainage infrastructure in their locality, while 39 % noted that the drainage infrastructure was in a poor condition owing to inadequate maintenance. Approximately one-quarter of the respondents (26 %) considered the condition of the stormwater and drainage system in their immediate environment as either fair or good.

4.2.3 Sensitivity

Different factors contribute to households' susceptibility or sensitivity to flood hazards. In terms of construction materials and structure, the building type is a factor in determining residents' vulnerability to a flood event, since this significantly influences the extent of the damage. Studies have shown that multi-storied buildings are less affected than single-story buildings (e.g., Kreibich and Thieken 2008). In the surveyed flood-affected areas, the majority (90 %) of buildings had one story and therefore were potentially at great risk of flood damage.

4.2.4 Flood Risk Awareness and Perceived Causes of the Flood Event

The majority (86.4 %) of the respondents indicated that no warning was received of the flood before the event. Those who received a warning indicated that this was based on forecasts (4.7 %) aired in the media and provided by the Nigerian Meteorological Agency (NIMET) and the National Emergency Management Agency (NEMA), on religious predictions (2 %), on personal experience and the intensity of the rainfall event (3.3 %), and on personal observations of rivers overflowing in the past (1 %). The respondents also noted that they did not consider the forecast of heavy rainfall for that day as a reason for alarm. They regarded this forecast just like every other weather forecast on the network news, as well as the print and electronic media, since they never expected the rain to wreak the havoc experienced. The respondents suggested various reasons for the August 2011 flood (Table 4). While 22 % of the respondents attributed the floods to heavy rainfall, 17 % attributed it to an act of God, or the anger of the river goddess due to the discharge of effluents and pollutants by Ibadan city dwellers. Others noted that the floods had been human-induced through the construction of bridges and buildings on or near river channels (9 %), and city dwellers' use of the rivers and drainage channels for solid waste disposal, which led to blockages in these channels (27.3 %). Other reasons suggested include the lack of or bad drainage and the government negligence in providing and maintaining adequate storm water drainage system in the city. The damming of the Eleyele River was also listed as a contributing factor to the flood. The respondents from Eleyele, Odo Ona Elewe, Odo Ona, and Apata – all located along the River Ona in Ibadan (see Fig. 3) – were of the opinion that the rainfall of August 26, 2011 was not heavy enough to generate a flood, but that the Eleyele dam opening without warning brought about the large volume of water that covered these areas downstream, leading to the loss of lives and properties. Others (3.3 %) blamed the government for not completing projects, such as the bridge on the River Ona at Odo Ona and Apata. The large percentage of

Table 4 Respondents' perceptions of the causes of the August 26, 2011 flood

Cause	Frequency (% of total)
Heavy rainfall	65 (21.7 %)
Act of God	51 (17 %)
Blocked drainage and building along river channel	28 (9.3 %)
Waste disposal into drainage	82 (27.3 %)
Lack of or bad drainage	41 (13.7 %)
Government negligence	10 (3.3 %)
Damming of Eleyele River	23 (7.7 %)
Total number of respondents	300

Source: author's field survey, September 2011

respondents' suggestion that the flood was linked to human influences (as inferred from personal observations listed above) corroborates the view that personal observation is the most trusted source of information concerning the causes of flooding, and that it plays a primary role in risk perception and response (Whitmarsh 2007, 19).

Bubeck et al. (2012a) define perceived risk as the combination of the perceived probability and perceived consequences of an event (e.g., a flood) or activity. The respondents' flood risk perception was assessed by asking: "Did you ever imagine that your locality could be affected by flooding?" Only 8.3 % answered positively. The majority (81 %) never thought that a flood could occur, while 4 % were undecided. Although 6.7 % indicated that they were aware that their locality could be affected by flooding owing to the presence of streams that could overflow, they never expected this to happen in the near future. These people said that they had watched the river channel's water level increase but never expected that it would get worse that day, because the river usually shows signs of overflowing without actually doing so.

Studies have found a relationship between individuals' demographic characteristics and their risk perceptions (e.g., Savage 1993; Grothmann and Reusswig 2006). People with lower education levels have been shown to have higher risk perception levels (Savage 1993), and age has been found to be positively correlated with the risk perception of a number of natural hazards (Grothmann and Reusswig 2006; Kellens et al. 2011). Home ownership has also been related to higher levels of perceived risks than renting a residence (Grothmann and Reusswig 2006).

The results of a Pearson Chi-square test and likelihood ratio tests show that the respondents' age, gender, and education were related to the flood risk perception level, although age was more strongly associated with high flood risk perception than gender and education. The association was also statistically significant ($p < 0.05$). Older respondents showed higher risk perception levels. The respondents' risk perception level was also associated with the perception of rainfall pattern changes (Table 5). The respondents were asked whether they would rate the rainfall as increasing, reducing, not changing, or variable. A positive association was observed between flood risk perception and the respondents' opinion of rainfall

Table 5 Respondents' characteristics and flood risk perceptions

	Age	Gender	Education	Past flood experience	Experience of Ogunpa 1980 flood	Perception of rainfall pattern	Perception of increasing rainfall intensity
Maximum likelihood ratio	.516	.005	.020	.423	.895	.339	.509
Pearson Chi-square value	.633*	.008*	.005*	.370	.892	.329*	.368*

*Significant at the 0.05 level (2-tail)

intensity changes and the total amount received. The association was stronger for those who felt that the rainfall intensity had been increasing in recent years. A note of caution should be added, however: an association between variables does not imply causality (i.e., the independent variable does not necessarily cause the dependent variable).

4.2.5 Flood Experience and Behavioral Response

Some studies show that flood experience strongly influences risk perception and is an important factor in motivating mitigation behavior (Siegrist and Gutscher 2008). However, past experience of flood events did not translate into higher flood risk perception among the sample. Although the majority of the respondents had no previous flood experience, 75.4 % of those with past experience of floods never imagined they could be affected by floods in their present location. This also means that past flooding experience did not translate into better preparedness for future flood events. Even though a strongly positive association was observed between flood risk perception and past experience of the 1980 Ogunpa flood, this was not statistically significant. Furthermore, although significant correlations have been reported between perceived personal risk and adjustment to natural hazards (e.g., Grothmann and Reusswig 2006), the flood event did not result in any personal or individual risk mitigation or reduction decisions for the majority (80 %) of respondents. This corroborates findings by Slovic (1987), who noted that perceived risk does not always contribute to taking preventive actions. Decisions by the few respondents who did take mitigative and preventive measures include relocation (12 %), household damage repairs (2 %), clearing and avoiding disposal of solid waste in drains and canals (4.3 %), and construction and/or clearing of nearby drainages (1.7 %). Varying degrees of positive associations were obtained between individual risk mitigation decisions and those respondents with experience of the flood resulting in loss of lives in the household (.922), loss of lives in the community (.482), and damage to residence due to the flood (.487); these were not statistically significant (Table 6). Brody and colleagues (2008), however, show that the number of fatalities resulting from natural hazards is the most significant predictor of heightened risk perception.

The flood event's severity had little effect on most respondents' preparedness plans concerning future floods. Responses to the question: "What do you think can be done in future to prevent/manage floods?" focused largely on public agents. The channeling of rivers and construction of better drainage was listed by 58 % of the respondents, while 10 % suggested the government should strictly enforce the planning laws. Other measures mentioned were the demolition of structures on the flood plains (8 %), proper solid waste management and drainage clearing (11 %), and raising public awareness of the flood risk and the factors contributing to an increased flood risk (2 %). In addition, the majority (82 %) was of the opinion that the government was the responsible agency for managing floods, 5 % felt it was the joint responsibility of citizens and the government, while only 5 % felt that

Table 6 Relationship between flood impact and individual risk mitigation decision

	Lives lost in household	Lives lost in the neighborhood	Residence affected
Maximum likelihood ratio	.922	.482	.487
Pearson Chi-square value	.923	.481	.485

citizens have a role to play. The above perception is a critical limitation of flood risk management in the city, given that private flood mitigation measures have become an integral component of contemporary flood risk management in developed societies (Bubeck et al. 2012b).

5 Conclusion

The effects of global environmental change are already manifesting and are, due to the current urbanization pattern, likely to be increasingly felt in West African cities and towns. The nature of West African urbanization and the consequences of urban growth under conditions of global environmental change are therefore of utmost importance in the context of the increasing frequency and severity of extreme events, including severe floods in the twenty first century. Most African cities are extremely vulnerable to flooding. Since human transformation of the natural environment is a significant factor in flood generation, and risk perception is an important moderator of human behavior, any attempt to manage flood risk and reduce people's vulnerability to this hazard requires an integrated approach. Besides the scientific knowledge of the hazardous event, there is the need for researchers to study the hazards' combined effects, previous hazardous experiences, sources of risk information, household risk perception characteristics, and their behavioral consequences (Lindell and Hwang 2008). An understanding of these is crucial for any social group or community. This article has examined these factors in the context of a major flood disaster in Ibadan. A holistic approach to flood risk management is important, given that physical vulnerability variables may not explain citizens' perception of risk as fully as socio-economic and attitudinal control variables (Brody et al. 2008). This may provide direction regarding addressing vulnerability to flood hazards arising from socio-cultural factors, as well as from environmental changes.

References

Adelekan, I. O. (2010). Vulnerability of poor urban coastal communities to flooding in Lagos, Nigeria. *Environment and Urbanization, 22*(2), 433–450.

Adelekan, I. (2011). Vulnerability assessment of an urban flood in Nigeria: Abeokuta flood 2007. *Natural Hazards, 56*(1), 215–231.

Adger, W. N. (1999). Social vulnerability to climate change and extremes in coastal Vietnam. *World Development, 27*(2), 249–269.

Agence Française de Développement (AFD). (2009). *AFRICAPOLIS: Urbanization trend 1950–2020: A geo-statistical approach. West Africa*. Agence Française de Développement (AFD). http://www.afd.fr/lang/en/home/publications/travaux-de-recherche/archives-anciennes-collections/NotesetEtudes/Africapolis. Accessed 10 Jan 2013.

Allen Consulting Group. (2005). *Climate change risk and vulnerability: Promoting an efficient adaptation response in Australia. Final report. Report to the Australian Greenhouse Office*. Melbourne: Allen Consulting Group. http://www.sfrpc.com/Climate%20Change/4.pdf. Accessed 4 Feb 2013.

Areola, O., & Akintola, F. O. (1980). Managing the urban environment in a developing country: The Ogunpa river channelization scheme in Ibadan City, Nigeria. *Environment International, 3* (3), 237–241.

Ayoade, J. O. (1979). A note on the recent flood in Ibadan. *Savanna, 8*(1), 62–65.

Ayoade, J. O., & Akintola, F. O. (1980). Public perception of flood hazard in two Nigerian cities. *Environment International, 4*(4), 277–280.

Bord, R. J., Fisher, A., & O'Connor, R. E. (1998). Public perceptions of global warming: United States and international perspectives. *Climate Research, 11*, 75–84.

Boulain, N., Cappelaere, B., Séguis, L., Gignoux, J., & Peugeot, C. (2006). Hydrologic and land use impacts on vegetation growth and NPP at the watershed scale in a semi-arid environment. *Regional Environmental Change, 6*(3), 147–156.

Brakenridge, G. R. (2013). *Global active archive of large flood events*. Archive excel file for all events, 1985-present, updated as the recent even Dartmouth Flood Observatory (DFO), University of Colorado. http://floodobservatory.colorado.edu/Archives/index.html. Accessed 5 Feb 2013.

Brilly, M., & Polic, M. (2005). Public perception of flood risks, flood forecasting and mitigation. *Natural Hazards and Earth System Sciences, 5*(3), 345–355.

Brody, S. D., Zahran, S., Vedlitz, A., & Grover, H. (2008). Examining the relationship between physical vulnerability and public perceptions of global climate change in the United States. *Environment and Behavior, 40*(1), 72–95.

Brown, J. D., & Damery, S. L. (2002). Managing flood risk in the UK: Towards an integration of social and technical perspectives. *Transactions of the Institute of British Geographers, 27*(4), 412–426.

Bubeck, P., Botzen, W. J. W., & Aerts, J. C. J. H. (2012a). A review of risk perceptions and other factors that influence flood mitigation behavior. *Risk Analysis, 32*(9), 1481–1495.

Bubeck, P., Botzen, W. J. W., Kreibich, H., & Aerts, J. C. J. H. (2012b). Long-term development and effectiveness of private flood mitigation measures: An analysis for the German part of the river Rhine. *Natural Hazards and Earth System Sciences. Special Issue: Costs of Natural Hazards, 12*(11), 3507–3518.

Cannon, T. (2008). *Reducing people's vulnerability to natural hazards: Communities and resilience* (UNU-WIDER research paper no. 2008/34). Helsinki: United Nations University World Institute for Development Economics Research (UNU-WIDER). http://www.wider.unu.edu/publications/working-papers/research-papers/2008/en_GB/rp2008-34/. Accessed 4 Feb 2013.

Carter, T. R., Jones, R. N., Lu, X., Bhadwal, S., Conde, C., Mearns, L. O., et al. (2007). New assessment methods and the characterisation of future conditions. In M. L. Parry, O. F. Canziani, J. P. Palutikof, P. J. van der Linden, & C. E. Hanson (Eds.), *Climate change 2007: Impacts, adaptation and vulnerability. Contribution of Working Group II to the fourth assessment report of the Intergovernmental Panel on Climate Change* (pp. 133–171). Cambridge: Cambridge University Press. http://www.ipcc.ch/pdf/assessment-report/ar4/wg2/ar4-wg2-chapter2.pdf. Accessed 4 Feb 2013.

Conway, D., Persechino, A., Ardoin-Bardin, S., Hamandawana, H., Dieulin, C., & Mahé, G. (2009). Rainfall and water resources variability in Sub-Saharan Africa during the twentieth century. *Journal of Hydrometeorology, 10*(1), 41–59.

Correia, F. N., Fordham, M., Da Graça Saraiva, M., & Bernardo, F. (1998). Flood hazard assessment and management: Interface with the public. *Water Resources Management, 12* (3), 209–227.

Crutzen, P. J., & Stoermer, E. F. (2000). The anthropocene. *International Geosphere-Biosphere Programme (IGBP) Global Change Newsletters, 41*, 17–18.

Cutter, S. L. (2003). The vulnerability of science and the science of vulnerability. *Annals of the Association of American Geographers, 93*(1), 1–12.

Cutter, S. L., Emrich, C. T., Webb, J. J., & Morath, D. (2009). *Social vulnerability to climate variability hazards: A review of the literature. Final report to Oxfam America*. Columbia: Hazards and Vulnerability Research Institute, University of South Carolina. http://adapt.oxfamamerica.org/resources/Literature_Review.pdf. Accessed 4 Feb 2013.

Descroix, L., Mahé, G., Lebel, T., Favreau, G., Galle, S., Gautier, E., et al. (2009). Spatio-temporal variability of hydrological regimes around the boundaries between Sahelian and Sudanian areas of West Africa: A synthesis. *Journal of Hydrology, 375*(1–2), 90–102.

Diagne, K. (2007). Governance and natural disasters: Addressing flooding in Saint Louis, Senegal. *Environment and Urbanization, 19*(2), 552–562.

Dodman, D., Francis, K., Hardoy, J., Johnson, C., & Satterthwaite, D. (2012). *Understanding the nature and scale of urban risk in low- and middle-income countries and its implications for humanitarian preparedness, planning and response*. London: International Institute for Environment and Development (IIED). http://www.dfid.gov.uk/r4d/Output/191454/Default.aspx. Accessed 5 Feb 2013.

Dow, K. (2005). *Vulnerability profile of West Africa*. Stockholm: Stockholm Environment Institute. http://sei-international.org/mediamanager/documents/Publications/Risk-livelihoods/SEI_Dow_Vulnerability_West_Africa_2005.pdf. Accessed 10 Jan 2013.

Drobot, S. D., Benight, C., & Gruntfest, E. C. (2007). Risk factors for driving into flooded roads. *Environmental Hazards, 7*(3), 227–234.

Few, R., Ahern, M., Matthies, F., & Kovats, S. (2004). *Floods, health and climate change: A strategic review* (Tyndall Centre working paper no. 63). Norwich: Tyndall Centre for Climate Change Research, University of East Anglia. http://www.tyndall.ac.uk/content/floods-health-and-climate-change-strategic-review. Accessed 4 Feb 2013.

Ford, J. D., & Smit, B. (2004). A framework for assessing the vulnerability of communities in the Canadian Arctic to risks associated with climate change. *Arctic. Special Issue: Human Dimensions of the Arctic System, 57*(4), 389–400.

Grothmann, T., & Pratt, A. (2003). *Adaptive capacity and human cognition*. Conference paper presented at the open meeting of the Global Environmental Change Research Community, 16–18 October 2003, Montreal, Canada. http://sedac.ciesin.columbia.edu/openmtg/docs/grothmann.pdf. Accessed 4 Feb 2013.

Grothmann, T., & Reusswig, F. (2006). People at risk of flooding: Why some residents take precautionary action while others do not. *Natural Hazards. Special Issue: German Research Network Natural Disasters: Towards an Integrated Risk Assessment – Case Studies in Germany, 38*(1–2), 101–120.

Harvatt, J., Petts, J., & Chilvers, J. (2011). Understanding householder responses to natural hazards: Flooding and sea-level rise comparisons. *Journal of Risk Research, 14*(1), 63–83.

Intergovernmental Panel on Climate Change (IPCC). (2007). *Climate change 2007: Contribution of Working Group II to the fourth assessment report*. In M. L. Parry, O. F. Canziani, J. P. Palutikof, P. J. van der Linden, & C. E. Hanson (Eds.), Cambridge: Cambridge University Press. http://www.ipcc.ch/publications_and_data/publications_ipcc_fourth_assessment_report_wg2_report_impacts_adaptation_and_vulnerability.htm. Accessed 4 Feb 2013.

Jha, K. A., Bloch, R., & Lamond, J. (2012). *Cities and flooding: A guide to integrated urban flood risk management for the 21st century*. Washington, DC: The World Bank.

Karley, N. K. (2009). Flooding and physical planning in urban areas in West Africa: Situational analysis of Accra, Ghana. *Theoretical and Empirical Researches in Urban Management, 4* (13), 25–41.

Kellens, W., Zaalberg, R., Neutens, T., Vanneuville, W., & De Maeyer, P. (2011). An analysis of the public perception of flood risk on the Belgian coast. *Risk Analysis, 31*(7), 1055–1068.

Kempton, W. (1993). Will public environmental concern lead to action on global warming? *Annual Review of Energy and the Environment, 18*(1), 217–245.

Kreibich, H., & Thieken, A. H. (2008). Assessment of damage caused by high groundwater inundation. *Water Resources Research, 44*(9). doi:10.1029/2007WR006621.

Lavell, A., Oppenheimer, M., Diop, C., Hess, J., Lempert, R., Li, J., et al. (2012). Climate change: New dimensions in disaster risk, exposure, vulnerability, and resilience. In C. B. Field, V. Barros, T. F. Stocker, D. Qin, D. J. Dokken, K. L. Ebi, et al. (Eds.), *Managing the risks of extreme events and disasters to advance climate change adaptation. A special report of Working Groups I and II of the Intergovernmental Panel on Climate Change* (pp. 25–64). Cambridge: Cambridge University Press. http://ipcc-wg2.gov/SREX/. Accessed 18 Feb 2013.

Lindell, M. K., & Hwang, S. N. (2008). Households' perceived personal risk and responses in a multihazard environment. *Risk Analysis, 28*(2), 539–556.

Mahé, G., & Paturel, J.-E. (2009). 1896–2006 Sahelian annual rainfall variability and runoff increase of Sahelian Rivers. *Comptes Rendus Geoscience, 341*(7), 538–546.

Mahé, G., Leduc, C., Amani, A., Paturel, J.-E., Girard, S., Servat, E., & Dezetter, A. (2003). Augmentation récente du ruissellement de surface en région soudano-sahélienne et impact sur les ressources en eau. In E. Servat, W. Najem, C. Leduc, & A. Shakeel (Eds.), *Hydrology of Mediterranean and semiarid regions. Papers selected for the international conference on hydrology of the Mediterranean and Semi-Arid Regions held in Montpellier, France, from 1 to 4 April 2003* (pp. 215–222). Wallingford: International Association of Hydrological Sciences (IAHS). http://iahs.info/redbooks/278.htm. Accessed 10 Jan 2013.

Mbow, C., Diop, A., Diaw, A. T., & Niang, C. I. (2008). Urban sprawl development and flooding at Yeumbeul suburb (Dakar-Senegal). *African Journal of Environmental Science and Technology, 2*(4), 075–088.

Merz, B., Kreibich, H., Schwarze, R., & Thieken, A. H. (2010). Review article "Assessment of economic flood damage". *Natural Hazards and Earth System Sciences. Special Issue: Risk Management of Extreme Flood Events, 10*(8), 1697–1724.

Messner, F., & Meyer, V. (2006). Flood damage, vulnerability and risk perception – Challenges for flood damage research. In J. Schanze, E. Zeman, & J. Marsalek (Eds.), *Flood risk management – Hazards, vulnerability and mitigation measures* (pp. 149–167). Berlin: Springer.

Miceli, R., Sotgiu, I., & Settanni, M. (2008). Disaster preparedness and perception of flood risk: A study in an alpine valley in Italy. *Journal of Environmental Psychology, 28*(2), 164–173.

Mileti, D. S. (1980). Human adjustment to the risk of environmental extremes. *Sociology and Social Research, 64*(3), 327–347.

Milly, P. C. D., Wetherald, R. T., Dunne, K. A., & Delworth, T. L. (2002). Increasing risk of great floods in a changing climate. *Nature, 415*(6871), 514–517.

Mitchell, J. K. (2003). European river floods in a changing world. *Risk Analysis, 23*(3), 567–574.

O'Connor, R. E., Bord, R. J., Yarnal, B., & Wiefek, N. (2002). Who wants to reduce greenhouse gas emissions? *Social Science Quarterly, 83*(1), 1–17.

Oguntala, A. B., & Oguntoyinbo, J. S. (1982). Urban flooding in Ibadan: A diagnosis of the problem. *Urban Ecology, 7*(1), 39–46.

Oluseyi, O. F. (2006). Urban land use change analysis of a traditional city from remote sensing data: The case of Ibadan metropolitan area, Nigeria. *Humanity & Social Sciences Journal, 1*(1), 42–64.

Oyo State. (2005). Ministry of Environment and Water Resources Environmental Regulations, 2004. *Supplement to Oyo State of Nigeria Gazette, 30*(5), March 2005- Part B.

Oyo State. (2011). *Report on the assessment of the 26th August 2011 flood disaster in Ibadan Metropolis*. Report of the Task Force on Flood Prevention and Mitigation, November 2011.

Peters, E., & Slovic, P. (1996). The role of affect and worldviews as orienting dispositions in the perception and acceptance of nuclear power. *Journal of Applied Social Psychology, 26*(16), 1427–1453.

Ruelland, D., Dezetter, A., Puech, C., & Ardoin-Bardin, S. (2008). Long-term monitoring of land cover changes based on Landsat imagery to improve hydrological modelling in West Africa. *International Journal of Remote Sensing, 29*(12), 3533–3551.

Satterthwaite, D. (2008). *Climate change and urbanization: Effects and implications for urban governance*. Paper presented at the United Nations Expert Meeting on Population Distribution, Urbanization, Internal Migration and Development, New York, NY, 21–23 January 2008. New York: United Nations Population Division. http://www.un.org/esa/population/meetings/EGM_PopDist/P16_Satterthwaite.pdf. Accessed 4 Feb 2012.

Savage, I. (1993). Demographic influences on risk perceptions. *Risk Analysis, 13*(4), 413–420.

Scheraga, J. D., & Grambsch, A. E. (1998). Risks, opportunities, and adaptation to climate change. *Climate Research, 10*, 85–95.

Séguis, L., Cappelaere, B., Peugeot, C., & Vieux, B. (2002). Impact on Sahelian runoff of stochastic and elevation-induced spatial distributions of soil parameters. *Hydrological Processes. Special Issue: The Future of Distributed Hydrological Modelling, 16*(2), 313–332.

Siegrist, M., & Gutscher, H. (2008). Natural hazards and motivation for mitigation behavior: People cannot predict the affect evoked by a severe flood. *Risk Analysis, 28*(3), 771–778.

Slovic, P. (1987). Perception of risk. *Science, 236*(4799), 280–285.

Smit, B., & Pilifosova, O. (2003). From adaptation to adaptive capacity and vulnerability reduction. In J. B. Smith, R. J. T. Klein, & S. Huq (Eds.), *Climate change, adaptive capacity and development* (pp. 9–28). London: Imperial College Press.

Smit, B., & Wandel, J. (2006). Adaptation, adaptive capacity and vulnerability. *Global Environmental Change, 16*(3), 282–292.

Steffen, W., Crutzen, P. J., & McNeill, J. R. (2007). The anthropocene: Are humans now overwhelming the great forces of nature? *Ambio, 36*(8), 614–621.

Tapsell, S. M., Penning-Rowsell, E. C., Tunstall, S. M., & Wilson, T. L. (2002). Vulnerability to flooding: Health and social dimensions. *Philosophical Transactions of the Royal Society of London, Series A: Mathematical, Physical and Engineering Sciences, 360*(1796), 1511–1525.

Tschakert, P., Sagoe, R., Ofori-Darko, G., & Codjoe, S. (2010). Floods in the Sahel: An analysis of anomalies, memory, and anticipatory learning. *Climatic Change, 103*(3–4), 471–502.

United Nations Development Programme (UNDP). (2007). *Human development report 2007/2008. Fighting climate change: Human solidarity in a divided world*. New York: Palgrave Macmillan for the United Nations Development Programme (UNDP). http://hdr.undp.org/en/reports/global/hdr2007-2008/. Accessed 10 Jan 2013.

United Nations Development Programme (UNDP). (2010). *Human development report 2010. The real wealth of nations: Pathways to human development. 20th anniversary edition*. New York: Palgrave Macmillan for the United Nations Development Programme (UNDP). http://hdr.undp.org/en/reports/global/hdr2010/. Accessed 10 Jan 2013.

United Nations Human Settlements Programme (UN-Habitat). (2007). Disaster risk: Conditions, trends and impacts. In UN-Habitat (Ed.), *Enhancing urban safety and security: Global report on human settlements 2007* (pp. 167–194). London: Earthscan. http://www.unhabitat.org/downloads/docs/GRHS2007.pdf. Accessed 4 Feb 2013.

United Nations Human Settlements Programme (UN-Habitat). (2010). *The state of African cities 2010: Governance, inequality and urban land markets*. Nairobi: United Nations. http://www.unhabitat.org/content.asp?cid=9141&catid=7&typeid=46. Accessed 10 Jan 2013.

United Nations International Strategy for Disaster Reduction (UN/ISDR). (2003). *Disaster reduction and sustainable development: Understanding the links between vulnerability and risk to disasters related to development and environment*. Geneva: United Nations Inter-Agency

Secretariat of the International Strategy for Disaster Reduction (UN/ISDR). http://www.osn.cz/soubory/dr-and-sd-english.pdf. Accessed 4 Feb 2013.

United Nations Platform for Space-based Information for Disaster Management and Emergency Response (UN-SPIDER). (2011). *UN-SPIDER Regional Support Office respond to Ibadan flood*. Abuja: UN-SPIDER. http://www.un-spider.nasrda.gov.ng/?q=node/1. Accessed 10 Dec 2012.

Wheeler, D. (2007). *Will the poor be flooded out? The IPCC's predicted flood disasters and their implications for development aid*. Policy notes of the Centre for Global Development, Washington, DC. http://www.cgdev.org/content/publications/detail/13452/. Accessed 4 Feb 2013.

Whitmarsh, L. (2007). Are flood victims more concerned about climate change than other people? The role of direct experience in risk perception and behavioral response. *Journal of Risk Research, 11*(3), 351–374.

Wisner, B., & Pelling, M. (2012). African cities of hope and risk. In M. Pelling & B. Wisner (Eds.), *Urban disaster risk reduction: Cases from urban Africa* (2nd ed., pp. 17–42). London: Earthscan.

Wisner, B., Blaikie, P., Cannon, T., & Davis, I. (2004). *At risk: Natural hazards, people's vulnerability, and disasters* (2nd ed.). London: Routledge.

An Integrated Approach to Environmental Management in Bangladesh

Raquib Ahmed and Chandan Roy

1 Introduction

Historically, Bangladesh is a densely populated country, mostly due to the large flows of migration from other parts of South and South-East Asia. The major pull factors are the fertile soil, the stable climate conditions, and easy agricultural practices. However, this rapid population growth, has degraded Bangladesh environmentally, which is perceived as an indirect threat to other parts of the world (United Nations Environment Programme 2001). It has a population of 150 million on 147,000 km^2, with an ever-increasing high person-land ratio – a potential threat to the land and, more generally, to the physical environment. Inappropriate managerial and governance skills have also contributed to aggravating poverty (Streatfield and Karar 2008).

To date, Bangladesh has developed social and political systems of governance and education that function, despite being burdened by their colonial legacy of blind bottom-down decision making – which is partly mismatched with the current, newer socio-political systems of South Asia (Reed 2008). Owing to intensive economic activities, new environmental problems are surfacing, which frequently pose significant challenges to policy makers and policy-implementing authorities. All issues are linked to each other. Addressing them separately often leads to unsustainable, short-term results, which indicates the need for a wider integration of different management approaches geared towards optimizing resource efficiency and policy effectiveness. Government agencies address the environmental problems in Bangladesh from spatial and sector-wise perspectives, which are partly

R. Ahmed (✉)
Institute of Environmental Science, University of Rajshahi, Rajshahi, Bangladesh
e-mail: raquib_ahmed@yahoo.com

C. Roy
Department of Computer and Information Science, Linköping University, Linköping, Sweden

B. Werlen (ed.), *Global Sustainability*, DOI 10.1007/978-3-319-16477-9_10

based on traditional mechanisms and nongovernmental organizations that use modern stakeholder participation methods (Reed 2008; Sultana and Thompson 2004).

2 Environmental Problems in Bangladesh

Environmental degradation and the depletion of natural resources are in Bangladesh are due to poverty, over-population, poor and/or limited access to education, and ineffective communication about the causes and consequences of environmental problems. They are manifested by deforestation, destruction of wetlands, and depletion of soil nutrients to name but a few. Natural calamities, such as floods, cyclones, and storm surges, also result in severe socio-economic and environmental damage. While major environmental problems are centered on a few aspects, in detail their range is almost unlimited. For example, the key environmental issues that the Ministry of Environment and Forest (MoEF) identified are (i) topsoil degradation, (ii) surface and groundwater contamination and toxicity, (iii) urban air pollution, (iv) coastal biodiversity loss, and (v) deforestation (MoEF 2005; Islam et al. 2009).

All these environmental problems are related to, or originate from, the major prevalent issue: topsoil degradation due to the continuous use of soil for crop production. This does not allow soil nutrients to be replenished, ultimately leading to a biochemical imbalance. Soil productivity therefore declines gradually, leading to food insecurity. The use of high-yield varieties of crops to support the rapidly growing population results in farmers becoming dependent on commercial seed and fertilizer providers. This in turn may set a vicious economic cycle in motion, with farmers being maneuvered into a position where they are forced to buy seeds and fertilizer (and perhaps also herbicides and pesticides) from large global players in the agro-industry. In addition, to reducing biodiversity by limiting crops to a small number of selected varieties, the cultivation of such high-yield varieties lead to other environmental problems. These include the need for irrigation, which causes shortages of surface water and a lowering of the groundwater table, resulting in widespread arsenic contamination. Population growth results in rapid urbanization and, hence, land use change, which reduces the land available for crop production at a rate of 1 % annually. Excessive pressure on land resources causes people to turn to marine resources to meet their nutritional needs. This again causes large-scale, unmanaged biodiversity loss. Around 27 % of Bangladesh's land area was covered with natural forest about 100 years ago. This has dropped to about 8–12 % today depending on the definition of "natural forest". Currently, natural forests are almost at a critical level, as their margins are being transformed into agricultural lands by illegal logging. The following is a list of environmentally stressed regions in Bangladesh (Rashid 1996):

- Mahanada basin: frequently flooded, also subject to frequent drought.
- West-central Barind: desiccated by improper land use. Low water table and poor soils affect crop agriculture.
- Middle Karatoa floodplain: affected by the drying up of the Karatoa river. Double cropping of High Yield Variety (HYV) rice lead to severe sulphur and zinc deficiencies.
- Brahmaputra-Jamuna floodplain: the entire stretch is affected by the Brahmaputra's right bank, which is breached four times in every 5 years.
- Chalan Beel: one of the richest wetlands that development activities and agriculture expansion have almost ruined.
- Atrai-Hurasagar drainage basin: drainage is impeded and water-logging has become a serious problem.
- South-west Jessore: the fresh water flow has decreased and salinity increased.
- Northern Khulna: large-scale shrimp farming has increased the salinity, conflicts among farmers has reduced the rice production.
- Khulna city and Mongla town: industrial pollution, oil spills from ships, and urban congestion have caused problems.
- Sundarbans: the increased salinity and the agriculture pressure from the vicinity cause forest degradation.
- Patuakhali-Bhola-Noakhali Char areas: water-logging, increased salinity, and diluvion. The excessive use of pesticides affects humans.
- Garo Hills in piedmont: erosion, flash floods, and the loss of tree cover have led to a decrease in the crop production.
- Tangail: the silting of rivers, more frequent flash floods.
- Madhupur tract: deforestation and improper use of sloping land have led to topsoil erosion.
- Shitalakshya River: industrial plants discharge toxic chemicals into the river, leading to a loss of fisheries and toxic fish, which are a public health hazard.
- Dhaka city: industrial pollution and urban expansion, which has destroyed the surrounding fertile agricultural and horticultural land.
- Haor basin: a reduction in fish spawning areas.
- South Sylhet: affected by deforestation, flash floods, and soil erosion.
- Lalmai range: deforestation, erosion and soil removal.
- Lower Meghna: affected by floods, erosion, stagnant productivity, loss of fisheries, population pressure.
- Central Noakhali: water-logging, lack of irrigation, salinity, decreasing crop production, and increasing population.
- Sandwip: dense population, erosion and accretion of land, land formation not consolidated, frequently affected by coastal cyclones and storm surges.
- Sitakunda range: deforestation, erosion, loss of productivity.
- Chittagong city and the port: industrial pollution, oil spills, ship-breaking industries, hill cutting.
- Chandraghona: industrial/chemical pollution dumped into the river, destroying its biodiversity.

- Hill Tracts: the ethnic local people's slash and burn cultivation and improper use of hill slopes have greatly increased erosion and the flooding of valleys, resulting in productivity loss. Serious and extensive deforestation.
- Chakaria Sunderban: a mangrove forest totally destroyed by shrimp farming, as the soil has become acidic.
- Chittagong coast – Kutubdia Island: severe bank erosion, loss of land, increase in salinity.
- Cox's Bazar: the tropical moist forest and unique biodiversity destroyed by deforestation and planting operations.
- St. Martin's Island: Coral reef ecosystem damaged by intensified tourism.

3 Government Policies, Rules, Acts, and Plans for Environmental Management

In recent years, the government has taken important steps towards conservation, the environmentally sound use of natural resources, and pollution control. These include the adoption of a National Environment Policy, the formulation of the National Conservation Strategy, and the National Environment Management Action Plan (NEMAP). It is hoped that these steps will help integrate environment and development actions into a coherent policy framework (Ministry of Planning 2010). Owing to the limited availability of land, virtually all land areas are utilized in one way or another for crop production, forestry, fisheries, and for urban and infrastructure development.

The distribution of land ownership is skewed towards comparatively wealthier citizens. Given this situation, small land owners, who are also repeatedly affected by disasters, become economically weaker and sooner or later have no alternative but to sell their land and migrate to other location, preferably bigger urban areas that offer marginal survival opportunities. This indicates that the poorer stakeholders have little resilience against natural disasters. Policy changes have so far contributed little to reverse this process. For several generations, the steady increase in the proportion of landless people has been compounded by natural disasters, land fragmentation, river meandering, and changes to river courses. Together, these factors have resulted in uneconomic farming that does not even provide the families concerned with a bare subsistence level.

Much land has also been lost due to river erosion due to seasonal floods and immediately after floods. The list of socio-ecological problems is very long, and this is only a partial description of the context within which strategic planning and policies have to be developed.

Basis for Laws and Policies As mentioned, there is currently no formal framework for policymaking, monitoring, and enforcement. The country lacks a comprehensive land use policy for appropriate and productive uses of the land. This refers to an optimum economic return from the land and the sustainable management of environmental conditions. Despite a general shortage of land in the country, there is sufficient abandoned and unused exists in urban and in rural areas. However, agricultural and other land is being lost due to:

- Land being converted into shrimp farming in the coastal regions, which has adversely affected agricultural production, has caused a loss of productive trees, and has led to the large-scale depletion of mangrove forests.
- Traditional land use patterns being changed with the introduction of HYV rice and irrigated agriculture.
- The expansion of urban settlements, sand and gravel mining, the development of the communication infrastructure, brick fields, and industrial developments.
- The erosion of agricultural land exacerbates the land consumption and rural poverty problems.
- Land fragmentation due to the vertical property inheritance system continues contributes to the number of landless families.
- Short-term leasing and the sharecropping system prevent poorer people from purchasing land from long-term absentee land owners, who can easily continue retain their ownership.
- Land tenure and ownership patterns are unclear in many areas, preventing long-term investment in land productivity.
- Inadequate land use planning based on an assessment of the land's capabilities constrains land productivity.
- Unregulated encroachment into forest lands leads to their unsustainable agricultural exploitation in the short term due to uncertainty of the future investments.

Given the above, laws are being prepared to address these. The following are a few landmark measures already taken (Disaster Management Bureau 2010; Government of Bangladesh 2010; MoP 2010).

- Environment Policy (1992) and Environment Action Plan (1992)
- National Environment Management Action Plan (NEMAP) (1995)
- Environmental Conservation Act (ECA) (1995)
- Environmental Conservation Rules (ECR) (1997)
- Environmental Courts Act (2000)

The key objectives of the Environment Policy (1992) and Environment Action Plan (1992) are:

- Maintaining an ecological balance and an overall development by protecting and improving the environment/
- Protecting the country against natural disasters.
- Identifying and regulating activities that pollute and degrade the environment.
- Ensuring environmentally sound development in all sectors of the economy.
- Ensuring the sustainable, long-term, and environmentally sound foundation of natural resources.
- Remaining actively associated with major international environmental initiatives.

The Environmental Conservation Act (ECA) of 1995 is the main legislative framework document relating to environmental protection in Bangladesh. This umbrella act includes laws to conserve the environment, improve the environmental standards, and to control and mitigate environmental pollution. The Act established the Department of Environment (DoE), and provided its Director General with the discretion to instigate inquiries, prevent accidents, advise the government, coordinate with other authorities or agencies, and collect and publish information about environmental pollution. According to the Act, no industrial unit or project should be established or undertaken without obtaining an Environmental Clearance Certificate (ECC) from the Director General of DoE.

The government of Bangladesh issued the Environment Conservation Rules (1997) which the Environment Conservation Act (Section 20) (1995) sanctioned. Under these Rules, the following aspects are also covered:

- Identification of ecologically critical areas.
- Classification of industries and projects into four categories (from “1 – highly polluting” to “4 – low environmental impact”) that indicate their contribution to environmental damage.
- Procedures for issuing the Environmental Clearance Certificate.
- Determination of environmental standards.

The Environment Court Act (2000) was enacted to establish environmental courts in each administrative division. Under this Act, the court has concurrent jurisdiction in civil and criminal cases. The basis for instituting a case is a violation of the Environment Conservation Act (ECA) (1995) and the rules following from this. The environment court is specifically empowered to:

- Impose penalties for violating court orders.
- Confiscate any object, equipment, or vehicle used in the execution of the offence.
- Pass any order or decree for compensation.
- Issue anyone violating the law with directions (a) not to repeat or continue the offence; (b) to take preventive, mitigative, or remedial measures with relation to any injury, damage, or harm caused, specifying the time limit and reporting to the DoE regarding the implementation of the directions.

4 Introduction to Land and Environmental Management

In Bangladesh only very limited areas of land are still available for further development and agriculture, or are suitable for land use change from an ecological perspective. Land areas under anthropogenic influence are already under intense pressure that adversely affects the carrying capacity of the local ecosystems and ultimately leads to the degradation of terrestrial ecosystems. This stress is expected to increase with the continued population growth. A key problem is the suboptimal and/or inappropriate use of land, such as government forested areas, coastal areas, and deciduous hill forests.

Agriculture is still the economic mainstay in Bangladesh involving a majority of the population. However, its contribution to the GDP has declined to 17.5 % as other sectors have grown significantly. For example, industry contributes 28.5 % to the GDP and the service sector 53.9 % (Central Intelligence Agency 2013). Given the government's lack of true commitment to policy modifications, this pattern is likely to continue with other sectors taking up land available for cropping. Consequently, the population pressure on land is still and will continue to be a crucial factor in the management of terrestrial natural resources. In the past, land use planning favored food crop production to the detriment of forest, cash crops, fish, cattle feed, etc. Almost all sectors compete for the use of land.

Issues like the continual erosion of land by rivers – the flood plain comprises a dense network of rivers, many of which have a tendency to change course, thus causing erosion – increasing landlessness and land fragmentation; the unclear status and pattern of land ownership, particularly with regard to new land; a low level of year-round land utilization; and the lack of land use planning need a decisive solution for sustainable economic growth.

The Fourth Five Year Plan (FFYP) outlines a number of policies with regard to land resources and land use planning. These are aimed at improving land availability for the rural population and improving its productive use (economically). Points worth mentioning are:

- A systematic and comprehensive analysis of optimal land use (in terms of long term productivity and related cost) is to be undertaken.
- Enforcing zoning laws to limit the loss of agricultural land to urban development.
- Possible further reductions in the land ownership ceiling, which is expected to lessen land fragmentation.
- Enforcing laws to enable share croppers to purchase land through the provision of long-term loans.
- Distributing Khas (public/government) lands.
- Improving land use based on an assessment of its optimal cropping pattern.
- Promoting the use of marginal and homestead land for intensive cultivation to contribute to the total productivity.
- Formulating and executing a land use policy to prevent wasteful resource use and conflicts (e.g. about shrimp and rice farming). The policy will take the physical and environmental aspects of conflict areas into account. Leasing of inland open waters will be increased to a minimum of 4 years. Higher production and resource conservation.
- Creating and maintaining a permanent green belt along the coast and participatory forestry (social forestry) along roadsides, etc.

The government also focuses on its Environmental Policy (1992). This environment policy makes four specific statements on land issues: (i) formulate an environmentally balanced and sound national land use policy and plan, (ii) prevent and reduce/mitigate soil erosion, (iii) preserve and increase soil fertility and protect reclaimed land and newly accreted land for environmentally sound management, and (iv) encourage the adoption of land use practices that are compatible with various ecosystems. The government also emphasizes the need to reduce the impact of salinity and alkalinity.

5 Agencies Working in Environmental Management Sectors

Those government agencies working in these sectors mostly apply government plans and often respond to a situation's immediate needs. Most of their efforts are supported by foreign donors. Some of the large donors are: the World Bank (WB), United Nations Development Programme (UNDP), United Nations Environment Programme (UNEP), World Health Organization (WHO), Asian Development Bank (ADB), U.S. Agency for International Development (USAID), the German government, Japan International Cooperation Agency (JICA), the government of The Netherlands, Norwegian Agency for Development Cooperation (NORAD), Canadian International Development Agency (CIDA), etc. These donors support local and collaborative (international) nongovernment organization (NGO) partners. Generally, NGOs are created for specialized sectoral and local issue-based activities. Sometimes, however, one NGO works simultaneously in various sectors and in different geographical regions. A paradigm shift has occurred because both the government and the NGOs are increasingly focusing on long term measures and working on conceptually complicated issues, such as global environmental change and climate change, nature and biodiversity conservation, sea-level rise, etc., within international partnerships. NGOs working on environmental issues in Bangladesh focus on the following:

- Program implementation (e.g. social forestry, crop diversification)
- Policy initiation or policy inputs (e.g. NEMAP)
- Providing assistance with international negotiations
- Research
- Supporting environmental movements (e.g. against genetically modified organisms)
- Taking a watchdog role (e.g. ship breaking)

The following sections provide an overview of the most important NGOs active in environmental management sectors in Bangladesh. Examples of their activities are given here to help gain an understanding of the nature, extent, and method of their work.

- The Bangladesh Center for Advanced Studies (BCAS) focuses on:

  (a) environment-development integration,

(b) good governance and people's participation,
(c) poverty alleviation and sustainable livelihoods, and
(d) economic growth and public-private partnership.

BCAS promotes people-centered sustainable development by applying and advancing scientific, technical, and local knowledge through research, by developing models, by giving demonstrations, policy advocacy, and project implementation. It is committed to developing southern perspectives, to develop the North-South Dialogue, and ensure environmental justice and access to resources and knowledge for the poor (BCAS 2011).

BCAS' specialization fields (BCAS 2013) include:

- Natural resource management (land, water, fisheries, forestry, agriculture, and biodiversity)
- Enhancing the resilience of natural and human systems
- Livelihood analysis, poverty reduction strategies, and food security
- Environmental Impact Assessment (EIA) and Social Impact Assessment (SIA)
- Clean technology and pollution management
- Energy (conventional and renewable)
- Environmental education and health
- Public-private partnership
- Global climate change, and the human dimension of global change
- Multilateral Environmental Agreements (MEAs)
- Trade, environment, and sustainable development

At the local and grassroots levels, BCAS works with communities on ecology-specific participatory management processes, such as at Chanda Beel on the Madhumati floodplain, where it helps the local communities' active participants develop natural resource management capabilities. In addition, BCAS aims to initiate the diffusion of renewable energy technology (solar photovoltaic systems) among the river Meghna's islands dwellers.

At the national level, BCAS has initiated activities along with others, including the NEMAP, the Sustainable Environment Management Program, National Conservation Strategy, New Fisheries Management Program, the Bangladesh State of Environment Report (2001), and the vulnerability assessment of Bangladesh to climate change and sea level rise. It also works closely with the Ministry of Environment and Forest, the Department of Environment, the Ministry of Science and Technology, the Ministry of Agriculture, the Bangladesh Agricultural Research Council, the Ministry of Energy and Mineral Resources, the local governments, and the engineering department.

At the regional level, BCAS is the secretariat for the Climate Action Network South Asia (CANSA) working on climate change related issues. BCAS is the international wing of Centre for Environment Education (CEE), it is also a member of Regional and International Networking Group (RING) in South Asia.

At the international level, BCAS is involved in a number of collaborative research projects, in policy analysis, and advocacy, including the North-South

Dialogue (NSD) on climate change, and the Global Compact Initiative under the auspices of the UN. BCAS participated in all the UNCED preparatory meetings in Rio and in WSSD in Johannesburg, where it also played a key role in advising government delegations. BCAS provides the Global Forum on Environment and Poverty (GFEP), which originated at the Earth Summit in Rio, with secretarial support. As a center of excellence in the south, BCAS has been involved in formulating and reporting process of all of UNEP's Global Environment Outlook Reports.

- The Bangladesh Environmental Lawyers' Association (BELA), an advocacy group of lawyers, was established in 1992 with the broad objective of promoting environmental justice and contributing to the development of sound environmental jurisprudence. Its efforts through public litigation have truly sensitized the concept of environmental justice, leading to special courts dealing with environmental offences.

BELA's Objectives and Strategies Are:

- Undertaking studies and research on the local, national, and international regulatory environment regimes
- Undertaking legislative advocacy
- Providing judicial or administrative help to ensure the implementation of existing environmental laws
- Resolving environmental disputes through court cases, alternative dispute resolution, mediation, and by other means
- Creating greater awareness of environmental laws and issues
- Providing legal assistance and supporting endeavors for the protection of the environment and human rights
- Developing networks with local, national, and international groups, bodies, or agencies that concentrate on environmental issues
- Developing a core group of environmental activists

- Proshika is an NGO that mostly concentrates on educating people on specific issues. A social forestry program is one of its major activities. This is a systematic intervention to enhance the planting, protecting, and regeneration of forest resources (Proshika 2005). It also provides disaster relief during natural calamities such as floods, cyclones, cold spells, etc. One of its projects on crop diversification cooperates closely with the Department of Agricultural Extension and the Agricultural Bank.
- The Bangladesh Rural Advancement Committee (BRAC), has two environment-related programs, one of which is concerned with natural disasters, the other with climate change (BRAC 2013). These programs seek to (i) promote better disaster preparation and management, (ii) increase awareness about climate change and its impacts, and (iii) undertake research to identify successful risk reduction and coping mechanisms.

  In order to provide communities with access to safe water, BRAC undertakes various activities, including developing a water safety plan, installing deep tube wells, doing water quality tests, and providing loans to construct tube well platforms. In arsenic and saline-affected areas, BRAC installs deep tube wells, constructs ponds, installs filters and arsenic removal filters, and installs piped water supply systems to provide safe water. BRAC raises awareness of sanitation issues, which creates community demand for facilities such as sanitary latrines. It installs water supply and sanitation facilities to improve people's health and also establishes schools in rural areas.

- The Wildlife Trust of Bangladesh (WTB) is a non-profit organization whose aim is to conserve the country's biological diversity. WTB's activities include (WTB 2012):
  - Research and monitoring
  - Institutional and policy development
  - Communication and education
  - Wildlife-human conflict mitigation
  - Legislation and law enforcement
- The Bangladesh Poribesh Andolon (BAPA) (Bangladesh Environment Movement) is a forum for citizens and organizations concerned about the country's natural environment. BAPA, which acts as a pressure group against any kind of environmental degradation, tries to create a broad-based citizens' movement for the protection and enhancement of the environment in Bangladesh. It organizes seminars, meetings, conferences, and workshops to draw attention to general and specific environmental problems, as well as to educate the public on such issues. It holds rallies and demonstrations to build public awareness and ensure the public's wide participation in environmental issues. Its objectives are (BAPA 2009):
  - To prevent further environmental degradation in Bangladesh,
  - To reverse, where possible, the damage that has already been done to the environment
  - To build a nationwide, united, civic movement to achieve its aims of preventing and reversing environmental degradation in Bangladesh.

6 Integrated Approach Towards a Sustainable Future

As discussed previously, various sectors are involved in various aspects of environmental policy. However, they have not as yet yielded appropriate sustainability results. This is obviously due to a lack of integration efforts. The National Environmental Management Action Plan (NEMAP) was actually a first integrated approach based on concrete programs and interventions aimed at promoting the better management of scarce resources, at reversing the present environmental degradation trends, and at increasing awareness of environmental issues more generally. NEMAP considered the commitments made under Agenda 21, and was the first initiative aimed at the preparation of a national Agenda 21, which the government of Bangladesh was supposed to present before 2005. It was an effort to bring together different perspectives within a common framework when addressing different types of issues. This exercise was also undertaken in the hope that it would provide decision makers with a comprehensive account of all the main environmental issues from different points of view, including those of stakeholders, implementers, and policy makers.

The government of Bangladesh was engaged in the preparation of a National Environmental Management Action Plan in cooperation with UNDP for some time. It proactively initiated a consultation process in association with NGOs, journalists, academics, and other segments of civil society. This was probably the first time that the government had finalized a national plan with the active participation of representatives from all sectors of the economy and all regions of the country. The NEMAP identified key environmental issues of concern to Bangladesh and the actions required to halt or reduce the rate of environmental degradation, improve the natural and human-made environment, conserve habitats and biodiversity, promote sustainable development, and improve the measurement accuracy of quality of life indicators.

Integration would occur at four levels, for example, (1) spatial activities were mainly addressed by local NGOs as they were accountable to local stakeholders; (2) sectoral issues were mostly addressed by local and international NGOs and supported by international donations from various external organizations and countries; (3) long-term policy and institutional issues were handled almost exclusively by the government and supported by international protocols that would ensure global integration; (4) government policy and laws to control internal activities under various ministries and departments. The formulation of NEMAP and the Bangladesh Climate Change Strategy and Action Plan (2009) fall under long term issues (MoEF 1995).

The NEMAP, formulated in response to current and projected environmental scenarios, required mitigating adverse environmental impacts. Environmental conditions and policies changed over time as progress was made towards sustainable development and the NEMAP was evolved in responses to these changes. The NEMAP identified the major environmental problems and their causes and suggested mitigation measures. NEMAP's full integration is shown in the following diagram (Fig. 1).

When NEMAP's mandate expired in 2008, the Bangladesh government started a new environmental management program to suit the new situation regarding global climate change and the new international protocols and responses. This program is known as the Bangladesh Climate Change Strategy and Action Plan (2009). It is a long-term working document and sets out the strategy for the next 25 years. Its major actions are to ensure (MoEF 2009):

- Food security, social protection, and health
- Comprehensive disaster management
- Infrastructure

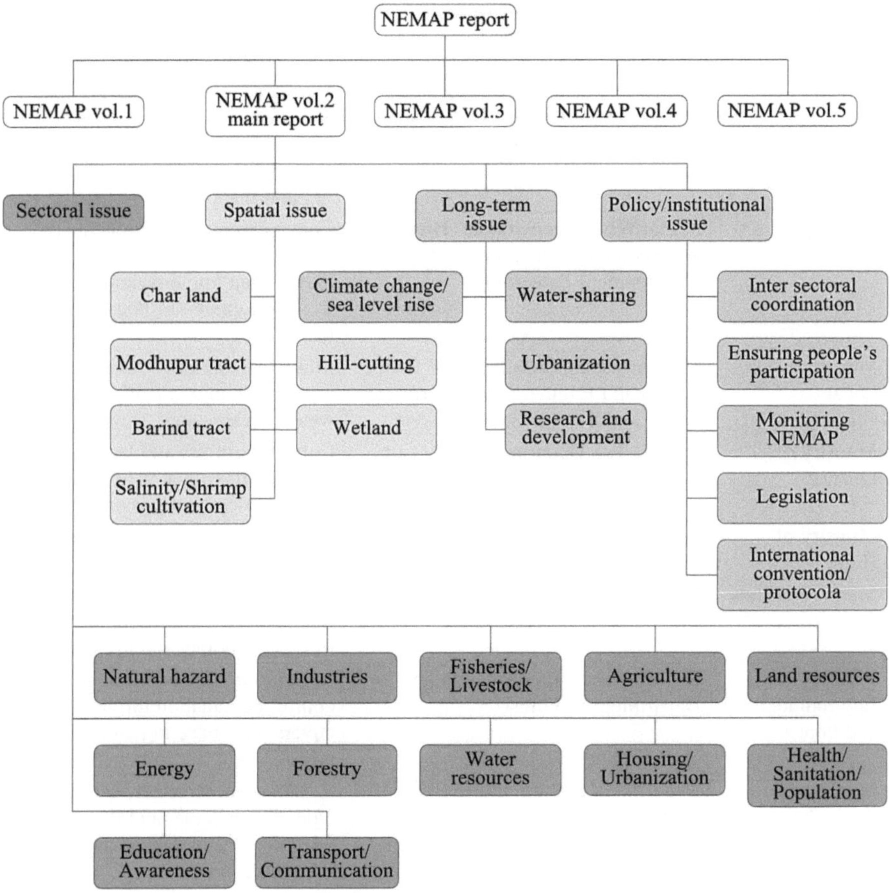

Fig. 1 Method of the NEMAP integration for environmental management in Bangladesh (Reproduced from MoEF 1995, p. 15)

- Research and knowledge management
- Low carbon development
- Capacity building and strengthening institutions

7 Conclusion

Approaches to the management of environmental issues are usually under control of the government, despite partial integration, NGOs have been tasked with responsibility for these issues. This is an experimental measure, as Bangladesh is a role model of NGOs participation in the national development. Many NGOs such as the Grameen Bank, BRAC etc. use this model, which is currently used in more than

Table 1 Management of environmental issues

Ministry	Authorities	Responsibilities
Ministry of Land	Directorate of Land Records and Survey, Land Reforms Board, Deputy Commissioners/Collectors of Land	Maintains records of rights, change of ownerships/transfer of land, collection of land tax based on use, settlement of government land, survey and mapping, land dispute, etc.
Ministry of Works	Public Works Department, Housing and Settlement Department, Urban Development Directorate	Construction of government structures, development of urban lands, master plan implementation authority, development of housing facilities
Ministry of Local Governments	Municipalities, Directorate of Public Health and Engineering Department, Water and Sewage Authority, etc.	Municipal facilities, rates and holding taxes, supply of safe drinking water, conservancy, sewage, etc.
Ministry of Finance	Income Tax Department, City Corporation Authorities	Collection of tax and income taxes on properties situated on urban land, fund development projects
Ministry of Home	Police and Border Protection Force (BGB)	Help implement government decisions, internal security
Ministry of Law	Judiciary	Legal coverage to plan implementation authorities, formulation of law to support policy and urban plans
Ministry of Environment and Forest	Department of Environment and Department of Forest	Monitor urban environment and control degradation through legal coverage, counseling and awareness development, forest management
Ministry of Agriculture	Department of Agriculture Extension, Agriculture Development, etc.	Ensure seed, fertilizer, and insecticide are available, and that knowledge dispersion and product marketing occur
Ministry of Industry	Chemical, textile, jute, iron and steel, etc.	Licensing industries, environmental monitoring
Ministry of Food and Disaster Management	Department of Food, Department of Disaster Management	Food management, storage, pricing, food security, disaster management, environment management due to disaster
Ministry of Chittagong Hill Tracts affairs	Chittagong Hill Tracts (CHT) Development Board, CHT Regional Council, etc.	Land, ethnic issues, forest and environment management in a specific region

Based on Ahmed (2002)

44 countries, including some European countries. Their activities have encouraged the transparent mass-participation of citizens in national development and are aimed at reducing the impact of non-democratic political power. Major land use conflicts arise from insufficiently coordinated action between the 11 Ministries and 24 agencies concerned with land management. However, the table below (Table 1) shows the weakness of this approach due to the government ministries' overlapping power and responsibilities regarding the management of environmental issues (Ahmed 2002).

The Key Constraints:

- Lack of inter-sectoral coordination
  - The management responsibilities of different environmental components are divided into different sectors and ministries. This hampers the efficient operation and execution of sustainable management.
- Contradictions in sectoral policies
  - Some policies sometimes contradict each other. Some provisions of the Fisheries Policy contradict that of the Land Policy and Industrial Policy, and the Environment Policy does not conform to the narrow objectives of the Export Policies.
- Lack of institutional capacities
  - The Ministry of Environment and Forest (MoEF) has yet to develop a strong functioning planning cell to support its work. It lacks essential baseline data on resources and areas of environmental concerns.
- Regulatory and institutional inadequacies for policy implementation
  - All the important policies, including the National Environment Policy of 1992, were formulated during the last decade and only a few of them seem to be coordinated well with the older legal instruments of the given sector.
- Outdated laws
  - A law passed two to five decades back cannot incorporate the concept of sustainable development or everyone's right to a healthy environment, which are the outcomes of very recent environment concerns. Such laws cannot play an effective role in combatting environment pollution in today's Bangladesh, where overpopulation, poverty, and illiteracy are aggravating this crisis every day.
- Non-punitive approach of laws
  - The existing laws do not have a punitive approach.

The challenge to sustainable development in Bangladesh hinges on three guiding principles:

- Achieving significant development for all citizens as demonstrated by human development indicators.
- Integrating environmental dimension in all development activities and at all levels: plans, policy, program, projects, community, and ecosystem.

(continued)

- A governance structure in which environmental concerns are central and the people, i.e. the stakeholders, are the ultimate custodians of the resource base and the source of all decision making.

It should also be noted that the actions indicated in the Bangladesh Climate Change Strategy and Action Plan (BCCSAP) 2009 are mostly adaptation and adjustments strategies. This appears to be a critical issue between sources of environment degradation, mitigation strategies, and adaptation. Only developing countries seem to be required to adapt, although they are not significant contributors of sources of problems. In the future, this might be a matter of debate regarding sustainability. Although the simultaneous achievement of all three of the above-mentioned objectives may appear an impossible and improbable task, there is an inherent synergy amongst the three objectives that could be the driving force to achieve sustainability in Bangladesh. The question is not only how to achieve sustainable development in Bangladesh, but – more appropriately – how to include the sustainability concept in the development of communities, ecosystems, projects, and programs. Thus, achieving sustainability must become a central objective, and the majority of the country's population, who are poor, must be brought into the development-environment nexus, be central in the decision making, and become contributors.

Acknowledgement The authors are thankful to Professor Gregor C. Falk of the Institute of Geography, University of Education, Freiburg, Germany for helping with an earlier draft of the paper.

References

Ahmed, R. (2002). Urban environmental issues in Bangladesh and government intervention. In A. Narman & K. Karunanayake (Eds.), *Towards a new regional and local development research agenda* (Meddelanden från Göteborgs universitets geografiska institutioner, Department of Geography, Series B No. 100, pp. 79–94). Gothenburg: University of Gothenburg.

Bangladesh Center for Advanced Studies (BCAS). (2011). Climate change adaptation and mitigation. http://www.bcas.net/about-bcas.php?id=7. Accessed 2 June 2012.

Bangladesh Center for Advanced Studies (BCAS). (2013). *Fields of specialization*. http://www.bcas.net/about-bcas.php?id=3. Accessed 2 June 2013.

Bangladesh Poribesh Andolon (BAPA). (2009). *Bangladesh Poribesh Andolon objectives*. http://www.bapa.org.bd/AboutBAPA.aspx. Accessed 2 June 2013.

Bangladesh Rural Advancement Committee (BRAC). (2013). *Disaster, environment and climate change programme*. http://decc.brac.net/. Accessed 2 June 2013.

Central Intelligence Agency (CIA). (2013). *The world fact book*. https://www.cia.gov/library/publications/the-world-factbook/geos/bg.html. Accessed 2 June 2013.

Disaster Management Bureau (DMB). (2010). National Plan for Disaster Management 2010–2015.

Government of Bangladesh (GoB). (2010). Environmental impact management framework Dhaka.

Islam, K. S., Xue, X., & Rahman, M. M. (2009). Successful Integrated Coastal Zone Management (ICZM) program model of a developing country (Xiamen, China) – implementation in Bangladesh perspective. *Journal of Wetlands Ecology, 2*(1), 35–41.

Ministry of Environment and Forest (MoEF). (1995). *National Environment Management Action Plan (NEMAP)* (Volume II: Main Report, Vol. 2 of 5). Dhaka: NEMAP Secretariat, Ministry of Environment and Forest.

Ministry of Environment and Forest (MoEF). (2005). National adaptation programme of action Dhaka.

Ministry of Environment and Forest (MoEF). (2009). Bangladesh climate change strategy and action plan 2009.

Ministry of Planning (MoP). (2010). Environmental management framework Dhaka.

Proshika. (2005). *Social forestry and natural forest protection*. http://www.proshika.org.bd/sfp_n.htm. Accessed 22 June 2013.

Rashid, H. (1996). *Geography of Bangladesh*. Dhaka: The University Press Ltd.

Reed, M. S. (2008). Stakeholder participation for environmental management: A literature review. *Biological Conservation, 141*(10), 2417–2431.

Streatfield, P. K., & Karar, Z. A. (2008). Population challenges for Bangladesh in the coming decades. *Journal of Health Population and Nutrition, 26*(3), 261–272.

Sultana, P., & Thompson, P. (2004). Methods of consensus building for community-based fisheries management in Bangladesh and the Mekong Delta. *Agricultural Systems, 82*(3), 327–353.

United Nations Environment Programme (UNEP). (2001). *Bangladesh state of environment report*. United Nations Environment Programme.

Wildlife Trust of Bangladesh (WTB). (2012). *The WildTeam mission*. http://wildlifetrustbd.org/our-mission. Accessed 22 June 2013.

Awareness of and Responses to the 2011 Flood Warnings Among Vulnerable Communities in Lagos, Nigeria

F. Olokesusi, F.B. Olorunfemi, A. Onwuemele, and M.O. Oke

1 Introduction

2007 marks a turning point in history: half of the world's population now lives in cities (UN-Habitat 2007). The total urban population is expected to double from two billion to four billion over the next 30–35 years (UN-Habitat 2006). An unwanted side effect of this rapid urbanization process is the increased susceptibility to flooding as a result of the concentration of people and assets in flood-prone areas, since many urbanized areas are located along major bodies of water. Furthermore, climate change may cause more frequent and more severe floods. This combination is likely to result in substantially larger flood impacts compared to former times, during which societies and environmental change drivers developed more slowly and societies continuously adapted to environmental changes (Zevenbergen et al. 2008).

The above accounts are corroborated by Sherbinin et al. (2007), according to whom cities are economic growth engines, centers of innovation for the global economy, and the hinterlands of their nations. The foundations the prosperity and prominence of most global cities are their longstanding commercial relationships with the rest of the world. Most cities are located on or near the coast, which has facilitated trade and contributed to their wealth. They are also often located in low-lying areas near the mouths of major rivers, which serve as conduits for commerce between interior agricultural and industrial regions and the rest of the world. Examples include Lagos, Marseille, Lisbon, Chennai, Shanghai, and New York. Such locations place cities at greater risk of current and projected climate hazards such as cyclones, high winds, flooding, coastal erosion and deposition, and a rising sea level.

F. Olokesusi (✉) • F.B. Olorunfemi • A. Onwuemele • M.O. Oke
Social and Governance Policy Research Department, Nigerian Institute of Social and Economic Research, Ojoo, Ibadan, Nigeria
e-mail: felixba2000@yahoo.com

B. Werlen (ed.), *Global Sustainability*, DOI 10.1007/978-3-319-16477-9_11

With the unprecedented rise in the number and severity of natural disasters, large urban settlements have become increasingly vulnerable. The concentration of substandard infrastructure and housing, material assets, and inherent socio-economic inequalities increases large cities and megacities' susceptibility to disasters. In addition to socio-economic and spatial vulnerabilities, these settlements' functions and geographical distributions make them especially prone to disaster risk (Sherbinin et al. 2007). This problem of high susceptibility to disaster risk is further compounded by megacities' extreme vulnerability to the impacts of climate change, especially if they are located in developing countries (Mehrotha et al. 2009; Adelekan et al. 2009; Satterthwaite et al. 2007). Most Nigerian cities are particularly vulnerable to disasters, especially floods, which have increased in frequency and intensity in recent years (Raheem et al. 2012; Gbadegesin et al. 2011; Olorunfemi 2008; Olorunfemi and Raheem 2007; Olokesusi 2011, 2004; Adelekan 2009). The vulnerability of Nigeria's cities to hazards is compounded by uncontrolled urbanization, widespread urban and rural poverty, degradation of the environment due to the mismanagement of natural resources, weak socio-economic infrastructure, and inefficient public policies (Raheem et al. 2012; Olorunfemi 2008; Olorunfemi and Raheem 2007; Olokesusi 2004).

Natural disasters that result from, or which are facilitated by, climate change can undermine decades of growth in urban regions in a single catastrophic event. Furthermore, low-lying cities situated near major rivers, deltas, coasts, and estuaries are especially vulnerable to the rising sea level (Stern 2006). As shown by Mortreux and Barnett – who analyze Funafuti, the main island of Tuvalu – floods badly impact the living conditions of local communities, who are forced to migrate in response to climate change (Mortreux and Barnett 2009). Lagos is both a megacity and a low-lying city located on the coast.

Recent global events indicate that floods are indeed increasing, both in terms of frequency and magnitude. Between 2001 and 2005, there were 818 flood disasters in the world, compared to 974 between 2006 and 2010. In addition, the estimated number of deaths from floods was 25,929 between 2001 and 2005, compared to 30,171 between 2006 and 2010 (International Federation of Red Cross and Red Crescent Societies (IFRC) 2011). Statistics also reveal the resulting economic costs of flood disaster damage (IFRC 2011; Stern 2006; Munich Re 2005; World Health Organization (WHO) 2002). Studies based on reported disasters (IFRC 2010) also suggest that there have been considerable shifts in flood event patterns and intensities, resulting in increased hazards for the world's growing urban population. Floods are also growing faster than non-climate-related hazards. It is predicted that increases in hazards due to climate change will impact different regions differently, but that increasing flood hazard is a shared future expectation. Under these circumstances, sustainable urban flood risk management is becoming an increasingly challenging task for urban communities and authorities. Unfortunately, many cities across the world are unable to keep up with the intensification of such challenges.

The flooding incidents in Lagos, too, have increased both in frequency and intensity and thus have had significant impacts (Mehrotra et al. 2009; Adelekan 2009; Gbadegesin et al. 2011). Perhaps in response to people's concerns and to avoid further calamity, as experienced in 2010, the Lagos State government issued several warnings in 2011 (most recently on June 10, 2011) to residents and property owners along river banks and flood-prone areas, urging them to move to higher ground (Vanguard 2011, 37; National Mirror 2011, 18). The government was responding to the Nigerian Meteorological Agency (NIMET) forecast of heavy rainfalls that could create severe flooding in the State. NIMET had forecast that heavy rainfalls, accompanied by tsunami-like flooding, were expected to start earlier and end later than usual in 2011.

NIMET forecast 1,200–1,700 mm rainfall in 2011 and specifically noted that communities in the northeastern part of the State, which is naturally susceptible to flooding, were expected to be worst hit during heavy rainfalls. The State listed the flood-prone communities: Ikosi-Ketu, Mile 12, Agiliti, Ajegunle, Thomas Laniyan Estate, Owode-Onirin, Agboyi, Owode-Elede, Mairan, and Isheri North Scheme. The official who issued the warning said that to avoid a repeat of the 2010 flooding, which displaced more than 1,500 residents of a community in Ikorodu (also in Lagos State), the residents of the affected communities had to move to higher ground from June to mid-September and October, and possibly to January. According to the State government official, "this became necessary because the capacity of most of the canals cannot contain the volume of runoff expected from the rainfall" (Vanguard 2011, 13). This study was prompted by the flood warnings issued to the abovementioned communities.

Our study had two objectives: to conduct a survey of the awareness of and responses to the flood warnings, as well as to offer a rapid assessment of the affected communities' physical vulnerabilities to flooding. This article is divided into six sections. After the introduction, Sect. 2 clarifies some conceptual issues. Sections 3 and 4 provide information on the study area and the methodology. In Sect. 5, we discuss the study findings, while the final section provides recommendations for the effective communication of flood warnings in general and for the reduction of these communities' vulnerabilities in particular.

2 Conceptual Considerations

Flooding is widely viewed as the most dangerous source of disaster risk. A wide range of literature spanning the insurance industry (Munich 2005) and international humanitarian and development agencies (IFRC 2011; United Nations Office for Disaster Risk Reduction (UNISDR) 2004) support this. There are two main discourses on flood disasters. The first – and dominant view – is that flood disasters are inherently a characteristic of natural hazards (Dixit 2003; Adger 1999). Disasters are inevitable when hazard magnitude is high. This contrasts with the alternative discourse, which sees flood disasters as being produced by the interaction of the physical hazard and the social vulnerabilities. The latter discourse identifies social relationships, structures, institutions, and governance in the quest to better understand flood disaster. It posits that flood disasters are not only the result of natural hazards, but also of socio-economic structures and political processes that render individuals, families, and communities vulnerable (Dixit 2003).

To fully understand urban flood risks, one should be familiar with the different components of risks. Risk is often understood only superficially as the occurrence of an extreme event or hazard (flood, drought, earthquake, storm, landslide, etc.) caused by natural forces or by a combination of natural forces and human influences. Although the occurrence of such a hazard is the primary precondition, it is only one component of risk. The second component of risk is that somebody or something is *at risk*; i.e., vulnerable to a hazard (World Meteorological Organization (WMO) and Global Water Partnership (GWP) 2008). This definition clarifies the basic structure of risks.

However, with reference to the term "vulnerability", a further distinction enhances our understanding of flood risk creation, since the notion of vulnerability in the abovementioned definition does not distinguish between physical exposure to hazards and the susceptibility of persons or things to hazards. This definition therefore does not address flood risk analysis or the question regarding which measures are most effective in reducing such risks. Hence, we adopt the definition of risk proposed by Crichton (1999), who defines risk as the probability of a loss. This loss depends on three elements: the magnitude of the hazard, vulnerability, and exposure. If any of these three elements in risk increases or decreases, the risk increases or decreases.

While exposure in the context of floods refers not only to the question whether or not people or assets are physically in the path of floodwaters, vulnerability may be defined as "[t]he conditions determined by physical, social, economic, and environmental factors or processes, which increase the susceptibility of a community to the impact of hazards" (UNISDR 2004, 16). Figure 1 shows the functional relationships between risks, hazard, exposure, and vulnerability. The UNISDR (2004, 17)

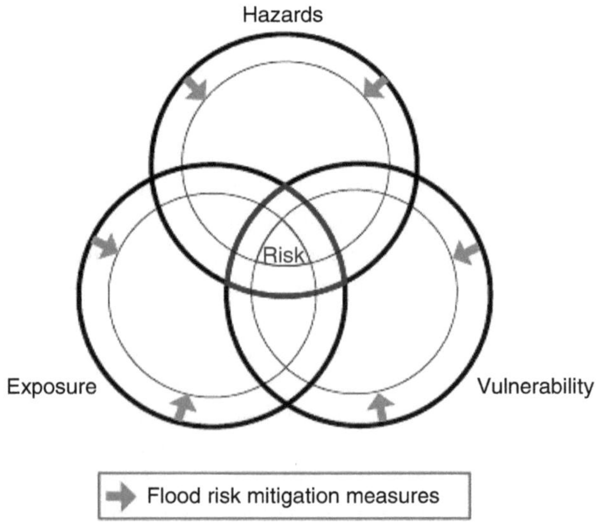

Fig. 1 Construct of urban flood risk and its reduction (WMO and GWP 2008, 3)

defines disaster as "a serious disruptions of the functioning of a community or society causing widespread human, material, economic or environmental losses which exceed the ability of the affected community or society to cope using its own resources".

This definition is complemented by a commonly accepted definition of flood risk, which defines it as a function of the probability of the flood hazard, of exposure to the flood hazard, and of the vulnerability of receptors of the flood hazard. There are many versions of such models for disasters generally (Thywissen 2006). Crichton (1999) formalizes this definition of risk with regard to floods in a risk triangle. In this triangle, hazard is a function not only of natural processes, but also of anthropogenic environmental changes that alter natural flow patterns and pathways to generate increased flood hazards from a similar magnitude weather event. In an urban context, this implies that upstream conveyance becomes part of the hazard that cities experience. This definition also encompasses the important notion of the difference in relative damage that those affected by a hazard experience. Such variations in the ability to cope with and to recover from a hazard are described either as vulnerability or as resilience to the hazard. Vulnerability and resilience are negative and positive measures of the same characteristic. Increases in the impacts and risks from flooding can, therefore, result from increases in the severity of hazards, from populations and their assets' exposure, or from the vulnerability of exposed populations and their assets to flooding.

Inherently, urban development creates larger risks, but those in higher income groups are better able to avoid or bear such risks than those with low incomes. There is a clear socio-spatial segregation concerning settlement locations' hazard exposure. Since urbanization is essentially an increase in population density, space becomes rare and expensive. Consequently, those who cannot afford to purchase or rent space in secure environments are forced to move to cheaper places. Given that the urban poor's livelihood often depends on their proximity to informal economies in the centers of large cities, many prefer to inhabit hazard-prone areas. Two further factors aggravate spatial marginalization. On the one hand, hazard-prone areas are often not privately owned and informal dwellers are therefore less likely to be displaced. On the other hand, however, many urban poor are migrants from rural areas who are not familiar with the various hazards and therefore tend to underestimate the risk of living in such exposed areas. This pattern is explained in Lagos's morphology by Adelekan (2009) and Olokesusi (2011).

3 The Study Area

Massive development to meet the socio-economic needs of emerging metropolises and megacities in Africa and Asia is usually carried out without adequate consideration of the hydrological environment (Actionaid 2006). Among the unresolved challenges faced as a result of this development are destructive flood incidents – even in regions previously considered safe. This is more prominent in the

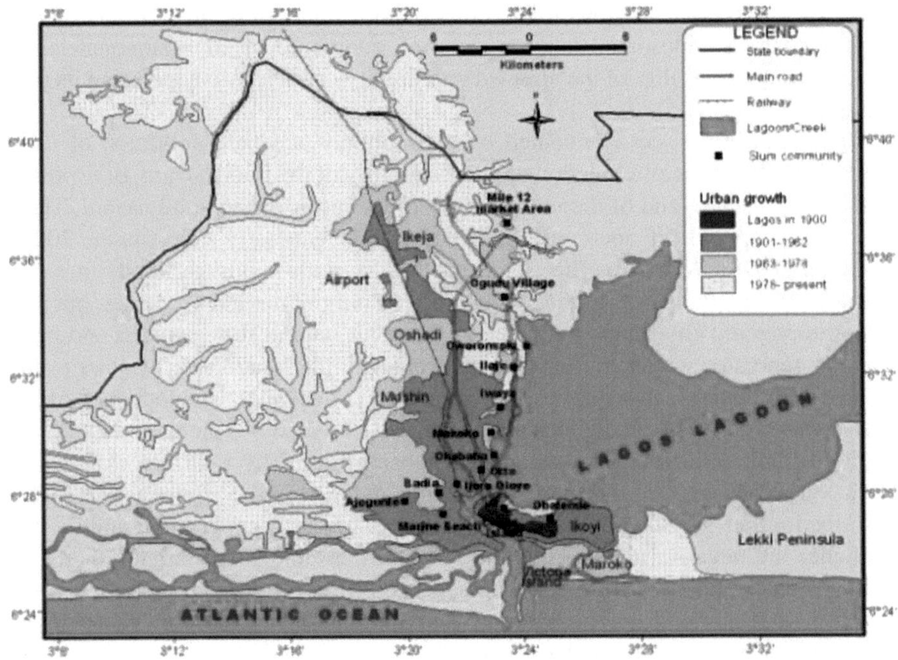

Fig. 2 Growth of metropolitan Lagos from 1900 to present (Source: Adelekan 2009, 8)

developing world, which is characterized by inadequate spatial arrangements in growing cities and in their socio-economic structures, few new investments in physical infrastructure, poor planning processes, and weak political and economic bases.

Interestingly, besides being the industrial and commercial hub of Nigeria and the West African subregion, metropolitan Lagos is also West Africa's largest city. It is expected to be one of the world's five largest cities by 2015 (Mehrotra et al. 2009). Lagos is Africa's second most populous city after Cairo. It grew explosively from 300,000 people in 1950 to an approximately 18 million in 2010, when it ranked as one of the world's ten largest cities. Lagos is home to many industries and has a large commercial infrastructure. It has greatly benefited from Nigeria's natural resources in the form of oil, natural gas, coal, wood, and water. For an overview of the state of Nigerian cities, see UN-Habitat (2004).

Figure 2 is a map of Lagos showing the city's growth from 1900 to the present. For this study, the metropolis Lagos is conceptualized as the city of Lagos within the city's official administrative area (in Lagos State), as well as the northeastern incursion or extension of the city into neighbouring Ogun State (as shown in Fig. 3). The city has a relatively large land area with a very diverse and fast-growing population. Thus, normal urbanization pressures and modern development forces have resulted in the city expanding to nearby suburbs such as northeast Lagos, which politically covers Lagos State and Ogun State. Northeast Lagos falls within

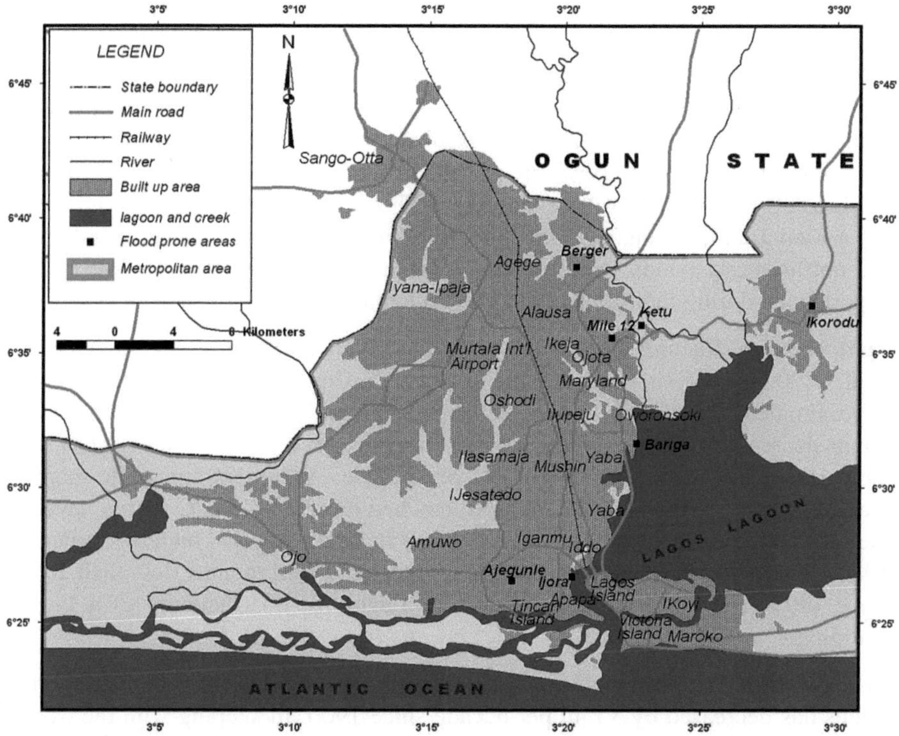

Fig. 3 Map of Lagos metropolis showing flood-prone areas (Composed by authors)

the lower part of the Ogun-Osun River Basin, with a larger portion of the region within the River Ogun floodplain. The region is part of Nigeria's Hydrological Area VI, with an annual runoff of $35.4 \times 109\ m^3$ (or a runoff depth[1] of 352 mm per year) and an average annual runoff increase of approximately 17 % (Federal Republic of Nigeria 2006).

The average population density of Lagos's local government areas is approximately 2,094 people per square kilometer, with a minimum of 164 people per square kilometer in Owode Obafemi Local Government Area (LGA) of Ogun State and a maximum of 55,939 people per square kilometer in Ajeromi-Ifelodun LGA in Lagos State. A better picture of the city's high population density can be inferred from a breakdown of each local government area's population density in built-up areas. The local government areas in Lagos State have an average population density of 13,194 people per square kilometer. Ajeromi-Ifelodun LGA, where most of the affected communities are located, has a staggering population density of 60,204 people per square kilometer (Mehrotra et al. 2009). This high population

[1] In hydrology, the runoff depth is determined by dividing the catchment's runoff volume by the catchment area. It provides a convenient way to compare runoff with precipitation in a given area and is usually expressed in depth units per unit time, usually as mm per hour (mm/h).

density has implications for flood disasters in terms of the number of people who would be directly affected. Figure 3 is a map of Lagos showing flood-prone areas.

Ocean and atmospheric interactions within and outside its environment, in which the Inter-Tropical Convergence Zone (ITCZ) is a controlling factor, affect Lagos's climate. The ITCZ movement is associated with a warm, humid maritime tropical air mass with southwesterly winds and the hot, dry continental air mass with dry northeasterly winds (Mehrotra et al. 2009). The maximum temperatures during the dry season are high, ranging from 28 to 33 °C when the region is dominated by the dry northeasterly winds. A minimum temperature of approximately 24–26 °C is experienced during the wet season (May to September).

Temperature Records from the two stations (Ikeja and Lagos) used in Mehrotra et al.'s (2009) analysis show that the monthly maximum temperature increased by approximately 0.1 °C per decade between 1952 and 2006, while monthly minimums decreased by approximately 0.5 °C per decade; since the 1900s, the average temperature has increased by 0.07 °C per decade. At the extremes, the monthly maximum temperatures for Lagos have reached more than 34 °C during seven of the past 200 years. The number of heat waves in Lagos has also increased since the 1980s. There were very few incidences of unusually cold months (less than 20 °C) since 1995. The temperature projected for Lagos for 2050 anticipates a 1–2 °C warming.

Precipitation According to historical records, the total annual precipitation in Lagos has decreased by 8 mm per decade since 1900. In keeping with the overall precipitation trends, most of Lagos has experienced rainfall decreases during the rainy season. For example, between 1950 and 1989, more than 20 months saw rainfalls over 400 mm. Between 1990 and 2006, however, only four rainy months recorded more than 400 mm. In the twenty-first century, precipitation in Lagos is expected to be less frequent but more intense. The projected precipitation for Lagos in 2050 anticipates an uncertain 5 % change in the mean (Mehrotra et al. 2009).

This megacity is sensitive to climate change owing to its flat topography and low elevation, location, high population, widespread poverty, and weak institutional structures. Many more vulnerabilities stem from these characteristics, including the high potential for water backups in drainage channels, inundation of roadways, and severe erosion (Mehrotra et al. 2009; Adelekan 2009; Awosika et al. 1993a, b). Thus, a large number of people are likely to feel the effects of any negative consequences of climate change and climate variability extremes – such as flooding – , especially the urban poor living in the city's marginal flood-prone areas.

4 Research Methodology

Since this study presents a brief assessment, primary data was mostly used. This data was obtained from a structured household survey questionnaire administered to the residents, as well as from focus group discussions (FGDs) with the key stakeholders. The research team also used on-the-spot assessment to determine the survey communities' physical vulnerabilities. These are the communities to whom NIMET and the Lagos State government issued flood warnings. These include Agiliti, Ikose-Ketu, Ajegunle, Owode Onirin, and River View Estate, all in northern Lagos along the Ogun River Basin. Forty questionnaires were administered in each community, complemented by FGDs with key groups that included men, women, and young people. Interviews were also conducted with key government officials in charge of flood management in the Lagos State Ministry of Environment. This was done to obtain information on the government's preparedness for floods in the affected communities.

The survey questionnaire elicited information on the respondents' socio-economic characteristics, their awareness of the flood warnings, and responses to them. In addition, information was asked regarding the sources of information, the neighborhood characteristics, and the characteristics of the buildings in which the respondents reside. These are important indicators of the level of vulnerability to flood disasters. Relevant secondary data was obtained from published and unpublished material, including the internet, to complement the primary data. Of the 200 questionnaires administered, 193 were retrieved and analyzable. The data collected from the questionnaire survey was analyzed using Statistical Package for the Social Sciences (SPSS).

5 Results and Discussions

Early warning signals are important elements of disaster risk reduction. They provide vulnerable communities with ample opportunities to prepare for disaster risk, which help reduce the impact of disasters when they occur. The following section presents the study results and a discussion of the vulnerable residents' awareness levels and responses to this flood warning by the government and its agencies.

5.1 Respondents' Socio-economic and Demographic Characteristics

Demographic and socio-economic characteristics have remarkable impacts on social geography and environmental behavior (Onokerhoraye 1994). As shown in Table 1, more than two-thirds of the respondents (72.5 %) were males, while the remaining 27.5 % were females. The high percentage of males is indicative of the interviews targeting the heads of households (usually males). Also, more than two-thirds of the respondents were married (71.5 %). The survey results further reveal that 43.0 % and 31.1 % of the respondents were between the ages of 26 and 35 years, as well as 36 and 45 years, while 45.1 % had attained secondary education

Table 1 Socio-economic characteristics (authors' analysis (2011))

Variables	Number of respondents	Percentage
Sex:		
Male	140	72.5
Female	53	275
Total	**193**	**100.0**
Age (years):		
18–25	19	9.8
26–35	83	43.0
36–45	60	31.1
46–55	18	9.3
56–65	7	3.6
Above 65	6	3.1
Total	**193**	**100.0**
Marital status:		
Single	51	26.4
Married	138	71.5
Separated	3	1.6
Divorced	1	0.5
Total	**193**	**100.0**
Educational qualification:		
No formal education	15	6.8
Primary education	10	5.1
Junior secondary	2	1.0
Senior secondary	81	42.0
Tertiary	87	45.1
Total	**193**	**100.0**
Monthly income:		
Less than NGN 7,500	12	6.2
NGN 7,501–15,000	27	14.0
NGN 15,001–20,000	56	29.0
NGN 20,001–45,000	16	8.3
NGN 45,001–60,000	8	4.1
GN 60,001–75,000	40	20.7
Above NGN 75,000	11	5.7
No response	23	11.9
Total	**193**	**100.0**
Household size:		
0–2	13	6.7
3–5	69	35.8
6–8	111	57.5
Total	**193**	**100.0**

and 42.0 % had attained tertiary education. Only 6.8 % of the respondents had no formal education. The high number of adults and their reasonable educational attainment are considered good for a study of this nature, because the respondents are old enough and sufficiently educated to be well informed about the issues at hand. Of the respondents, 43.0 % were traders, 12.0 % artisans, 12.0 % public servants, 10.0 % active in the transport sector, while 1.0 % were farmers. The high proportion engaged in trading further lends credence to Lagos's position as Nigeria's commercial engine.

Large household sizes have implications for disaster impacts in the sense that when a disaster strikes, large numbers of people per household are likely to be affected. More than half of the respondents (58.0 %) had between 6 and 8 children, 35.8 % between 3 and 5 children, and 6.7 % between 0 and 2 children. By inference, the respondents had an average of 4.4 children per household. The rather large household size and the large number of children have implications for disaster risk in that high absolute numbers of casualties during disasters may be compounded by an increased need for disaster relief for survivors.

Analysis of the respondents' incomes indicates that 29.0 % of the respondents earn between NGN 15,000 and 20,000 per month.[2] Another 20.7 % of the respondents earn between NGN 60,000 and 75,000 per month. However, 7.1 % of the respondents earn less than NGN 7,500 per month. With absolute poverty defined globally below USD 1 per capita per day and given an average household of 4.4 people, the average family will need to earn NGN 19,800 per month to live above that poverty line. The implication is that one of every five of the respondents lives below the absolute poverty line. An individual's income also largely determines his or her capability to cope with and recover from disasters. Poverty makes urban residents vulnerable to natural disasters such as flooding.

The next section focuses on the impact that the demographic and socio-economic data has on the respondents' awareness levels and responses to flood warnings in Lagos.

5.2 Respondents' Awareness of Flood Warnings

The respondents in five of the communities were asked if they were aware of the flood warnings that the Lagos State government issued. The results indicate that more than two-thirds of the respondents (77.2 %) were aware of the flood warnings, while 22.8 % indicated they were not. Further analysis presented in Table 2 shows a strong correlation between the levels of educational qualification and flood warning awareness. In other words, a larger proportion of the respondents with higher educational qualifications was aware of the flood warnings compared to the less educated respondents. Specifically, 85.2 % and 69.0 % of respondents with

[2] During the study period, the currency exchange rate was USD 1 to NGN 155.

Table 2 Respondents' educational qualification and awareness of flood warnings (authors' analysis (2011))

Educational qualification	Awareness of flood warnings: Yes	No	Total
No formal education	2	2	4
	50.0 %	50.0 %	100.0 %
Primary education not completed	7	2	9
	77.8 %	22.2 %	100.0 %
Primary school	10	0	10
	100.0 %	0.0 %	100.0 %
Junior secondary school	1	1	2
	50.0 %	50.0 %	100.0 %
Senior secondary school	69	12	81
	85.2 %	14.8 %	100.0 %
Tertiary education	60	27	87
	69.0 %	31.0 %	100.0 %
Total	149	44	193
	77.2 %	22.8 %	100.0 %

secondary and tertiary qualifications were respectively aware of the flood warnings. A *t*-test analysis of the relationship between the two variables shows a calculated χ^2 value of 11.8, significant at 0.05, impliying a significant difference in the level of education and the awareness of flood warnings.

5.3 Sources of Information on Flood Warnings

We sought to establish the media through which the respondents received their information. The survey results reveal that 36.0 % of the respondents received their information via the radio, 32.2 % were informed by others, 20.0 % via television, and 9.0 % via newspaper. Although 77.2 % of the respondents were aware of the flood warnings, the large number of respondents who received the information via other people implies that the state agency in question should make an effort to disseminate information better by implanting different media – such as town criers – in affected communities and should work with community development associations (CDAs), of which there are many in many of the affected communities. The result of this analysis is presented in Table 3.

5.4 Responses to the Flood Warnings

While awareness of flood warnings is important, individual response patterns to such information is crucial, because the actions taken determine the extent to which

Table 3 Flood warning information sources (authors' analysis (2011))

Information source	Frequency	Percentage
Radio	53	35.6
Television	30	20.1
People	48	32.2
Newspaper	13	8.7
Other means	5	3.4
Total	**193**	**100.0**

Table 4 Response to flood warnings (authors' analysis (2011))

Response	Number of respondents	Percentage
Planning relocation	12	6.2
Clearing drainage channels	44	22.8
Sand-filling building surroundings	40	20.7
No action	97	50.3
Total	193	100.0

people suffered from the disaster impact. Hence, the study also identified the actions that individuals took in response to the warning.

Table 4 shows that the majority (73.4 %) of the respondents claimed that they had not made any emergency plans. Only 26.6 % stated that they had undertaken some preparations in anticipation of the flood. People's lack of preparation undermined the purpose of the flood warning.

As indicated in Table 4, 50.3 % of the respondents took no action at all, depending solely on "God and the government" for help. On the other hand, the actions taken by more than 40 % of the respondents involved clearing drainage channels and sand-filling the surroundings of their homes. These measures do not necessarily mean an improvement in these areas' infrastructures in that the drainage channels were not expanded. As a result, these measures may not prevent flood waters or mitigate the disaster impact. Despite the government's key measures to prevent a loss of life – which emphasized relocation, among others – , only 6.2 % of the respondents were willing to relocate.

During the focus group discussions, the respondents noted that the government flood warnings were an "attempt by government to acquire their land like it did in Maroko [a slum that was upgraded and allocated to the rich in another suburb of Lagos]". They also noted that the relocation camp that the government opened in Agbowa (for the victims of 2010 flooding) was "more or less like a psychiatric home." One respondent from the Ajegunle neighborhood noted that flooding is a regular phenomenon in their community, that it comes and goes and they therefore did not contemplate relocation. Others maintained that the relocation cost was beyond their reach. However, many believed that because the flooding would have no effect, they made no plans to relocate. They maintained that past predictions had led to nothing. Others noted that they are homeowners and therefore could not abandon their homes. The government's failure to properly inform the residents that were at risk and to provide alternative safe havens or temporary shelters is an indication of the general poor readiness for the impending disaster.

5.5 Vulnerability Assessment of the Studied Communities

The state of the city infrastructure is a major factor that determines a city's vulnerability to flooding. The significance of infrastructure in the normal functioning of any city cannot be over-emphasized. For instance, the Intergovernmental Panel on Climate Change (IPCC 2007) notes that risks from flooding are greatly reduced by well-maintained flood control, sanitation infrastructure, and public health measures. Similarly, Olokesusi (2004) has noted that, in Nigeria, settlements' vulnerability to frequent environmental emergencies is largely determined by the following two variables:

- The vulnerability of the elements within them that are at risk, i.e., the ability of the built physical environment – the buildings, site improvement, and infrastructure – to withstand the stress of natural hazards.
- The hazards of locations of settlements; i.e., the extent to which they are subjected to environmental emergencies.

It is against this background that we analyze the state of the study area's infrastructure and environmental characteristics with regard to vulnerability. This analysis focuses on building characteristics, building locations, availability of drainage facilities, and waste disposal methods. In Table 5, we present and analyze the type, age, location, and condition of the housing that the respondents occupied.

37.3 % of the buildings are above 30 years of age, while 23.8 % are between 21 and 30 years old. Only 4.6 % of the buildings are less than 5 years old. The high percentage of buildings older than 20 years (61.1 %) shows that these city buildings are highly vulnerable to flooding, given that many of the houses are poorly maintained. This is revealed in the results of the analysis, which indicate that 75.0 % of the buildings need minor or major repairs. Only 24.9 % of the respondents indicated that their houses were physically sound. The implication of this is that, in the event of flooding, most of the houses might not withstand the flood's impact.

A physical assessment of some of the communities shows some coping measures in place due to past flood incidents. Many residents use wooden pedestrian bridges to reach their homes (see Figs. 4 and 5). Some of these measures actually strengthened the residents' resolve not to relocate.

A building's location plays an important role in increasing or decreasing its vulnerability to flooding. Buildings located close to floodplains are more vulnerable to flooding than those on higher ground. An analysis of the study data shows that 34.7 % of the buildings are located on floodplains or marshy land (see Fig. 6). On-the-spot assessment found that many dilapidated houses are located in unsafe places (Fig. 7).

Table 5 Characteristics of respondents' houses and neighborhoods (authors' analysis (2011))

Building characteristics	Number of respondents	Percentage
Age of buildings:		
Less than 5 years	9	4.6
6–15 years	49	25.4
15–20 years	14	7.3
21–30 years	46	23.8
Above 30 years		37.3
No response	3	1.5
Total	**193**	**100.0**
Building condition:		
Structurally sound	48	24.9
Poor, in need of minor repairs	67	34.7
Poor, in need of major repairs	78	40.4
Total	**193**	**10.0**
Location of buildings:		
Hilly terrain	9	4.7
Sloping terrain	19	9.8
Flat terrain	68	35.2
Very close to river or stream	30	15.5
On floodplain or marshy land	67	34.7
Total	**193**	**100.0**
Building construction materials:		
Laterite or mud	1	0.5
Laterite or clay blocks	8	4.1
Cement and sand concrete	176	91.2
Corrugated iron sheets	3	1.6
Wooden materials	4	2.1
Others	1	0.5
Total	**193**	**100.0**
Solid waste disposal method:		
Dump it in nearby stream	28	14.5
Put it in refuse bin or bag	34	17.6
Burn it	27	13.9
Dump it on the street or compound	69	35.8
Drop it in refuse dumps	35	18.1
Total	**193**	**100.0**

Of the homes, 15.5 % are located close to a stream or river, and 9.8 % on slopes (Figs. 8 and 9).

According to the analysis results, almost all the buildings (91.2 %) are constructed from cement or sand concrete. There is poor solid waste management in the affected communities. Of the respondents, 14.5 % dump their waste in the nearby bodies of water, 17.6 % use refuse bins, and 13.9 % burn their refuse, 35.8 %

Figs. 4 and 5 Wooden pedestrian bridges to houses in the Ajiliti community (Source: authors, June 8, 2011)

Fig. 6 A House in marshy land at Owode Onirin

dump their waste in the streets, and 18.1 % in refuse dumps. All the wastes in the streets and refuse dumps eventually find their way into the drainage systems, which often results in blocked drains. Most buildings in the study area are not sound and are therefore very vulnerable to flood disasters owing to their locations and structures.

Henderson (2004) stresses inadequate physical infrastructure with respect to the risks and vulnerability associated with climate change impacts, especially flooding. We paid attention to the availability and state of an access road, as well as the availability and nature of the drainage facilities. Table 6 reflects the respondents' opinions of the state of these facilities.

Fig. 7 A dilapidated house at Ikosi (Source: authors, June 8, 2011)

Fig. 8 A house at Ajiliti located on slopes very close to a stream

Table 6 shows that 60.6 % of the respondents indicated that there is an access road to their neighbourhood. However, the majority (54.7 %) indicated that these are not asphalted and are in poor condition. The importance of access roads cannot be over-emphasized. Poor access roads increase people's vulnerability to flooding and hinder rescue operations in cases of flood disaster. Similarly, 56.5 % of the respondents stated that there were drainage facilities in their neighborhoods. However, 94.8 % of the drainage facilities are open, making them dangerous during flooding. In addition, 75.1 % of the respondents noted that the drainage is already blocked. More than half the respondents noted that the drainage facilities are poor and not maintained. Well-kept drainage facilities are important, especially in areas susceptible to flooding, such as the study area.

Fig. 9 A similar structure at Ikosi-Ketu (Source: authors, June 8, 2011)

Table 6 State of infrastructure in the study area (authors' analysis (2011)

Infrastructure	Number of respondents	Percentage
Availability of access road:		
Yes	117	60.6
No	76	39.4
Total	**193**	**100.0**
State of access road:		
Not asphalted but good	52	27.1
Not asphalted and bad	105	54.7
Not asphalted but motorable	25	13.0
Asphalted	11	5.2
Total	**193**	**100.0**
Availability of drainage:		
Yes	109	56.5
No	84	43.5
Total	**193**	**100.0**
Nature of drainage:		
Open drain	183	94.8
Covered drain	10	5.2
Total	193	**100.0**
Condition of drainage:		
Blocked	145	75.1
Cleared	48	24.9
Total	**193**	**100.0**
Frequency of drainage clearance:		
Daily	9	4.7
Weekly	18	9.3
Monthly	30	15.5
No maintenance	136	70.5
Total	**193**	**100.0**

6 Summary and Conclusion

The study findings show that the affected communities are very vulnerable to flood disasters. The location of the buildings and the state of the available infrastructural facilities in these communities contributed immensely to their vulnerability to floods. The affected communities' responses to the flood warnings were poor, thereby rendering people vulnerable to flood disasters. Whereas the Lagos State government asked people who were vulnerable to flooding to relocate to higher ground, it failed to provide temporary shelter for them, thereby discouraging relocation. Based on these findings, we make the following recommendations.

- There is an urgent need for the government to continuously provide advocacy campaigns on the dangers of building on floodplains and indiscriminate refuse dumping.
- Lagos should increase its efforts to create awareness of climate change to the grassroots and to promote and sustain adaptation strategies to build the capacities of institutions and communities.
- The government should enforce building regulations to ensure people do not build on the flood plains.
- The government should explore new ways, particularly in terms of stakeholder involvement, to mitigate floods' impact.
- The government should utilize diverse means (including town criers) to disseminate flood warnings to communities. This will ultimately increase awareness levels in the various communities.
- The social and political aspects of flood management should be strongly emphasized to complement the hydraulic and engineering aspects that the government is currently addressing.
- Proper temporary accommodation should be provided for people at risk of flooding and for displaced persons to motivate them to relocate from flood-prone areas.
- Clearing of blocked drains should be an ongoing activity, rather than only occurring after the onset of rains.
- Removable concrete covers should be placed on open drains to reduce dumping of solid waste and to permit regular clearing. In addition, solid waste receptacles should be provided at strategic locations in the city, and households should be encouraged to practice sustainable waste management.

Acknowledgement We thank our home institution, the Nigerian Institute of Social and Economic Research (NISER), Ibadan, for financial support to carry out this research at very short notice. NISER is Nigeria's premier policy think tank and a member of the Think Tank Initiative (TTI). TTI is funded by the International Development Research Council (IDRC), Canada. The views expressed are entirely those of the authors.

References

Actionaid. (2006). *Climate change, urban flooding and the rights of the urban poor in Africa: Key findings from six African cities*. London: ActionAid International. http://www.actionaid.org.uk/doc_lib/urban_flooding_africa_report.pdf. Accessed 24 May 2010.

Adelekan, I. O. (2009). *Vulnerability of poor urban coastal communities to climate change in Lagos*. Paper presented at the Fifth Urban Research Symposium 2009, Marseille. http://www.urs2009.net/docs/papers/Adelekan.pdf. Accessed 5 Feb 2012.

Adger, N. W. (1999). Social vulnerability to climate change and extremes in Coastal Vietnam. *World Development, 27*(2), 249–269.

Awosika, L. F., French, G. T., Nicholls, R. J., Ibe, C. E. (1993a, March 14–19). *Impacts of sea level rise on Nigeria*. Proceedings from IPCC symposium: *The rising challenge of the sea*, Margarita.

Awosika, L. F., Ojo, O., & Ajayi, T. A. (1993b). *Implications of climate changes and sea level rise on the Niger delta, Nigeria. Phase 1 report*. Nairobi: UNEP.

Crichton, D. (1999). The risk triangle. In J. Ingleton (Ed.), *Natural disaster management* (pp. 102–103). London: Tudor Rose.

Dixit, A. (2003). Floods and vulnerability: Need to rethink flood management. *Natural Hazards, 28*(1), 155–179.

Federal Republic of Nigeria. (2006). *Report of the presidential committee on redevelopment of Lagos mega-city region*. Lagos: Lagos Megacity Project.

Gbadegesin, A. S., Olorunfemi, F. B., & Raheem, U. A. (2011). Urban vulnerability to climate change and natural hazards in Nigeria. In H. G. Brauch, U. Oswald Spring, C. Mesjasz, J. Grin, P. Kameri-Mbote, L. Chourou, et al. (Eds.), *Coping with global environmental change, disasters and security – Threats, challenges, vulnerabilities and risks* (Hexagon book series on human and environmental security and peace, Vol. 5, pp. 669–688). Berlin: Springer. doi:10.1007/978-3-642-17776-7.

Henderson, L. J. (2004). Emergency and disaster: Pervasive risks and public bureaucracy in developing nations. *Public Organization Review: A Global Journal, 4*, 103–119.

Intergovernmental Panel on Climate Change (IPCC). (2007). Summary for policy makers. In: S. Solomon, D. Qin & M. Manning (Eds.), *Climate change 2007: The physical science basis*. Contribution of Working Group I to the Fourth Assessment Report of the Intergovernmental Panel on Climate Change.

International Federation of Red Cross and Red Crescent Societies (IFRC). (2010). *World disaster report (focus on urban risk) 2011*. Geneva: International Federation of Red Cross and Red Crescent Societies.

International Federation of Red Cross and Red Crescent Societies (IFRC). (2011). *World disaster report: 2011*. Geneva: International Federation of Red Cross and Red Crescent Societies.

Mehrotra, S., Natenzon, C.E., Omojola, A., Folorunsho, R., Gilbride, J., & Rosenzweig, C. (2009). *Framework for city climate risk assessment: Buenos Aires, Delhi, Lagos, and New York*. World Bank Commissioned Research Fifth Urban Research Symposium Cities and Climate Change: Responding to an Urgent Agenda Marseille, France.

Mortreux, C., & Barnett, J. (2009). Climate change, migration and adaptation in Funafuti, Tuvalu. *Global Environmental Change, 19*(1), 105–112.

Munich Re. (2005). Megacities – Megarisks: Trends and challenges for insurances and risk management. http://www.munichre.com/publications/302-04271_en.pdf. Accessed 6 Jun 2009.

National Mirror (2011). http://nationalmirroronline.net/new/. Accessed 12 Jul 2011.

Olokesusi, F. (2004, July 6–7). *The vulnerability situation and emergency management in Nigeria*. Paper presented at the Maiden National Conference of Directors and Heads of Disaster Management Organisations in Nigeria organised by NEMA, held in Abuja.

Olokesusi, F. (2011). *Lagos: The challenges and opportunities of an emergent African Megacity*. Paper presented at the NISER Seminar Series on 5th Jul, Ibadan.

Olorunfemi, F. B. (2008). Disaster incidence and management in Nigeria. *Research Review, 24*(2), 1–23.
Olorunfemi, F. B., & Raheem, U. A. (2007). Urban development and environmental implications: The challenge of urban sustainability in Nigeria. *Ibadan Journal of the Social Sciences, 6*(1), 69–78.
Onokerhoraye, A. G. (1994). *Human geography for Africa. The Benin social series for Africa.* Benin City: University of Benin.
Raheem, U. A., Olorunfemi, F. B., Sulyman, A. O., & Awotayo, G. P. (2012). Multiple vulnerabilities and urban health challenges from extreme weather events in Ilorin, Nigeria. *Human Settlements and Urban Development Journal, 3*(1), 45–57.
Satterthwaite, D., Saleemul, H., Pelling, M., Reid, H., & Lankao, P. R. (2007). *Adapting to climate change in urban areas the possibilities and constraints in low- and middle-income nations.* Human Settlements Discussion Paper Series, IIED.
Sherbinin, A., Schiller, A., & Pulsipher, A. (2007). The vulnerability of global cities to climate hazards. *Environment and Urbanization, 19*(1), 39–64.
Stern, N. (2006). *The stern review on the economic effects of climate change*. Report to the British Government. Cambridge: Cambridge University Press.
Thywissen, K. (2006). *Components of risk: A comparative glossary*. Bonn: United Nations University, Institute for Environment and Human Security (UNU-EHS).
UN-Habitat. (2004). *The state of the world cities*. Nairobi: UN-HABITAT.
UN-Habitat. (2006, June 19–23). *A tale of two cities*. In World Urban Forum III, An International UN-Habitat Event: *Urban sustainability*, Vancouver. Nairobi: UN-HABITAT.
UN-Habitat. (2007, April 16–20). *Sustainable urbanization: Local actions for urban poverty reduction, emphasis on finance and planning*. 21st session of the Governance Council, Nairobi.
United Nations Office for Disaster Risk Reduction (UNISDR). (2004). Living with risk: A global review of disaster reduction initiatives. In *United Nations Inter-Agency Secretariat of the International Strategy for Disaster Reduction (UN/ISDR)*. Geneva: United Nations.
Vanguard Newspaper Nigeria Limited. (2011, May 29).
World Health Organization (WHO). (2002). *Floods: Climate change and adaptation strategies for human health*. London: WHO.
World Meteorological Organisation (WMO), & Global Water Partnership (GWP). (2008). *Urban flood risk management: A tool for integrated flood management*. Associated Programme on Flood Management (APFM) Technical Document 11, Flood Management Tool Series.
Zevenbergen, C., Veerbeek, W., Gersonius, B., & van Herk, S. (2008). Challenges in urban flood management: Travelling across spatial and temporal scales. *Journal Flood Risk Management, 1*, 81–88.

Part IV
Interfaces Science Policy

Solution-Based Spatial Planning for Disaster Risk Reduction and Climate Change Adaptation in Taiwan

Yu-Fang Lin

1 Introduction

The results of a global risk analysis of natural disaster hotspots indicate that Taiwan is one of the most hazard-exposed countries (Dilley et al. 2005). It suffers frequent and multiple hazards (three to five comprising earthquakes, landslides, flooding, etc.) affecting large areas of the population and total land area (both around 73.1 %) (Dilley et al. 2005, Table 1.1 on p. 4 and Table 1.2 on p. 8). Consequently, Taiwan is the country with the highest relative mortality risk from three or more hazards (Dilley et al. 2005, Table 1.2 on p. 8) and also faces the highest economic risk from these hazards. Owing to urbanization and rapid development, natural hazards have frequently caused significant harm and damage to people and property over the past decades. The number of casualties and substantial economic losses in heavily populated hazardous areas is increasing. Moreover, climate change is likely to affect the variability of rainfall patterns and will therefore influence the occurrence of hydro-meteorological hazards (e.g., extreme rainfall events and subsequent flooding). This, in turn, is expected to exacerbate uncertainty and adversely affect the population's resilience in disaster prone areas (United Nations Office for Disaster Risk Reduction (UNISDR) 2008). Therefore, environmental planning in Taiwan has to focus on minimizing exposure and preventing vulnerability to natural hazards. Spatial planning, as an instrument to promote sustainable development, plays an important role in disaster risk reduction (DRR) and climate change adaptation (CCA) in the long term.

In this article, spatial planning is used as a generic term to refer to different kinds of planning practices that aim to influence the spatial patterns of human life and

Y.-F. Lin (✉)
Department of Landscape Architecture, National Science and Technology Center for Disaster Reduction, Chung Yuan Christian University, CYCU, No. 200, Chung-Pei Road, Chungli City, Taoyuan County 32023, Taiwan
e-mail: v.yflin@gmail.com

B. Werlen (ed.), *Global Sustainability*, DOI 10.1007/978-3-319-16477-9_12

social organization, for instance, the location and distribution of different activities in space. This helps create a more rational territorial organization of land uses and zoning. In disaster risk management, spatial planning aims to shift the focus from an ad hoc disaster response to carefully planned prevention. It is designed as a systematic development model, which involves risk assessment, scenario development, and the monitoring of the disaster prevention measures' implementation. Hazard maps are important for determining key vulnerable areas and potential hazards resulting from environmental and climate changes. The assessment results can contribute to risk communication with all the stakeholders and can help decision makers with adapting to climate change and reducing disaster risks by means of spatial planning. Solution-based spatial planning, which focuses on disaster risk management and climate change adaptation, draws on a wide range of disciplines and requires competences from a variety of professions (e.g., law, engineering, policy strategy, technology, and environmental planning) and their different areas of expertise (e.g., database development, climate modeling, and the monitoring and enforcement of policies and regulations). While the framework presented in this article is tailored to the case study context, it aims to highlight general issues and insights that are useful in other multi-level and cross-sectoral governance contexts.

2 Disaster Risk Management

Disaster risk management is an important practice in the context of sustainable development. Risk is commonly defined as the function of hazards (probability), exposure, and vulnerability (extent of the damage) (Asian Disaster Reduction Center (ADRC) 2005; Schmidt-Thomé 2006). The intensity of hazards, the potential extent of damage, and the ability of cities (regions) to adjust to these changes and to minimize their effects determine the disaster risk (Schmidt-Thomé 2006). Since natural hazards cannot be avoided, increased efforts have been made to reduce vulnerability, minimize exposure, and adapt to hazards in order to reduce losses (Fig. 1; ADRC 2005; Dilley et al. 2005; Schmidt-Thomé 2006). From this viewpoint, a systematic effort to analyze and reduce disasters' causes is the key function of disaster risk management (UNISDR 2008). Disaster risk identification and assessment should therefore be the first steps in specifying the objectives of disaster risk management.

2.1 *Risk Identification and Assessment*

Risk identification is a process of combining scientific knowledge with factual data and normative elements. It helps determine which natural hazards impacts are potentially key and what is regarded as "dangerous" in natural hazards' dynamic

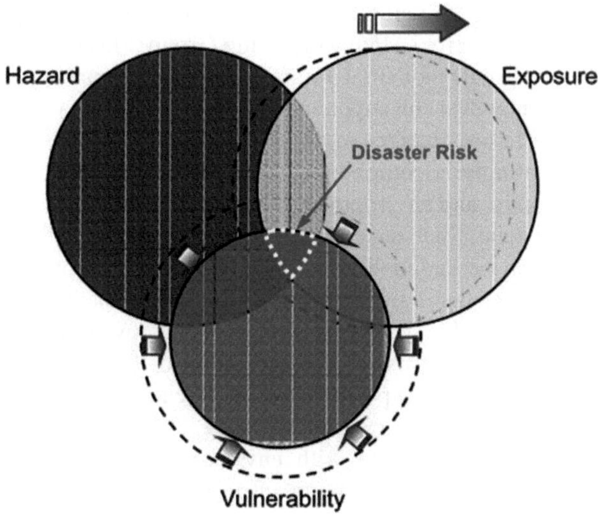

Fig. 1 Mechanism of natural risk reduction (ADRC 2005, 6)

processes (Schneider et al. 2007). Natural hazards have always posed threats to human beings and their assets. Extreme events occur on different timescales and comprise natural hazards that disturb, disrupt, or change relatively constant situations or processes within a matter of seconds (e.g., an earthquake or tsunami), days (e.g., a flood or a storm surge), months (e.g., extreme temperatures), or years (e.g., a drought), after which the initial, or a new "normal" state is reached (Schmidt-Thomé 2006). However, human settlements in different geographical, geological, and climatic conditions have been exposed to different natural hazards of varying intensities and frequencies. These natural hazards have led to different levels of risk perception, which are based on different natural, social, and cultural contexts and frameworks of understanding. Risk perception is an important factor that leads to different short-term risk responses and long-term adaptation strategies in different cultures. Therefore, disaster risk identification is essential to highlight the range of possible extreme events and their potential impacts in a site-specific and culture-specific context (Carter et al. 2007). It is thus necessary to identify all possible environmental (e.g., erosion), social (e.g., population and economic growth), and climatic (e.g. rising temperature) changes. To this end, the intensity and frequency of extreme events, along with the probability of their occurrence, should be determined and recorded over time. These records will reveal uncertainties in the changes, in the potential impact (including exposure and sensitivity), and in the adaptation capacity (Schneider et al. 2007). A comprehensive database is a valuable foundation for disaster risk assessment.

Disaster risk assessment is used to estimate the quantitative damage resulting from natural hazards and the socio-economic impacts (ADRC 2005). From the beginning of the vulnerability assessment process, multiple sources of risk, historical hazard data, forecasting, risk perceptions, and the uncertainty of environmental changes (regarding the direction and magnitude of the changes) should be considered. Vulnerability is a function of the potential damage and the population's coping capacity (Allen Consulting Group 2005). Historical records on hazards can help estimate potential future threats to humans and their welfare. These record must, however, be supplemented with recent data and, ideally, also with estimates of the future changes in the hazard frequency and severity due to climate change. Accordingly, a comprehensive database, including geophysical, biological, socio-economic, and environmental information, should form the basis of the disaster risk assessment. Such data records will help determine the system characteristics, which can then be included in the hazard data. The resulting maps will serve to identify potential hazard areas and key vulnerabilities. These maps have proven to be an effective tool in disaster risk assessment, developing policy strategies, and decision making. They provide a visualization of the "invisible" risks and vulnerabilities, and are therefore a powerful communication tool for planners, decision makers, and stakeholders alike (Schmidt-Thomé 2006). While the information on maps that combine hazard data with environmental and climate change impacts is very complex, it provides significant insights into the severity of the threat and characteristics of the affected territory, as well as into potential response strategies through spatial planning (Schmidt-Thomé 2006; Schneider et al. 2007).

2.2 *Spatial Planning in Disaster Risk Reduction and Climate Change Adaptation*

As an instrument for promoting sustainable environmental planning, spatial planning can be used to develop an approach to reduce exposure to hazards, to lessen the vulnerability of people and property, to wisely manage land use and land use change, and to strengthen preparedness for adverse natural extreme events (Brackhahn and Kärkkäinen 2001; Schneider et al. 2007; UNISDR 2008). Although particular issues regarding spatial planning are very specific to the local context, it is also necessary to consider multi-level and cross-sectoral governance issues. Spatial planners should identify, analyze, and communicate information on natural hazards to all stakeholders, which include public authorities, decision makers, landowners, affected residents, and the wider public, as part of a long-term disaster management strategy. Spatial planning may involve a combination of proactive, strategic, and forward-looking disaster management dimensions (Davoudi and Layard 2001; Davoudi 2010).

In the past decades, there has been a particular focus on integrating climate change's potential effects on natural extreme (weather) events (Schmidt-Thomé 2006) because the increasing frequency and magnitude of extreme weather events may increase disaster vulnerability. Spatial planning is the interface between key vulnerabilities and the strategies to protect the environment, to reduce disaster vulnerability and to mitigate the increasing impacts of natural hazards (including the potential impacts of climate change) on people and on the built environment.

The core spatial planning concepts regarding disaster risk reduction and climate change adaptation, i.e., protection, adaptation, and retreating, involve the following main steps: (1) managing new developments, or developing away from the disaster-prone areas; (2) designing buildings and regulating land uses, or legal zoning plans; (3) drawing up an energy-efficient infrastructure plan (e.g., regarding energy, water, transport, waste, and communications); (4) managing sustainable resource utilization; and (5) creating effective emergency preparedness planning to minimize the effects of natural hazards (Schmidt-Thomé 2006; Sutanta et al. 2009; Davoudi 2010).

A discussion platform is essential for sharing a common understanding of the entire planning process, for example, estimating disaster frequency and intensity, projecting and modeling climate changes, developing land use plans, improving decision making, etc. On this platform, spatial planners and decision makers can provide all stakeholders with information and can communicate effectively with the private and public sectors (Schmidt-Thomé 2006). The information on the discussion platform is evidence-based knowledge, which ensures that all processes take place appropriately.

This article focuses on using planning practices as an approach to reduce the risk of disasters associated with climate change. Figure 2 shows the framework for solution-based spatial planning used in this article. The phases comprise: (1) identify problems and determine objectives, (2) gather information and available data for analysis, (3) establish decision making criteria, (4) assess risks, (5) identify options, (6) appraise options, (7) develop strategies and make decisions, (8) implement decisions, and (9) evaluate, monitor, and review the planning process.

3 Study Materials

3.1 A Risky and Diverse Environment in Taiwan

Taiwan is an island state in East Asia (120°58′55.2886″ 23°58′25.9486″, TWD97[1]). The main island lies on the Tropic of Cancer, and has a subtropical and tropical climate. The island has a diverse landscape of mountains, plains, and hills. Owing to Taiwan's geographic location and geological characteristics, environmental changes (loss of certain species' habitat) and social phenomena that influence the environment (e.g., urbanization) often cause natural hazards. The increase in the frequency and intensity of extreme events also increases the frequency and intensity of disasters (Hsu et al. 2011). According to Dilley and colleagues' (2005) study of global natural disaster hotspots, Taiwan is the country most exposed to multiple hazards, as 73.1 % of its land area is subject to up to four different natural hazards (Dilley et al. 2005, Table 1.1 on p. 4). Moreover, natural hazards affect about

[1] TWD97 is the coordinate system of the Taiwan geodetic datum 1997.

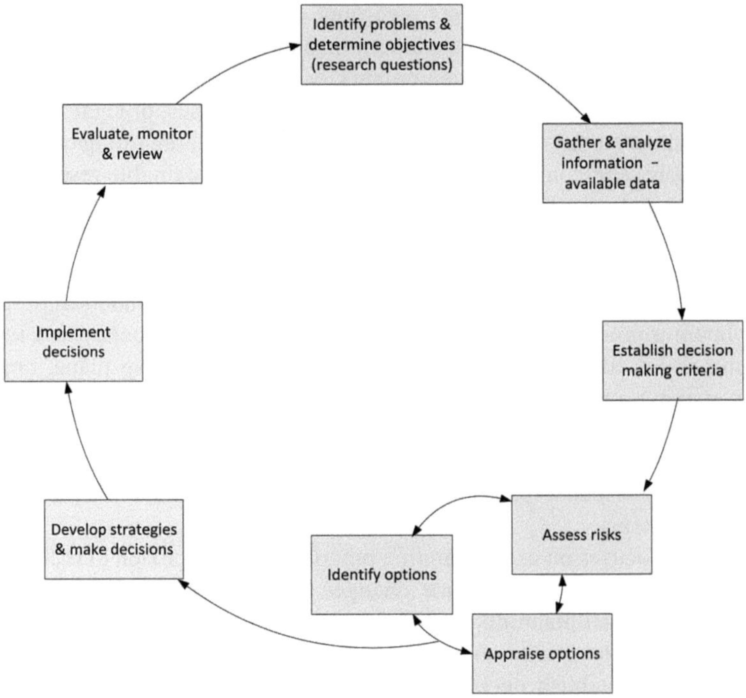

Fig. 2 A framework to support decision making for CCA and DRR (Adapted from Willows and Connell (Eds.) 2003, 7)

73.1 % of Taiwan's population (Dilley et al. 2005, Table 1.1 on p. 4). Taiwan is ranked as the country with the highest relative mortality risk from three or more hazards, with 95.1 % of the total population at risk of natural hazards (Dilley et al. 2005, Table 1.2 on p. 8). Similarly, Taiwan faces the highest relative economic risk from three or more hazards, with 96.5 % of the country's GDP generated in hazardous areas (Dilley et al. 2005, Table 7.2 on p. 89). It is thus essential to develop strategies to reduce the risks from natural disasters sustainably and adaptively. Spatial planning is, therefore, a useful tool to reduce disaster risks and adapt to climate change.

The following questions were considered in this article: (1) What are the potential natural disaster risks in Taiwan, and how can the potential impacts of climate change be addressed in the proposed integrated approach? (2) What are appropriate criteria for assessing and displaying natural hazards and risks? (3) What are the main challenges of and opportunities for spatial planning regarding adapting to the increased frequency and severity of natural hazards? (4) What are appropriate processes for risk communication with stakeholders and how should the information that is provided in the discussion platform messages be framed? (5) What are the core issues and appropriate strategies of DRR and CCA?

This article suggests that solution-based spatial planning is an effective strategy for facing the potential challenges of DRR and CCA, and answering the abovementioned questions.

3.2 *Risk Assessment in Taiwan*

We consider the most frequent and most destructive hazards related to climate change, namely landslides, floods, and coastal hazards. Figure 3 shows the risk of these three different hazards on separate maps. The central mountain areas, encompassing eight different townships (Wufong in Hsinchu, Taian in Miaoli, Renai in Nantou, and Alishan in Chiayi, among others; for details, see the red spots on the left map in Fig. 3) are most at risk of landslides. Chianan Plain, Tainan-Kaohsiung Metropolitan, and the coastal township in the Pingtung Plain (center map in Fig. 3) are at risk of floods and inundation. Owing to climate change, the sea level is projected to rise 1.4 m by 2100. Consequently, seven counties, Tainan, Chiayi, Kaohsiung, Pingtun, Yilan, Changhua, and Yunlin, and the estuary of the Tamsui River will face coastal risks (an area of ca. 770 km^2; for details, see the right map in Fig. 3). Each risk map shows different risk levels for the specific hazard: from no hazards to a high level of risk (computed by multiplying the probability of the hazard occurring and its consequences).

Overlaying the above hazard maps with maps of other environmental and social characteristics can help identify at-risk areas and their possible vulnerabilities. Corresponding strategies for spatial planning can then be designed and implemented on different scales (regional, local, or site-specific) for specific perspectives (e.g., integrated river management) in accordance with local conditions and planning contexts. For instance, the Chianan Plain is at high risk of inundation due to storm and coastal storm surges. This area's main strategies for disaster risk reduction and climate change adaptation are thus included in integrated river management, industrial relocation, land use management, and education. Integrated river management should involve building intercepting facilities (e.g., dams) in order to improve water retention and prevent flooding. Industrial relocation and land use management should prevent the creation of impervious surfaces and land subsidence. In addition, education should be geared toward reducing social vulnerability and raising environmental awareness.

4 Discussion and Conclusion

This article introduced the processual concept of solution-based spatial planning for disaster risk reduction and climate change adaptation on a national scale. We presented risk maps of landslides, flooding, and coastal hazards. Risk maps provide

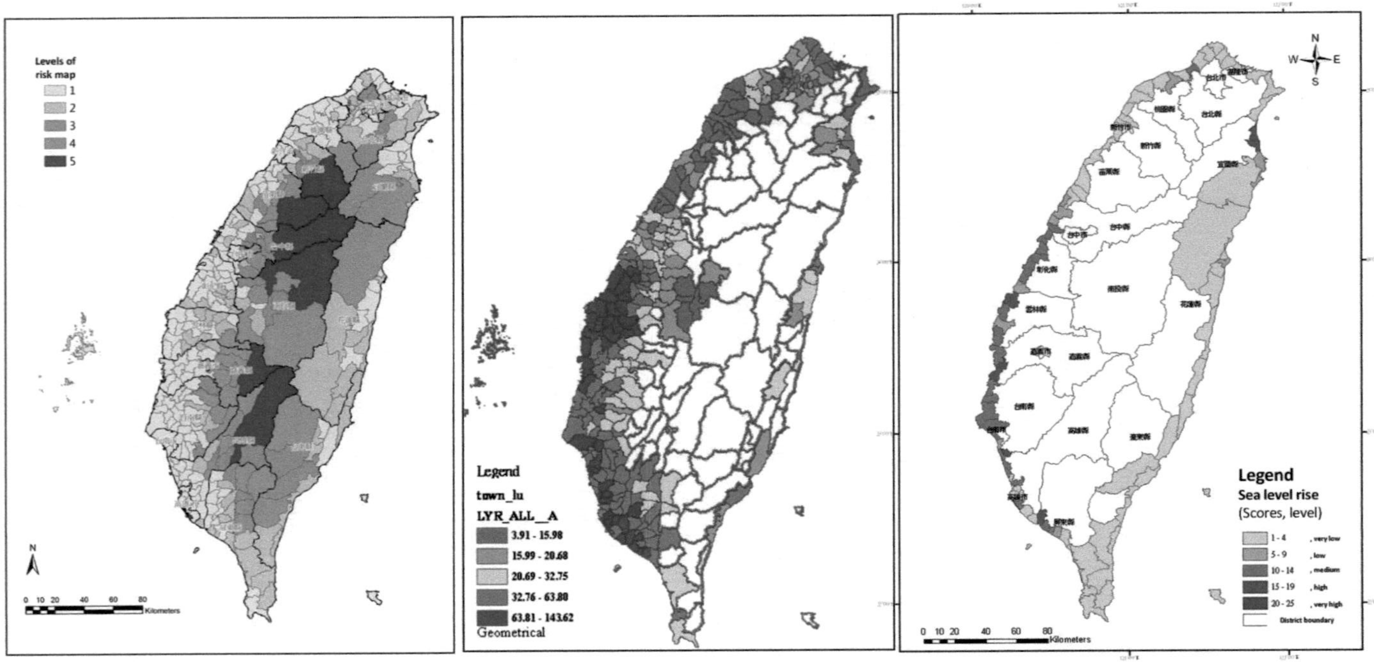

Fig. 3 Risk maps of Taiwan: landslide risk map, inundation risk map, and coastal risk map (From *left* to *right*: Lin et al. 2011, 19; Lu et al. 2010, 6; Chang et al. 2011, 35)

a visualization of potential threats to spatial development. We also described the corresponding strategies for the Chianan Plain as a study case.

While the risk maps can be combined into a map comprising all hazards in disaster-prone areas, effective and appropriate strategies should be devised to adapt to the local or regional characteristics and conditions in the planning contexts.

The methods applied in this article have certain limitations. First, since we downscaled the climate data from the Global Circulation Models (GCMs) further, their projections were subject to increased uncertainty regarding the spring and the plum rain season.[2] Second, it is difficult to estimate the uncertainty regarding climate projections and environmental changes. Third, integrated spatial planning for disaster risk reduction and climate change adaptation should be considered in different timescales (short term and long term), and accordingly, the planning measures should be implemented in multiple phases. Fourth, the case study spatial planners and decision makers in the case study had difficulties balancing the need for environmental protection with economic development in their adaptation strategies. Future amendments to spatial planning strategies and processes in Taiwan should consider the following challenges:

- Available data: Risk maps represent different aspects and functions, while introducing new indicators to the risk assessment. The fundamental information fed into the risk assessment (e.g., system scope, performance challenges, stakeholder expectations, implementation challenges, historical hazard data) should be as detailed and as high-resolution as possible to ensure the effectiveness of spatial planning. Risk maps should, moreover, mitigate the inaccuracy of the available information on climate risks and uncertainties resulting from the downscaling of climate data.
- Legislation: There is a need for a comprehensive legislative system to support spatial planning for disaster risk reduction and climate change adaptation in Taiwan. The adaptation strategies should be implemented in a legal framework that goes beyond accounting for the building regulations. A truly integrated approach would involve appropriate land use relocation, especially in disaster prone and environmentally sensitive areas.
- Timescale difference and time lags: Some natural hazards are difficult to recognize or are misjudged on a human timescale that may mislead decision makers when they develop strategies and policies for disaster risk reduction and climate change adaptation. For instance, hydro-meteorological hazards are difficult to estimate due to large uncertainty ranges in the calculation of recurrence intervals. However, immediate response actions can be taken with regard to certain natural hazard events (e.g., storms and typhoons). It is difficult to develop a comprehensive strategy that is applicable to all types of hazards when using spatial planning. The implementation of spatial planning strategies and environmental policies may fail to immediately reflect the core environmental issues. In

[2] The plum rain season refers to the East Asian rainy season of nearly 2 months of continuous precipitation (from the late spring to the early summer).

general, disaster risk reduction and climate change adaptation have different timescales in spatial planning strategies. The estimation and the corresponding adaptation strategies should consider the different timescales and time lags.
- Environmental awareness: An important part of risk communication is encouraging individuals and businesses to take actions to increase their adaptation capacity regarding disaster risk and climate change. Education, training, and media platforms may be helpful in providing people with information on short and long-term risks and on how to avoid or prevent them. In addition, such platforms may provide guidelines on how to react when emergencies are imminent.
- Uncertainty of changes: Climate change adaptation is based on the projections of possible future climate conditions. However, no one can prove the accuracy of projections. The environment is dynamic and continuously changing due to natural processes, as well as human activities. Designing and implementing solution-based spatial planning should be based on a combination of recent and historical data and future projections in order to minimize the impact of change uncertainties. In the planning contexts, specific adaptation strategies should solve existing problems and prevent similar or new catastrophes in the future.

The progress and effectiveness of solution-based spatial planning should be continuously reviewed to ensure that the approach is comprehensive and minimizes the gap between adaptation strategies and their actual implementation in multi-level and cross-sectoral governance. In this context, the results of a vulnerability assessment may provide important insights into decision making, mitigating disaster risks, and adapting to climate change.

Acknowledgements My thanks go to the team of the environment and climate change project at National Science and Technology Center for Disaster Reduction (NCDR) in Taiwan who provided the relevant data for this case study.

References

Allen Consulting Group. (2005). *Climate change risk and vulnerability: Promoting an efficient adaptation response in Australia. Final report. Report to the Australian Greenhouse Office.* Melbourne: Allen Consulting Group. http://www.sfrpc.com/Climate%20Change/4.pdf. Accessed 4 Feb 2013.

Asian Disaster Reduction Center (ADRC). (2005). *Total disaster risk management: Good practices 2005*. Handbook for the UN World Conference on Disaster Reduction, Kobe.

Brackhahn, B., & Kärkkäinen, R. (Eds.). (2001). *Spatial planning as an instrument for promoting sustainable development in the Nordic countries: Action programme for 2001–2004*. Denmark: The Ministry of the Environment.

Carter, T. R., Jones, R. N., Lu, X., Bhadwal, S., Conde, C., Mearns, L. O., O'Neill, B. C., et al. (2007). New assessment methods and the characterisation of future conditions. In M. L. Parry, O. F. Canziani, J. P. Palutikof, P. J. van der Linden, & C. E. Hanson (Eds.), *Climate change 2007: Impacts, adaptation and vulnerability. Contribution of working group II to the fourth assessment report of the intergovernmental panel on climate change* (pp. 133–171). Cambridge: Cambridge University Press. http://www.ipcc.ch/pdf/assessment-report/ar4/wg2/ar4-wg2-chapter2.pdf. Accessed 4 Feb 2013.

Chang, C.-H., Lin, Y.-C., Chen, Y.-M., Chou, S.-Y., & Lee, Y.-N. (2011). *Coastal disaster risk map (I)* (Technical Report). Taiwan: NCDR.

Davoudi, S. (2010). *The national country reports: United Kingdom*. Newcastle: Newcastle University. http://www.bbsr.bund.de/BBSR/DE/FP/MORO/Studien/2009/NatStrategienKlimawandel/LaenderstudieUK,templateId=raw,property=publicationFile.pdf/LaenderstudieUK.pdf. Accessed 15 Feb 2013.

Davoudi, S., & Layard, A. (2001). Sustainable development and planning. In A. Layard, S. Davoudi, & S. Batty (Eds.), *Planning for a sustainable future* (pp. 7–18). London: Spon.

Dilley, M., Chen, R. S., Deichmann, U., Lerner-Lam, A. L., & Arnold, M. (2005). *Natural disaster hotspots: A global risk analysis* (Disaster risk management series, Vol. 5). Washington, DC: World Bank.

Hsu, H.-H., Chen, J.-D., Lu, M.-M., Chen, J.-M., Chou, J., & Wu, Y.-C. (2011). *Scientific report on climate change in Taiwan 2011*. Taiwan: National Science Council.

Lin, Y.-C., Chang, C.-H., Chen, Y.-J., Lee, H.-C., Chen, Y.-M., & Huang, B.-C. (2011). *Landslide risk map (I)* (Technical Report). Taiwan: NCDR.

Lu, J.-C., Chen, Y.-M., Chang, C.-H. Kuo, & Y.-L. (2010). *Evaluation of inundation risk under climate and environmental changes – an analytical framework for township level evaluation in Taiwan*. 2010 Conference for Disaster Management in Taiwan, Taipei.

Schmidt-Thomé, P. (2006). *Integration of natural hazards, risk and climate change into spatial planning practices*. Ph.D. thesis, Department of Geology, University of Helsinki.

Schneider, S. H., Semenov, A., Patwardha, I., Burton, C. H. D., Magadza, M., Oppenheimer, A. B., Pittock, A., et al. (2007). Assessing key vulnerabilities and the risk from climate change. In M. L. Parry, O. F. Canziani, J. P. Palutikof, P. J. van der Linden, & C. E. Hanson (Eds.), *Climate change 2007: Impacts, adaptation and vulnerability. Contribution of working group II to the fourth assessment report of the intergovernmental panel on climate change* (pp. 779–810). Cambridge: Cambridge University Press.

Sutanta, H., Rajabifard, A., & Bishop, I. D. (2009). An integrated approach for disaster risk reduction using spatial planning and SDI platform. In B. Ostendorf, P. Baldock, D. Bruce, M. Burdett, & P. Cornoran (Eds.), *Proceedings of the surveying & spatial sciences institute biennial international conference* (pp. 341–351). Adelaide: Surveying & Spatial Sciences Institute.

United Nations Office for Disaster Risk Reduction (UNISDR). (2008). *Briefing note 01: Climate change and disaster risk reduction: Weather, climate and climate change*. Geneva: United Nations Office for Disaster Risk Reduction.

Willows, R., & Connell, R. (Eds.) (2003). *Climate adaptation: Risk, uncertainty and decision-making*. UKCIP Technical Report. Oxford: UKCIP.

Institutions and Planning: A Reflection from Disaster Management Planning in Indonesia

Hendricus A. Simarmata and Raka W. Suryandaru

Abbreviations

Bakornas PB	Badan Koordinasi Nasional Penanggulangan Bencana or National Coordinating Agency for Disaster Management
Bappeda	Badan Perencanaan Pembangunan Daerah or Local Development Planning Agency
BNPB	Badan Nasional Penanggulangan Bencana or National Agency for Disaster Management
BPBD	Badan Penanggulangan Bencana Daerah or Local Agency for Disaster Management
CSR	Corporate social responsibility
DIY	Daerah Istimewa Yogyakarta province
DRR	Disaster risk reduction
FGD	Focus group discussion
HFA	Hyogo framework for action
NGO	Nongovernmental organization
NTB	Nusa Tenggara Barat province
NTT	Nusa Tenggara Timur province
Perka BNPB	Peraturan Kepala BNPB or Head Regulation of BNPB
Perpres	Peraturan Presiden or President Regulation
PP	Peraturan Pemerintah or Government Regulation
RA-PB	Rencana Aksi Penanggulangan Bencana or Disaster Management Action Planning
Renstra	Rencana Strategis or Strategic Plan

H.A. Simarmata (✉) • R.W. Suryandaru
Urban Development Studies Postgraduate Program, University of Indonesia, Gedung Pascasarjana 4th floor Jalan Salemba Raya 4, Jakarta, Indonesia
e-mail: simarmata.andy@yahoo.com; raka.suryandaru@gmail.com

B. Werlen (ed.), *Global Sustainability*, DOI 10.1007/978-3-319-16477-9_13

RKP	Rencana Kerja Pemerintah or Short-term Development Plan
RPB	Rencana Penanggulangan Bencana or Disaster Management Planning
RPJM	Rencana Pembangunan Jangka Menengah or Mid-term Development Plan
RPJP	Rencana Pembangunan Jangka Panjang or Long-term Development Plan
RTRW	Rencana Tata Ruang Wilayah or Spatial Planning
UNDP	United Nations Development Programme
UNISDR	United Nations Office for Disaster Risk Reduction

1 Introduction

The nexus between institutions and planning has only recently started receiving attention in the academic literature. This is all the more surprising given that institutions play a strong role in influencing planning processes and decisions. For the purpose of this article, *institution* refers to formal and informal institutions as well as tangible and intangible ones. This definition allows one to consider a broad range of institutions in this study, for instance, governments, markets, and cultural norms. Institutions provide frameworks within which planning operates. A lack of institutional support is likely to adversely affect the legitimacy, social acceptability, and political feasibility of planning processes, actual plans made, and their implementation.

A review of the literature shows that various planning scholars have attempted to have the institutional dimension feature more prominently in planning discourses. From an economic perspective, for instance, Alexander (2007) highlights "transaction cost as a means of reconciling the competing ideas of market and planning or those of private and public planning" (Alexander 2007, 7). From a policy perspective, Healey (1997) argues that planning has frequently been justified as a form of government intervention to limit excessive freedoms. The planning literature thus recognizes the manifold ways in which institutions shape the planning system.

Planning systems can be defined as "systems of law and procedure that set the ground rules for planning practice" (Healey 1997, 72). It is argued that, in the face of the growing complexity of current societies, planning systems cannot be seen as an independent phenomenon but more as a product of wider institutional forces (Booth 2005), which are influenced not only by their internal functions, but also by external factors. These external factors mainly result from globalization processes that are facilitating a freer transfer of policy ideas – including planning approaches – across nations (Dolowitz and Marsh 1996). That is, planning approaches from overseas influence the form and structure of domestic planning systems (Healey and Williams 1993; Sanyal 2005). A planning system therefore describes a dynamic phenomenon whose progression and changes should be seen in the context of both local and global institutional changes.

Conceptually, "disaster management includes the sum total of all activities, programs and measures which can be taken before, during and after a disaster for the purpose of avoiding a disaster, reducing its impact or recovering from its losses" (Khan et al. 2008, 46). Garatwa and Bollin (2002) suggest that disaster management should focus on risk management, which is a series of actions (programs, projects, and/or measures) and instruments expressly aimed at reducing disaster risk in endangered regions and mitigating the extent of disasters in the first place. Therefore, disaster management planning is a tool that also prevents economic losses, since there are indications that the risk of economic losses owing to disasters continues to increase across all regions – and seriously threatens the economies of low-income countries (United Nations Office for Disaster Risk Reduction (UNISDR) 2011).

In Indonesia, disaster management planning is known as Rencana Penanggulangan Bencana (RPB) (RPB is the bahasa acronym for Disaster Management Planning), which is governed by BNPB (bahasa acronym for Badan Nasional Penanggulangan Bencana or National Agency for Disaster Management). BNPB is the ministry-level organization established through Act no. 24/2007 regarding Disaster Management (Government of Indonesia 2007a). This organization formulated and endorsed the guidance to conduct RPB through Perka BNPB (bahasa acronym for Peraturan Kepala BNPB or Head Regulation of BNPB) no. 4/2008 regarding the Guideline for Disaster Management Planning (Government of Indonesia 2008c). This regulation covers the definition, procedures, and methods to conduct RPB. RPB consists of planning for disaster risk management, contingency and mitigation, and public awareness (pre-disaster phase), planning for emergency responses (disaster phase), and planning for recovery and reconstruction (post-disaster phase).

This article argues that RPB emerges from a rather recent paradigm shift in disaster management. Previous approaches in disaster management have tended to focus on coping and dealing with disasters and their consequences; that was a reactive rather than a preventive approach. Following a paradigm shift, this sequence has been turned around, such that disaster management moved from a focus on disaster cleanup to a more comprehensive, preventive approach that mainly sought to reduce the probability of natural events and phenomena turning into disasters. Disaster risk management usually seeks to involve various stakeholder groups (Sperling and Szekely 2005). However, disaster management planning in general still faces the complexity of the causes and consequences of disasters, as well as the uncertainty of time between disasters, and kind of actions to be taken. This is why planning scholars recommending robust planning processes for disaster management.

In practice, the integration of disaster management planning into development planning also requires examination because every state has a different development system. The position of RPB in the Indonesian development planning system, how it is influenced by, and how it influences planning institutions, as well as how

Indonesia's development planning system is coping with the current change towards a more democratic and decentralized structure, are the key questions this article is concerned with. Indonesia's "decentralized governance", as defined by the United Nations Development Programme (UNDP 2004), has involved a shift to a power-sharing situation between central and local governments based on subsidiarity concepts and the involvement of the private sector and civil society (Work 2002). According to Mayntz (2003), governance has been used for modeling "a more co-operative mode where state and non-state actors participate in mixed public/private networks" (Mayntz 2003, 27). Dirk Messner, a German Sociologist, argues that the shifting governance paradigm from 'strong state' to 'cooperative state' or the recent model 'moderating state' (Brunnengraber et al. 2006, 5) would change the domain and framework for planning processes, including RPB.

The authors are not aware of any empirical study that examines the relationships between governance institutions in the disaster management planning process. As a relational resource in management planning processes, institutions have the potential to contribute to the improvement of governance capacity. Healey (1997) has argued that the planning environment has shifted its focus from government to governance. This trend has been particularly strong in Indonesia over the past decade as well as in other developing countries. The simultaneous reformation and decentralization in the late 1990s have provided the mandate to adopt good governance as the primary principle of development. Therefore, planning practices – including RPB processes – should be examined in the context of decentralized governance.

It is argued that institutions will continuously evolve owing to social reproduction by many institutional actors (Giddens 1984). Before the reformation era, the planning process in Indonesia was technocratic rather than inclusive and relied on the technical assistance of planning professionals. Nowadays, planning structures provide an enabling environment for an inclusive process. The role of nonstate actors, especially of those with a prominent social and economic status and with considerable influence on the planning process is presented in more detail below. Concerning planning implementation, institution building becomes a primary concern since the transformation process often requires collective actions of the actors who have a shared meaning and commitment (Gualini 2001).

According to Giddens (1984), social systems can be understood as an interactive and mutually constitutive relationship between institutions (structure) and actions (agency). The RPB regulation, which is represented by Perka no. 4/2008 (Government of Indonesia 2008c), has constrained the planning team in organizing the short-term planning process at the sub-national level. The planning team is simply the 'repository of the role' (see Miller 2011), which is meant by Perka no/4/2008. Whereas planners have their own value and role which is defined by their professions. Therefore, planning practices of the RPB process at the sub-national level, which is structured by Perka 4/2008 and implemented by planners, define planning institutions for disaster management. The authors will use three main factors in the institutional analysis that can be used to examine the institutions of RPB as follows.

First, present national regulations or rules that enable disaster management planning at the subnational level; second, planning actors and agencies who operationalize the planning process; and third, resources that have been provided to support the planning practices.

This article seeks to understand the institutions of disaster management planning at subnational levels through case studies of RPB processes in nine Indonesian provinces. The discussion is organized into three parts. Part 1 discusses the institution of RPB and reviews the hierarchy of Indonesia's planning system and RPB's position within it. Part 2 presents case studies on the implementation of disaster management planning in three provinces. It identifies limitations and challenges through three institutional factors: regulation, actors, and resources. Part 3 draws conclusions on the emergence of institutional arrangements that create an enabling environment for disaster management planning.

This study uses qualitative data that was gathered by the authors during their time as planning consultants in an RPB project for nine provinces between August and December 2011. However, for this article, the authors select three provinces: Daerah Istimewa Yogyakarta (DIY), Central Java, and Bali as case study areas, since the authors were directly involved as part of a planning team. In addition, these provinces provide different characteristics of governmental and nongovernmental actors and their interactions. These three case studies explore the RPB planning process in order to find underlying practices that were taken during the planning period. The authors also applied participant observation of serial planning activities, such as a consultation meeting in Jakarta, a focus group discussion (FGD) in Yogyakarta, Semarang, and Denpasar, and a national seminar in Jakarta attended by nine provincial governments. Such participant observation was important to provide meaning from those institutional actors' interaction from an insider perspective. The authors also conducted nonstructured interviews in a conversational style with three different sub-groups of the planning team: three reviewers, three writers, and three planning drafters. Those interviews were used to identify the roles, expectations, and problems experienced in operationalizing RPB at subnational levels. Literature reviews of planning policies in Indonesia were also examined to provide the institutional context in which RPB operates.

2 RPB as Shifting Institution: From Reactive to Preventive, and from Government to Governance

RPB is the first disaster management planning concept in Indonesia to be regulated and released by the BNPB via Perka BNPB no. 4/2008 (Government of Indonesia 2008c), which orders each level of government to have RPB for managing disaster issues in their region. The presence of RPB today cannot be removed from the development of the BNPB, which was founded in 2008 as a result of Act no. 24/2007 regarding Disaster Management (Government of Indonesia 2007a).

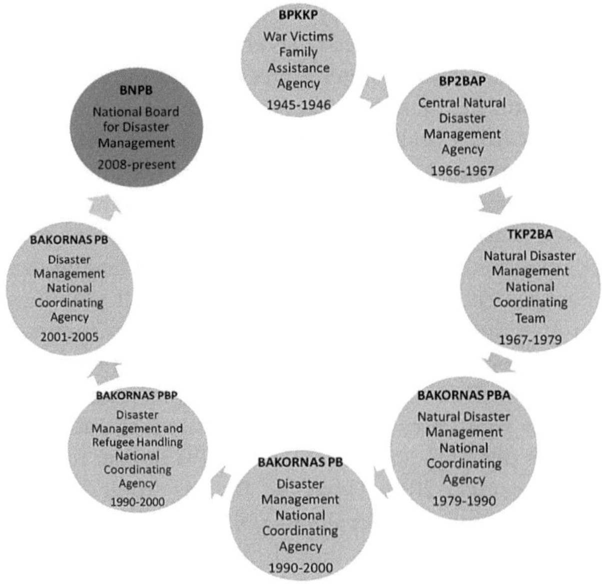

Fig. 1 Evolution of the Indonesian Disaster Management Agency (Badan Nasional Penanggulangan Bencana (own diagram, 2011))

The history of the institutional setting of disaster management in Indonesia dates back to the independence declaration of 1945. A special agency dedicated to assist war victim families was established at the time; it has evolved six times, most recently into a special agency at ministry level, but it has not (yet) been included into the presidential cabinet (see Fig. 1). The authority and organization of BNPB, which has survived several political regimes, governments, and changes of power, was regulated and established by the Perpres (bahasa acronym for Peraturan Presiden or President Regulation) no. 8/2008 regarding National Board for Disaster Management (Government of Indonesia 2008a).

The BNPB is more powerful than its previous organization, since the Head of BNPB's reports directly to the President. The greatest difference between the era of *orde lama* (authoritarian regime) of 1945–1966 and the *order baru* (centralized regime) during 1966–1998 is the agency's working area. The previous agency, Bakornas PB (bahasa acronym for Badan Koordinasi Nasional Penanggulangan Bencana or National Coordinating Agency for Disaster Management), was supposedly only a coordinating agency whose members were central government ministries and public agencies. It was headed by the Vice President and supported by an executive agency that ensured the agency's day-to-day operations. Bakornas PB was arguably no longer fit for its purpose, given the increasingly more sophisticated challenges of disaster management. These new challenges include the requirement of allocating responsibility for disaster management to a dedicated agency, as opposed to adding this responsibility to a section of an existing ministry. This is particularly important, since the existing ministries seldom share the sense of urgency, and as a result, do not prioritize disaster risk management as part of their duties. Therefore, a separate and appropriately resourced agency is required – named BNPB under Act no. 24/2007 regarding Disaster Management (Government of Indonesia 2007a).

BNPB is established to meet the need for an agency which can provide guidance and direction for disaster management efforts in a fair and equal manner to both national and local government in the context of decentralized governance. At national level, to maintain and strengthen relationships with the various ministries and public agencies that previously were members of Bakornas PB, these ministries and public agencies are included in the BNPB Advisory Board. It is argued that by applying this organizational structure, BNPB can maintain the coordination function so that the disaster risk management can be implemented in an integrative manner. At province and district/city level, BNPB establishes BPBD (bahasa acronym for Badan Penanggulangan Bencana Daerah or Local Disaster Management Agency) which is part of the organization of local government, and is equal in status to other local government agencies. Therefore, Act no. 24/2007, which was initiated by Parliament in response to calls from nonstate actors in 2005 (Lassa 2011), has given new power to BNPB and BPBD to manage the disaster management affairs in their areas.

The founding of BNPB through Act no. 24/2007 has shown that the role of nonstate actors, supported by international organizations, is crucial in shifting Indonesia's disaster management paradigm. Lassa (2011) argued that, since 2008, more nonstate actors have been involved compared to previous periods, including community-based organizations, private firms, external funders, academia, and civil society (Lassa 2011, 124). Therefore, disaster management governance in Indonesia has actually followed global disaster risk governance discourse by following the Hyogo Framework for Action[1] (HFA), which contains the disaster risk reduction paradigm and acknowledgement of international cooperation as the keys of disaster management via PP (bahasa acronym for Peraturan Pemerintah or Government Regulation) no. 21/2008 regarding Operation of Disaster Management (Government of Indonesia 2008b).

RPB is acknowledged as a required disaster management tool. Act no. 24/2007 prescribes this as one of the BNPB's key tasks (Articles 35 and 36 in Government of Indonesia 2007a). Perpres no. 21/2008 addresses the RPB as a part of development planning (Article 6 in Government of Indonesia 2008b). The Act enabled a twofold shift in focus: from recovery to prevention, and from government to governance. According to this Act, disaster management action in the pre-disaster phase consists of prevention, mitigation, integration into development planning, disaster risk analysis, implementation of spatial plans, education and training, and implementation of disaster risk management technical standards. The Act requires the various levels of government (national, provincial, and district/city governments) to adopt RPB as the guiding principle for disaster risk management.

---

[1] Hyogo Framework for Action is a worldwide agreement on disaster risk reduction whose aim is to promote a strategic and systematic approach to reducing vulnerabilities and risks to hazards. It is signed at The World Conference on Disaster Reduction in Kobe, Hyogo, Japan which was held from 18 to 22 January 2005.

RPB essentially provides a master plan with appropriate actions for the four cycles of the disaster risk cycle: a conceptual plan for mitigation, a contingency plan, an operational plan, and a recovery plan. RPB's focus is on reducing people's vulnerability and increasing institutional capacity in dealing with disasters and their short-term and long-term consequences. The BNPB strongly believes that the most effective way to protect people from disasters is to build their coping capacity. The idea is that sufficient knowledge on disaster risk management will lead them to choose lifestyles that enhance their resilience and lead them to take preventative and curative actions well in advance of an actual hazard.

The Perka BNPB no. 4/2008 (Government of Indonesia 2008c) prescribes the RPB planning procedures for local government. It divides the planning process into two main phases: risk mapping and planning conception – within 26 weeks (6.5 months) (Fig. 2). As shown in Fig. 2, the RPB process timeline is only 26 weeks, which apparently do not provide adequate time for the technocratic process and the participatory process. In comparison, the timeline for conducting RTRW (bahasa acronym for Rencana Tata Ruang Wilayah or Spatial Planning) is 24 months, according to PP no. 15/2010 regarding Implementation of Spatial Planning (Government of Indonesia 2010). The RPB process design and its tight timeframe have actually severely constrained the inclusive process that should have legitimized the planning product.

RPB is conducted by the planning team, which consists of the disaster management expert team (which was only involved in the risk mapping phase) and the planners (planning drafters, writers, and reviewers). According to the guideline, the risk mapping should use particular scientific formula based on disaster risk data in the various areas. In contrast, the planning conception process should use participatory planning approaches through a nine-step strategy to formulate the disaster management policies, programs, and activities as well as the mechanism for monitoring and evaluating the plans. Risk mapping is the only source of information used to develop plans. It means that programs and actions are developed after establishing a big picture of what causes the risks. Therefore, high-quality risk assessment is urgently required, since its results determine the quality of RPB.

Therefore, in the era of decentralized governance, where local and provincial governments have been empowered, the national government should be more creative and provide benefits for the local institutions in addressing the disaster management affairs. The shifting of institutions from government to governance has also driven the planning process to become more inclusive – involving more nonstate actors. Meanwhile, in terms of disaster management, the shifting paradigm from reaction to prevention through the disaster risk reduction (DRR) approach has been recognized and has become the main approach of RPB. Hence, the RPB's planning substance is now about managing the risks of disasters in decentralized governance. The authors will now address the practices of RPB, especially at subnational level, within shifting institutions of disaster management.

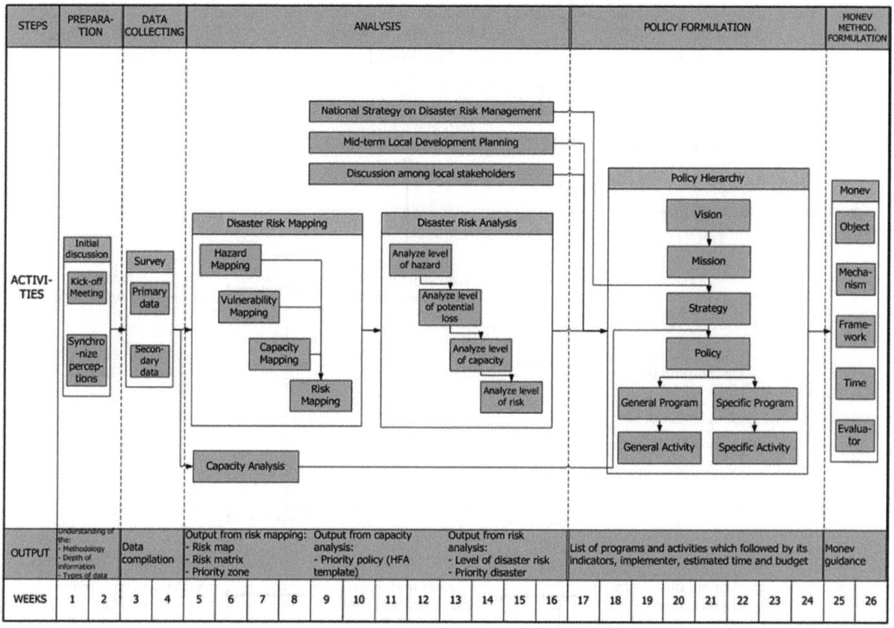

Fig. 2 Framework of Indonesian disaster management planning (own diagram, 2011)

3 Translating RPB into Statutory Planning: A Missing Converter

To gain access to government development budget and spatial confirmation, the authors argue that RPB should be integrated into Indonesia' planning system. There are two kinds of statutory planning laws: (1) development planning, regulated by Act 25/2004 regarding national development planning system (Government of Indonesia 2004), and (2) spatial planning, regulated by Act 26/2007 regarding Spatial Management (Government of Indonesia 2007b). Both laws consider decentralized governance to ensure that both planning processes are conducted at all levels of government (national, provincial, local) in an inclusive and participatory way, within the timeframes of short-term, mid-term, and long-term periods, and considered intersectoral and cross-cutting issues. RPB's position in the planning system, as interpreted by Mr. I., BNPB's expert staff, shows that interaction between RPB and other planning documents is conceptually established (Fig. 3).

Based on the above planning framework (Fig. 3), there are four nexuses that can be conceptually examined, (1) the strategic value of Province RPB, (2) Province RPB's role in Province RA-PB (bahasa acronym for Rencana Aksi Penanggulangan Bencana or Disaster Management Action Planning), (3) the relationship between Province RPB and Province RPJM (bahasa acronym for Rencana Pembangunan Jangka Menengah or Mid-Term Development Plan), and (4) the relationship

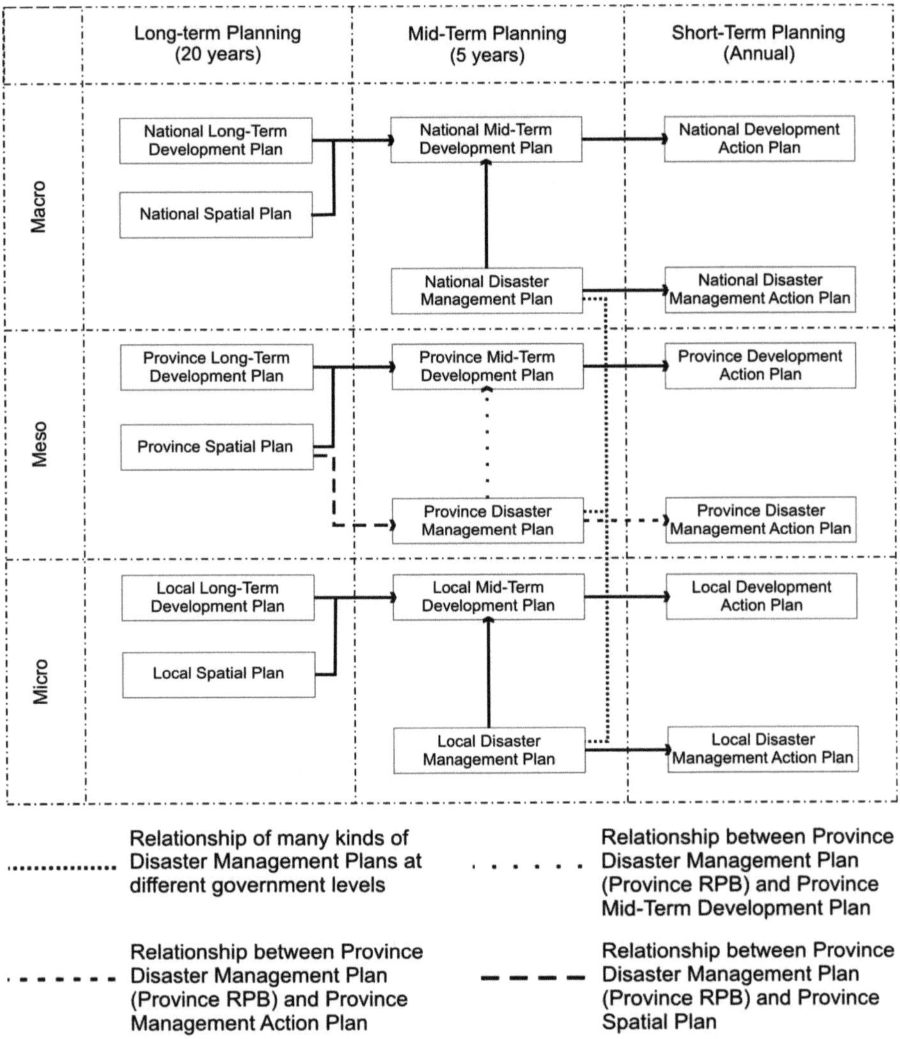

Fig. 3 RPB's Position in Indonesia's National Development Planning System (Modified from Irawan 2011, 4)

between Province RPB and Province Spatial Plan. The authors found that RPB has met difficulties in addressing the substance of RPB in development planning, especially in translating planning language.

First, as stated in Act no. 24/2007 regarding Disaster Management (Government of Indonesia 2007a), the national, provincial, and local governments have the mandate to develop an RPB at a certain point in time. The national-level RPB is used as a reference for RPB at the provincial level, which in turn is used by district or city governments to develop their RPB at the local level. In the actual planning process, however, each government level must be equally represented. This means the national-level RPB process must invite all provincial governments and that the RPB at provincial level should invite the districts and city governments within its administrative area. Based on this mechanism, the RPB at provincial level (the meso-level in Fig. 3) is of strategic importance because it should accommodate national and local interests. Therefore, ensuring the complementary aspects of RPBs at the intergovernmental level is the key task when developing an RPB.

Second, the relationship between RPB and the RA-PB should be strong. These documents are inextricably linked, like the two sides of a coin. RPB acts as a master plan, complemented by RA-PB as an operational plan for each of the levels. The concept of RA-PB is used to reexamine an RPB program by considering each governmental level's de facto institutional capabilities and to withdraw contributions from actors other than government institutions (Mr. I., BNPB, expert staff, interview on October 20, 2011). Therefore, instead of being conducted exclusively by experts, an RPB program should also be formulated inclusively in order to make an RA-PB effective concerning design and easier to implement.

Third, the link between RPB and RPJM should be strengthened. According to Act no. 25/2004 regarding the National Development Planning System, all development sectors or functional areas of government should be integrated in the RPJM in order to gain financial support from the national and/or local budget. Thus, based on PP no. 38/2007 regarding the Distribution of Governmental Affairs among Central Government, Provincial Government, and Local Government (Government of Indonesia 2007c), disaster management is defined as one of the development sectors in one government unit. The RPJM is organized by national or local planning agencies and it is obligatory for all levels of government to have it, along with RPJP (bahasa acronym for Rencana Pembangunan Jangka Panjang or Long-term Development Plan) and RKP (bahasa acronym for Rencana Kerja Pemerintah or Short-term Development Plan). BNPB emphasizes that RPB is part of RPJM and not a stand-alone planning document (Mr. R., BNPB's expert staff, interview on October 21, 2011). The RPB should inspire the development of disaster-adaptive RPJM. Therefore, RPB and RPJM are supposed to be interconnected and synchronized.

The authors also argue that RPB's role to enrich RPJM by introducing disaster management approach into development planning can be accomplished by compiling a working paper that sets out the efficient use of resources in this process. By creating a special technical paper that adds to and complements RPB, the drafting of yet another planning document can be avoided. That is, the need for new guidelines on RPB procedures arises from the inefficiencies and illegitimacy of current RPB processes.

Fourth, the relationship between RPB and RTRW should be reciprocal. The authors assert that the RPB must be synchronized with the spatial planning. The disaster-prone area identified by the RPB should be recognized by the RTRW in order for it to have special zoning regulation and/or alternatives for disaster mitigation. The RPB will provide useful information for the RTRW on which risk a region is exposed to, at what level, and which factors are causing the risk. By understanding the levels of physical, social, economic, and institutional vulnerability, planners can formulate land use policy to reduce these vulnerabilities. On the other hand, the RTRW's structure and land use plan can be utilized by the RPB to promote soft or nonstructural mitigation measures. Nonstructural measures include adaptation through managed retreats, such as coastal setbacks, density restrictions, phased out development, and rolling easement (Intergovernmental Panel on Climate Change (IPCC) 2001). Planners can come together with forward-thinking policy in order to create cities that will be places of environmental equilibrium, mitigation against climate change, and healthy places for people to live (AECOM 2010).

Results of the above framework analysis suggest that there are three practical problems in the RPB process, namely the complexity, the exclusiveness, and the lack of embeddedness problem. As for the complexity problem, the multiple levels of government involved in the planning process have different priorities and interests in prioritizing the program in the RPB scheme. An intersectoral approach that sought to solve complexity does not take place because the disaster management is only part of sectors (the functional areas of government). As a new agency in the local bureaucracy, BPBD faces difficulties in negotiating the disaster risk as the point of departure of envisioning the planning goal. The RPJM's vision derives from the elected governor, who is not aware of the notion of disaster risk reduction. In addition, various planning products by other agencies with different timeframes and scales were hard to synchronize, because they had their own laws that were not mainstreamed by the disaster risk reduction.

Second, the exclusiveness problem occurs at the provincial level, because the process was dominated by government agencies, without significant numbers of other actors. The RPB process could not invite a wide range and numbers of other stakeholders owing to limited financial resources and time repetition that has been designed for procedural compliance. One or two nongovernmental organizations (NGOs) and academia were invited, but not within a participatory planning framework, in which all participants should have the same deliberation in negotiating the issues. The inclusiveness of disaster management planning is important in order to recognize the experience and knowledge by other stakeholders and to invite their contributions and commitment to the disaster management plan in later planning process stages. Third, there was a lack of embeddedness, despite the Perka BNPB 4/2008 stating that the RPB is supposed to be part of the overall development plan (Government of Indonesia 2008c). However, the RPB process was subject to different temporal and spatial scales than the development planning or spatial planning process.

In terms of planning communication, based on experiences from the 2011 provincial-level RPB process, the authors recognize that the planning team consisted of 10–12 government staff members from various development sectors. Their assurance to integrate the RPB program in their sectors was, however, unconvincing. The most common argument brought forward was that the terminology used in RPB could not be automatically used in the development planning term. The RPB program terminology would need to be converted prior to being used in a development program. The case from Central Java shows that they preferred using names and codes of RPJM codification to ensure access to government funding. Meanwhile, the RPB still used the program names that derived from the HFA Action assessment. RPB programs need a converter in order to get financial support from government, if not, it is just a compilation of programs.

> In a consultation meeting with BNPB's experts, Mr. T., BPBD of Central Java province, stated that RPB of Central Java Province has followed development program terminology and was confident that the program recommendation can be financed through regular government budgeting. He implicitly argued that other RPBs that did not follow development program terminology may not be funded by the government (author meeting notes, December 2, 2011)

In addition, the authors highlight the transaction costs for the coordination activities to the planning agency in terms of time and human resources. Within the same process, it is also realized that the absence of the private sector in the planning process impacted on the possibilities for implementing specific RPB programs. For instance, in coordinating the flood infrastructures program in the disaster-prone area of Jakarta province, where there are many industrial and warehouse estates and property developers who would benefit from the program, these actors or private sector representations were not invited, whereas they could have contributed to the program through corporate social responsibility (CSR), for instance. That is, the planning process should be inclusive; it should consider a variety of stakeholders and the likely costs and benefits that would accrue to them under different implementation scenarios.

The authors also highlight that the large number of planning instruments used in Indonesia should be better integrated and coordinated. Planning institutions in Indonesia are always evolving, depending on development needs and political interests. The mainstreaming of ideas of disaster management, climate change adaptation, green development, etc. into development has eventually created ineffectiveness in development planning processes at the subnational and local levels. Therefore, it is important to embed the RPB process into development planning as well as to make this an inclusive process that actively seeks participation from a variety of stakeholders.

The authors argue that the complexity, exclusiveness, and lack of an RPB's embeddedness, which was also exacerbated by poor communication skills and lack of coordination of the RPB team concerning limited resources and time, are the main issues that would lead an RPB to have difficulties in settling itself harmoniously in the planning system. The practices of RPB at subnational level provide a lesson that on the one hand, RPB needs robust methods, and on the other hand, that development planning institutions need to be adapted to the new evolving paradigm, such as disaster risk reduction. A planning process that adopts a procedural approach cannot be used in isolation. It should be transformed into a more collaborative way to reduce the complexity, to be more inclusive, and to be embedded in the development planning process.

4 Unsupported Regulation to Promote Inclusive Planning

Based on the case studies of RPB processes in East Java, Central Java, and Bali, there are three key findings that relate to the regulation environment in the RPB-making process: (i) uncoordinated planning practices of disaster management, (ii) the use of strict planning methodology with little attention to the different characteristics of each province, and (iii) the obligation to provide tangible products by the end of the year.

First, the authors argue that fruitful organization of disaster management can only be achieved if all stakeholders participate in a collaborative manner. The regulations regarding coordinative tasks in disaster management have already existed before the inception of the BNPB. The high expectation through Perpres no. 8/2008 regarding National Board for Disaster Management (Government of Indonesia 2008a), which clearly assigned BNPB to act as a coordinating planning agency in disaster management, is not yet satisfied. The difficulties of coordination activities not only happened in coordinating state and nonstate actors, but also coordinating the multiple governance levels, as well within government actors. The authors found that the difficulties were also due to the egotistical sectoral culture in governmental bureaucracy and the fairly recent establishment of BPBD Provinces, under the Governor. As stated by Mr. T. (2011) from BPBD of Central Java Province, the reluctance of other government agencies to join the plan-drafting discussion and to share their thoughts with the team occurred because BPBD was just perceived as the same-hierarchical agency with other sectors and not a strategic agency like Bappeda (bahasa acronym for Badan Perencanaan Pembangunan Daerah or Local Development Planning Agency) (Mr. T., interview on December 2, 2011). Ms. D. from BPBD of DIY Province also noted that no one of her colleagues from the Social Service agency share their program details with the plan drafters because they perceive the RPB as unnecessary and they already had their own disaster rehabilitation and reconstruction program (Ms. D., interview on November 15, 2011).

The authors argue that the young age of BPBD Provinces has put them in an uncomfortable situation in which they need more time to establish themselves as well as gain respect and trust from the other long-established agencies. These challenges became clear when BPBD of Bali Province invited the province's government agencies to attend the FGD on capacity assessment. Instead of sending senior public officials, the meeting attendees were low-ranking officers. In DIY province, problems emerged when BPBD of DIY province appointed the planning drafters. Ultimately, the plan drafters consisted only of BPBD staff members, with no representatives from other agencies. This was in part due to the ignorance of stakeholders other than BPBD and partly to the insufficient efforts by the BPBD to promoting these events. However, this problem was not present in Central Java's RPB planning process since the head of BPBD, a former high officer of Bappeda, who had a vast network among provincial government agencies so that he could unofficially call his colleagues from the other agencies to attend the plan-drafting meeting.

The second issue relates to the use of strict methodology of RPB. BNPB has specified the methodology and data that would be used in RPB at the sub-national (i.e. province) level through Perka BNPB no. 4/2008 (Government of Indonesia 2008c). BNPB applied the certain parameters in the methodology in order to integrate all the provinces into one single database, which apparently not comply the expectation of BPBD provinces. The general data and simplification in determining the risk levels are the key shortcomings here. For instance, the methods did not include the tidal floods that severely inundated large areas of Java's North Coast and the cold lava floods of the previous Mount Merapi eruption in DIY and Jawa Tengah province risk assessment.

Indonesia is a tropical archipelago with very different characteristics from one island to another. One of the results of this circumstance is that the provinces' disaster risks can differ significantly. The authors argue that, to get a clear profile of individual provinces's disaster risk, every province should have the flexibility to develop its own risk assessment methodology that suits their own localities and resources. Provinces that have an island group formation, like Nusa Tenggara Timur (NTT) province, are certainly different than provinces that lie inside big islands, like DIY province. While DIY province is more urbanized and land based, NTT province features archipelagic and rural areas. The generalization thus unduly homogenizes the diversity of risk levels among the provinces. Therefore, the authors argue that the guidelines for RPB should only provide the basic minimum requirements rather than strict methods and data.

The third issue relates to the obligation to provide tangible products by the end of the year. Since the RPB activity was under the BNPB's program, thus all kind of activities must be completed within one fiscal year. It is common that the government program just started in the second semester of the year due to procurement and fiscal administration and must be ended at the end of the year. So, the available time is between 6 and 8 months, which severely certainly constraints the RPB process. Having completed the RPB, the authors argue that the planning time of RPB should be extended. The time allocated to the project (6 months) was very limited, and was insufficient for conducting disaster risk mapping and disaster risk management planning concomitantly. Besides, an inclusive and collaborative RPB process is urgently needed. The planning team of DIY province and Bali expressed their limitation in organizing intense serial meetings to conduct the RPB process. They expected that there was additional time for socialization and internalization of government bodies. In comparison, PP no. 15/2010 regarding Implementation of Spatial Planning requires 24 months to conduct a spatial planning process (Government of Indonesia 2010).

5 Knowledge Gaps in RPB Teams

The BNPB set up three groups of actors in the RPB making process: (i) the plan drafters, (ii) the writers, and (iii) the reviewers (Fig. 4). The plan drafters consist of local stakeholders from a variety of institutions, including the provincial government agency, NGOs, private sector, and academia. The writers consist of

Fig. 4 RPB actors (own diagram 2011)

consultants hired for the project, such as urban planners (the authors' role), human geographers, and environmentalists. Their task was to moderate the plan-drafting discussion and to compile the results in the RPB's final report. The writers could also provide further insights to enrich the quality of the discussion as well as data and literature. The reviewers are the BNPB's special task force for reviewing the RPB planning draft. This team assesses whether the document meets the requirements stated in Perka no. 4/2008.

Based on the case studies of RPB process in DIY, Central Java, and Bali provinces, there are three key findings relating to the interaction of actors involved in the RPB-making process: (i) high dependency on writers, (ii) reluctances and time constraints of plan drafters to actively become part of the planning team, (iii) knowledge gaps among actors in risk assessment and in planning conception owing to different backgrounds and perspectives.

First, interactions within the planning team are dominated by the writers, who reside in Jakarta. In fact, the planning drafters should be the main actors – as required by the BNPB. However, the scale and pattern of interaction among the actors was not designed in detail; thus, some provinces chose to establish the interaction mechanism so as to be able to achieve the target in time. Central Java Province stakeholders preferred to give the data only to the writers and gave them the authority to use those data to produce the planning report. DIY province stakeholders ordered the writers to use the study done by NGOs. Bali Province stakeholders gave the writers almost no data and asked them to use the national data. These three case studies imply the same expectation that the plan should be delivered on time, rather than emphasizing dialogue among actors so as to have better argumentation in prioritizing the planning policies.

These interactions do not reveal the specific needs of local stakeholders with experience of their own region's disaster profiles. But if participatory planning is applied – in which case disaster risk can be fully assessed by local people and effectively reduced through a range of activities – the planning would have substance, would be more focused, and would relate to the specific needs of local people (Venton and Hansford 2006). This high dependency meant that the opportunity to enhance local stakeholders' capacity to understand the disaster management planning process and to share their expectations through this mechanism were lost. The other disadvantages of such types of practices are low acceptance of provincial government agencies to use the RPB as their platform in arranging programs and budgeting. It also hindered the opportunity to involve other stakeholders (i.e. the private sector) to contribute their resources in the implementation phase.

However, the writers sought to involve the other members more actively, with informal talks after the FGD, which was sometimes more effective for gathering information. For instance, from the discussion with Ms. D., BPDB DIY Province, the authors could identify their sector priorities relating to disaster management. Ms. D. is a high-ranking officer who has been serving in her current position for 5 years. She stated that the cold lava flood disaster from the Mount Merapi eruption was a tough challenge for her institution. She had recommended mitigation actions in the RPJM but found that her initiative could not yet be accommodated. She expressed the hope that RPB can be an alternative channel to promote mitigation, but doubted that RPB could achieve this, since it had no legal basis. Based on the interview, the authors understand why the plan drafters have different motivations to participate in the planning conception process.

Second, the plan drafters and the writers are unable to smooth over their differences, not least because there is little interaction between these two teams. This led to both teams being reluctant to take full responsibility for their respective contributions to the RPB process. The BNPB had developed a nine-step strategy for the planning conception process, which required intensive discussions within and between the plan drafters and the writers. However, participation in RPB was not mandatory for government officials; for example, those involved in the plan drafter team took a fairly passive stance. They asked to be thoroughly briefed before agreeing to actively contribute. They voiced their reluctance to participate and to take responsibility should they find the information provided in the brief to be unconvincing. Since the plan drafters showed their reluctance in formulating the RPB, the writers took over this responsibility and drafted the plan.

Internal quarrels culminated in the embarrassment of government officials when the planning drafters were due to present the draft RPB to reviewers. When the time came to submit the draft RPB, the reviewers asked the plan drafters, not the writers, to present the result in a plenary meeting attended by all actors involved in the RPB in Jakarta. Since the plan drafters had been dismissive of the entire RPB process, they were unaware of the contents of the draft RPB prepared by the writers. Instead of expressing remorse or regret and being apologetic for the low quality of the RPB, however, the plan drafters merely pointed at the writers, arguing that the draft RPB was poorly documented by this team, who should now be held accountable.

Third, there was a knowledge gap in both the risk assessment and planning conception phases among planning team members. In the risk assessment phase, most of the planning drafters had difficulties in using the risk map. Recall from the group discussions of Bali province that the plan drafters criticized the draft risk map that had been prepared by the writers because the spatial information was too general and did not highlight some sub-district which they believe were high risk areas. The same case also happened in DIY province: Most of the stakeholders

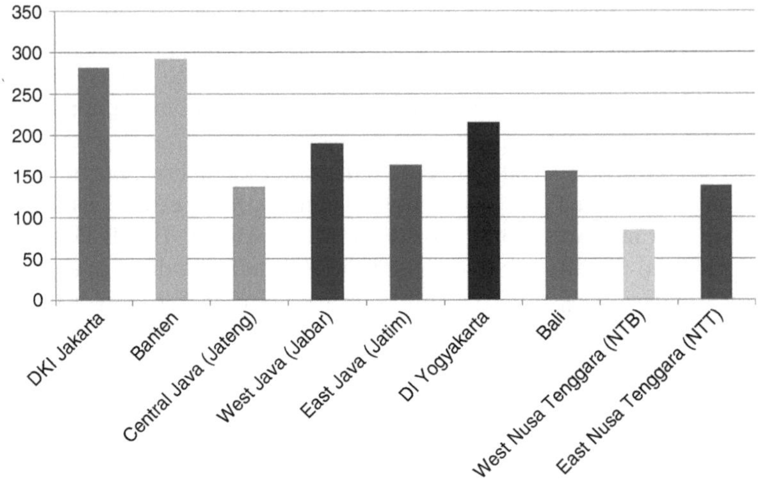

Fig. 5 Total number of programs per province (own diagram 2011)

strongly objected the risk map since the plan drafters had their own version, which was more detailed. Both cases demonstrate that the perceptions among planning teams are divergent in understanding and using the risk map for planning purposes. It was defined in the first place that the map scale at the sub national level is 1:250,000 and that this is used for program coordination purposes only. Yet, it was not prepared for operational purposes in most cases.

In the planning phase, the gap occurred in relation to the understanding of the HFA. Recall from the FGD event to fill in the HFA form, which was held in Semarang, the participants (government staff) faced the difficulties in translating the notions contained in the HFA into their daily work terminology. The limited time of debriefing session also influenced their level of understanding of the HFA modules. Based on the authors' observation, most of the participants had difficulties in answering the five main topic questions on the availability of local policy and/or a disaster management program; the availability of disaster risk assessment and a monitoring system; the use of knowledge, innovation, and education to build a culture of safety and resilience at all levels of governance; the underlying risk factors; and the presence of a disaster preparedness support system for effective response at all levels. The results of the nine case studies suggest that the program recommendations still do not reflect adequate risk assessment despite planning team capacities having increased. The reason for this may be that all RPBs contain a large number of programs (more than 100), except NTB province (Fig. 5). It is more than what could be considered effective and efficient due to the rather low institutional capacity of BPBD. This is reflected in Fig. 6 which shows that only small proportions of BPBD are taking on the role of the sole responsible actor. As new the organization in the provincial government structure, it would be difficult to delegate the implementation of those programs to other agencies.

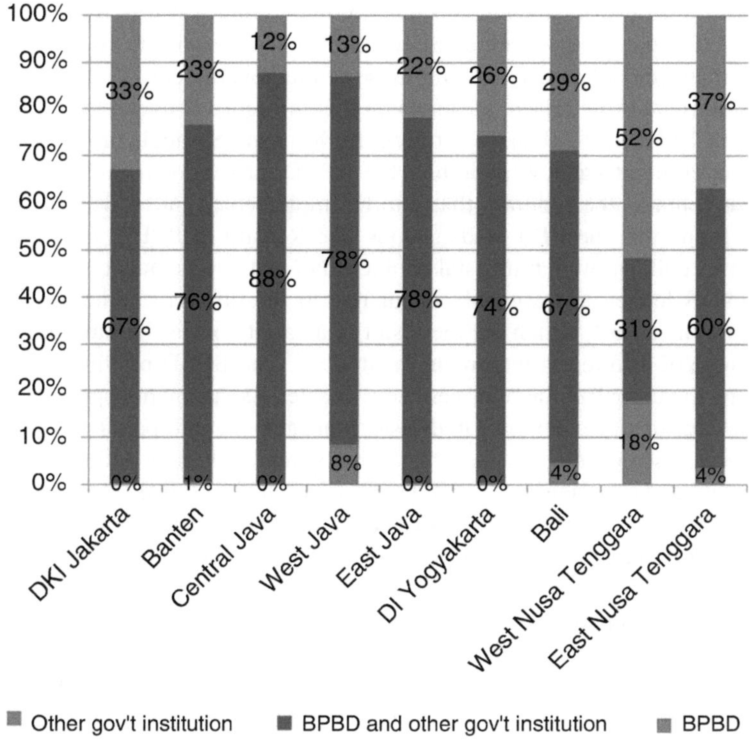

Fig. 6 BPBD as actor in RPB (own diagram 2011)

As shown in Fig. 5, the number of programs in the nine RPBs ranges from 70 to 290, a number that is considered too high. In Banten Province's RPB, for instance, there are 290 disaster management programs recommended for the next 5 years. DKI Jakarta Province also had an ambitious target of 280 disaster management programs in its RPB. Ms. D., BPBD of DIY province, stated that all these programs are needed to develop a resilient society, and insisted that the Provincial Government of DIY should implement all these programs. The authors doubt that the BPBD will succeed in implementing all these programs within 5 years, not least owing to its limited financial and human resources. The RPB should promote fewer but more strategic programs rather than developing a large number of ineffective programs.

A literature review on institutions responsible for implementing the nine RPBs showed that almost all programs proposed by the RPB involved the BPBD as one of the implementing institutions and that only a few programs were implemented without the BPBD's involvement. In Central Java's RPB, 120 (88 %) of 137 proposed programs are under BPBD's responsibility. In West Java's RPB, 165 of 190 proposed programs (86 %) are also managed by the BPBD (Fig. 6). Mr. T., BPBD of Central Java Province, argued that the role of disaster management is owned by BPBD and that it should therefore be responsible for implementing the RPB's programs. He doubted that other government agencies are legally allowed to implement a disaster management-related program, since it was BPBD's task. He also doubted that nongovernmental stakeholders were willing to integrate their disaster management initiatives with those of the government.

This dominant role raised questions on the difference between the RPB and the Renstra of BPBD itself (bahasa acronym for Rencana Strategis or Strategic Plan). Every government agency, including the BPBD, has its own mid-term strategic plan that contains the program that will be implemented in the next 5 years. The authors found that the RPB tends to look like Renstra of BPBD, and not like the innovative, collaborative multi-stakeholder plan that it was meant to be.

The knowledge gaps arguably occur due to the different competences of the BPBD province staff, which are the result of a variety of disciplinary backgrounds and a range of different government institutions. The BPBD province needs sufficient time to internalize the relatively new disaster management approach. The lack of knowledge of and experience in disaster management and spatial planning can be understood based on the fact that this notion is relatively new and the educational background and experience of BPBD staff are limited to this disaster management and planning field. Therefore, the BPBD are left with the task to build its capacity in a very short time, since disasters occur continually and require quick and planned responses from local authorities.

In addition, the plan drafters's technical capacity also needs to be addressed. At the meso-level, the plan drafters who conduct RPB must have the capabilities to employ available data, which vary from region to region. They should also seek to integrate risk mapping and risk assessment into the planning conception. Research results presented in this article hint at some general issues that are similar across all the provinces the authors studied because all case studies used the same data sources and methodology. The plan drafters should not limit their analysis to two-dimensional maps and spatial information, but should also pay attention to other dimensions, such as culture, because human and social capital are the most important factors in implementing RPB programs.

6 The Lack of Data Resources

Planning, including RPB, always deals with uncertainty. The critical step in conducting planning is the projection of future conditions as it provides the basis for scenario building. The key challenge for disaster management planning is that the frequency and magnitude of past disaster events cannot always be used for future hazard projections. For hydro-meteorological disasters, a combination of past flood event data, climate change data, and precipitation variability data can be adequate for projecting future scenarios for disaster-prone areas. For geophysical disasters, however, the availability of past earthquake data alone is not sufficient to predict the location of future earthquakes. To date, the most feasible approach is to identify variables that contribute to disaster risk – especially vulnerability and capacity – and that people can influence so as to improve the variables. Therefore, data availability on all variables becomes a significant determinant for the ability to adequately assess individual risk levels.

Based on the case studies of RPB processes in DIY, Central Java, and Bali provinces, there are five key findings that relate to the data issues in the RPB process: (i) inadequate natural hazard assessment owing to the use of historical disaster data as the only data source, (ii) inadequate vulnerability assessment owing to generalization of potential loss calculation, (iii) inadequate capacity assessment owing to the use of the HFA questionnaire as the only data source, (iv) inadequate disaster planning owing to poor risk assessment result, and (v) the unavailability of financial resource data.

First, hazard assessment is performed only by utilizing the historical records of events. BNPB provided the writing team with various past disaster event maps taken from various government organizations with the authority to issue such data. For instance, past volcanic eruption data was acquired from the Center for Volcanology and Geological Hazard Mitigation, and data on past floods was acquired from the Meteorology, Climatology and Geophysics Agency. These data was then used to build a hazard level typology for a province. Although this approach is acceptable, the authors hold that undertaking a study of natural hazards prior to disaster management planning will benefit the understanding of the province's individual hazard profile. The authors argue that, without support from other relevant data, it is impossible to identify disaster-prone area beyond those areas that have previously suffered disasters.

Second, vulnerability assessment is conducted using generalized potential loss data. To assess the level of vulnerability, three variables were used: population; the number of basic facilities, critical facilities, and housing; and the conservation area. The higher the quantity, the more vulnerable the location. Potential loss is analyzed by multiplying the quantity of each variable with the monetary value (in rupiah) per unit variable. The shortcoming is that the money value per unit variable is considered the same for all disasters and for all provinces, although in fact it is totally different. For instance, the potential loss calculated could not be used for reference purposes since the damage caused by earthquakes and droughts is very different. This shortcoming also pertains to the potential loss value per facility in Jakarta and Nusa Tenggara Barat. Since the price of materials used to build a house in Jakarta and Nusa Tenggara Barat differs, the potential loss of a house in Jakarta and a house in Nusa Tenggara Barat is different. The authors argue that more specific data on the monetary value per variable should be employed, based on previous actual disasters in each province.

Third, capacity assessment is undertaken by using the HFA institutional capacity model. However, there are shortcomings regarding the substance of the HFA and the data acquisition procedure. The HFA questionnaire does not adequately consider local government concerns since they have no opportunity to express their knowledge and experience regarding local institutional capacity other than what is asked in the questionnaire. Therefore, the capacity assessment developed based on this questionnaire cannot show local information, which plays a significant role in determining the institutional capacity level.

Concerning the data acquisition procedure, one FGD was held in each province to gather information about local disaster management capacity. Participants were asked to fill in the questionnaire on the spot. The problem is that the FGD participants came from government sectors only and thus could hardly provide comprehensive information about local capacity. The authors argue that the discussion should also have included stakeholders from academia, NGOs, the private sector, and the community such that a variety of viewpoints could have been elicited.

Fourth, the limited data used in the risk assessment resulted in poor assessment output. Assessment output determines the quality of the planning conception since the latter is heavily reliant on risk assessment information. The authors conclude that the risk assessment output did not fit the planning conception requirements, especially concerning the lack of local information at the province level. The information was so general that the planning drafters had difficulties identifying the regions's physical and social attributes. For instance, from the vulnerability assessment for earthquakes in DIY province, the information obtained was limited to the potential number of people impacted; potential loss of basic facilities, critical facilities and housing; and conservation area. It missed critical information, such as the socio-economic profile of the community, their habitual day-to-day practices, and their perspective in facing disaster.

In addition to a paucity of socioeconomic data, data on financial resources is also limited. For the planning process, it is important to have clear information on funding sources and the available budget that can be spent on measures suggested in the plans. In the RPB's case, there are shortcomings in terms of the exclusion of financial sources from nongovernmental actors and the unavailability of precise figures on the number of actors that can be dedicated to the disaster management program. A literature review on financial resources in the nine RPBs revealed that the RPBs only consider local government budgets as the sole funding source for disaster management. It excludes the potential funding from the community, donors, and the private sector. In West Nusa Tenggara Province, for instance, the BPBD actually had the right to ask Newmont Nusa Tenggara (a major mining company) to fund some RPB programs through its CSR program. Another example is DKI Jakarta's BPBD, which did not include its cooperation with the World Bank to implement the Jakarta Emergency Dredging Initiative, a program to improve Jakarta's flood management system by dredging selected floodways, canals, and retention basins.

It is hard to gauge local government financial capacity in funding RPB disaster management programs since data on their financial resources is not provided for the planning process. Therefore, when the plan drafters had to budget for the various programs, they simply assumed the ideal case, i.e. that the program could be funded at any cost. For instance, the East Java Province's RPB was estimated to cost 105 trillion rupiah (about 10.8 million US$) over 5 years. However, since this figure had not been and could not be compared to the local government's available total budget, the authors argue that this results in large uncertainties concerning the feasibility of East Java Province's RPB.

7 The Emergence of Institutional Arrangements of RPB

Analyses of the case study material presented in this article suggest that future RPB projects would benefit from an institutional arrangement that has been prepared in coordination with the development of planning methods. The results of this research further indicate that a well-designed planning environment – as defined by regulation, actors, and resources, and their interplay – is more likely to lead to effective, efficient, politically legitimate, and publicly accepted planning processes and planning outcomes. In other words, the findings of this study suggest that improvements concerning regulation, actors, resources, and – especially – their interplay are likely to enhance the overall planning system's success.

Based on the authors's analysis of the institutional arrangements for RPB, five improvements concerning the regulation of RPB are suggested, namely in areas of stakeholders participation, authorities of BPBD, cooperation between Bappeda and BPBD, regulation on project duration, and regulation on the RPB Guideline. The first issue is related to the lack of coordination and the ineffective action in current RPB. The authors suggest that a regulation that mandates the participation of multiple stakeholders (other than government institutions) should be established. Planning conducted with a collaborative planning approach under the guidance of a strong coordinating body will help local governments produce better plans. The second issue relates to small authorities of BPBD in consolidating other departments within provincial government body. Irrespective of these difficulties, the BPBD should maintain its efforts and should also work towards gaining respect and trust from colleagues in other government bodies.

The third issue is related to the lack of cooperation between Bappeda and BPBD in the RPB-making process. The present regulations, which allocate the responsibility for coordinating RPB to the BPBD province despite it being insufficiently resourced in terms of both finance and staff, need to be reviewed. The authors suggest that both agencies should develop better cooperation between them in order to tackle the issue. In particular, a need for capacity building among BPBD staff was diagnosed in order to enhance the ability to exercise the required coordinating action. In addition, Bappeda, which has its own language of planning, programming, and budgeting, need to be properly informed as well. These complexities merit an evaluation of the authority of BPBD provinces in providing RPB process.

The fourth issue is related to the obligation to provide a tangible product at the end of the year, as required by the national government's budgeting regulations. This obligation caused the BPBD to do RPB planning in only 6 months during one fiscal year, which is an insufficient timeframe. The authors suggest that an assessment on the regulation concerning project duration should be conducted to solve this issue. The last issue is related to poor guidelines for RPB. The authors suggest to review the Perka BNPB 4/2008 regarding the Guideline for Disaster Management Planning (Government of Indonesia 2008c), which is considered a rigid, yet ineffective methodology. The guideline for RPB should only contain the minimum requirements of planning substances and protocols, rather than requiring fixed data and certain methods. This would make the RPB process more customizable, flexible and adaptable to local needs and capacities within specific limits.

The main issue concerning the actors in the institutional arrangements for RPBs is the lack of capacity of the government staff involved in the RPB planning process. This is a general problem, and its rectification may help improve three actor-related issues, namely variety of involved stakeholders, sense of belonging on the part of the plan drafters, and variety of knowledge among different planning actors. The first issue is related to the domination of government agencies in the plan drafters's team. This issue correlates with the issue of regulations that limits participation in the planning process to governmental institutions. It is recommended that other stakeholders are included in the planning process, such as local community groups, NGOs, the private sector, and academia.

The second issue is related to a low sense of identification with the RPB's purpose and process on the part of the plan drafters. Members of the plan drafters's teams appeared to be disaffected, ignorant, disengaged, and alienated from the task of RPB, and were therefore unwilling to take full responsibility for the plan drafting. Providing the drafting teams with a clearer and more persuasive explanation of the background to and the bigger picture of disaster risk management, the role of planning within it, and an indication of where their contribution would fit in may help convince team members of the relevance of their task in the context of the whole disaster risk management process. This approach should be combined with the invitation to participate from an early stage. Involving the plan drafters from the outset of a project is expected to raise their interest and thus will get them to contribute.

Knowledge gaps among different planning actors are arguably the biggest issue in relation to actors's capacities. This phenomenon has three dimensions: (1) differences in the level of understanding of the HFA, (2) deficiencies in map-reading skills, and (3) an inappropriate RPB program. The inappropriate RPB program in particular correlates with the issue of regulation by the coordinating government agency. Among the approaches suggested to remedy these shortcomings are (i) intensive training aimed at capacity building among BPBD staff; (ii) seeking external expertise (e.g., from a consultancy) to assist the BPBD in assessing local disaster risk and in developing various disaster management strategies; and/or (iii) shifting the planning coordination function from BPBD to Bappeda Provinces.

In three of the nine studied provinces resource assessments highlighted a less-than-adequate supply of both financial and human resources. Mitigating this issue would enhance the overall performance of the planning process and would help rectify two specific issues, namely insufficient data for risk assessment and the unavailability of financial resource data. The first issue is related to insufficient data, or data at a too high aggregation level. For instance, hazard maps derived solely from historical records (which exclude climate change) cannot project future risks and potential hazards. Ideally, hazard maps should not only consider historical records but also modeling results that quantify the magnitude and frequency of potential future disaster events. Therefore, more specific data, along with the utilization of qualitative socioeconomic information from and about affected communities is required in addition to the standard data that is currently being used to generate hazard maps.

The second issue is related to the unavailability of financial resource data. Incorporation of potential funding from the community, individual donors, and the private sector can be an alternative source of funding, despite the program relying mainly on financing through government budgets. Data on local government budgets should be made available since it is important for determining a realistic number and size of individual programs. An initiative to co-host one particular program for one particular location by several government agencies is suggested here in order to address the lack of communication and coordination among different agencies.

8 Conclusion

Using disaster management planning practices in nine Indonesian provinces as case studies, this article describes the interplay between regulations, resources, and actors in forming and maintaining institutions. Applying Giddens's (1984) theoretical lens on social systems as an interactive and mutually constitutive relationship between institutions (structure) and actions (agency), the article points out that the shift in the planning paradigm prevalent in Indonesian institutions – from a reactive to a preventive approach – resulted in a new tool for provincial government to manage disaster events and disaster risk in their region: RPB. Even though the RPB terminology does not use the notion of risk, the RPB planning methods have departed from conventional risk assessment, and instead use risk maps to recommend planning policies and programs. The RPB terminology is derived from Act no. 24/2007 regarding Disaster Management (Government of Indonesia 2007a), which according to Lassa (2011) was influenced by the disaster risk approach introduced by the international donors and nongovernmental organizations. UNDP, as one of the active donors, is still facilitating the institutional capacity building of RPB through providing expert teams for implementing RPB.

The authors found that by implementing RPB at the sub-national level, Indonesia has practiced the shifting paradigms of disaster management, from reactive to preventive action. The results from RPB in three provinces suggest that RPB still faces difficulties concerning its integration into development planning due to the differences in planning terms between two systems. Development planning already has its own terms and mechanisms for producing planning products. So, even though disaster management affairs have been added as a development sector, on a practical level, the substance of RPB still needs to be synchronized with the other related planning sectors and mainstreamed into the spatial planning process, especially in defining disaster-risk areas. The authors argue that although at the normative level the promotion of disaster risk reduction was successful and has added new institutions to the national development system, on a practical level the changes – for instance at sub-national level – are slow. The unsupported regulations, knowledge gaps among planning teams, and lack of data are the reasons why RPB planning institutions need converter kits in order to be harmonized with the development planning process. Internally, the institution of RPB itself is not yet adequate to deliver collaborative action among stakeholders, as required by disaster management legislation.

According to Giddens (1984), structure only exists through agency that agents possess, and agents have rules and resources that will enable or undermine their actions to define structure. The planning institution of RPB can only be reconstituted if the local and national governments are willing to use their authority and resources in the planning process, invite more stakeholders, and provide more specific data. By applying the technical regulation regarding RPB, BNPB has already initiated the routinized activities from provincial government in exercising disaster management planning. In practice, however, the BPBD province along with other team members innovated and adapted the planning process. They asked BNPB to accommodate their specific local needs in the making of RPB. Based on empirical research of RPB practices at the sub-national level, it is argued here that there is an opportunity to strenghthen the planning institutions of BNPB by intensifying communication among government, planning professionals, academia, and NGOs. The authors suggest that Perka no. 4/2008 (Government of Indonesia 2008c) regarding disaster management planning should be reviewed to provide for an inclusive planning process, an effective link to development planning, and a flexible and adaptive approach to provincial RPB.

References

AECOM. (2010). *Climate design: Design and planning for the age of climate change*. Singapore: ORO.

Alexander, E. A. (2007). Institutionalist perspectives on planning: Why? Where? How? In N. Verma (Ed.), *Institutions and planning* (Series: Current research in urban and regional studies, pp. 37–60). Amsterdam: Elsevier.

Booth, P. (2005). The nature of difference: Traditions of law and government and their effects on planning in Britain and France. In B. Sanyal (Ed.), *Comparative planning cultures* (pp. 259–284). New York: Routledge.

Brunnengraeber, A., Dietz, K., Hirschl, B., & Walk, H. (2006). *Interdisciplinarity in governance research* (GARNET Working Paper: No 08/06). Berlin: Freie Universität Berlin. Available from http://www2.warwick.ac.uk/fac/soc/garnet/workingpapers/0806.pdf

Dolowitz, D., & Marsh, D. (1996). Who learns what from whom: A review of the policy transfer literature. *Political Studies, 44*(2), 343–357. doi:10.1111/j.1467-9248.1996.tb00334.x.

Garatwa, W., & Bollin, C. (2002). *Disaster risk management: Working concept*. Eschborn: GTZ. Available from http://www2.gtz.de/dokumente/bib/02-5001.pdf.

Giddens, A. (1984). *The constitution of society: Outline of the theory of structuration*. Cambridge, UK: Polity Press.

Government of Indonesia. (2004). *Act no. 25/2004 regarding national development planning system*. Jakarta: Government of Indonesia.

Government of Indonesia. (2007a). *Act no. 24/2007 regarding disaster risk management*. Jakarta: Government of Indonesia.

Government of Indonesia. (2007b). *Act no. 26/2007 regarding spatial management*. Jakarta: Government of Indonesia.
Government of Indonesia. (2007c). *Government regulation no. 38/2007 regarding distribution of governmental affairs between central government, provincial government and local government*. Jakarta: Government of Indonesia.
Government of Indonesia. (2008a). *President regulation no. 8/2008 regarding National Board for Disaster Management*. Jakarta: Government of Indonesia.
Government of Indonesia. (2008b). *Government regulation no. 21/2008 regarding operation of disaster management*. Jakarta: Government of Indonesia.
Government of Indonesia. (2008c). *Regulation of BNPB Chief no. 4/2008 regarding the guideline for disaster management planning*. Jakarta: Government of Indonesia.
Government of Indonesia. (2010). *Government regulation no. 15/2010 regarding implementation of spatial planning*. Jakarta: Government of Indonesia.
Gualini, E. (2001). *Planning and the intelligence of institutions: Interactive approaches to territorial policy-making between institutional design and institution-building*. Aldershot: Ashgate.
Healey, P. (1997). *Collaborative planning: Shaping places in fragmented societies*. London: Macmillan.
Healey, P., & Williams, R. (1993). European urban planning systems: Diversity and convergence. *Urban Studies, 30*(4–5), 701–720. doi:10.1080/00420989320081881.
Intergovernmental Panel on Climate Change (IPCC) Working Group II. (2001). *Climate change 2001: Impacts, adaptation and vulnerability*. GRID-Arendal. http://www.grida.no/publications/other/ipcc_tar/?src=/climate/ipcc_tar/?wg2/301.htm. Accessed 11 Sept 2012.
Irawan. (2011). *Disaster management plan and mid-term development plan: Position of RPB*. Presented on formulation of disaster management planning workshop at Millennium Hotel, Jakarta, October 28, 2011.
Khan, H., Vasilescu, L. G., & Khan, A. (2008). Disaster management cycle – A theoretical approach. *Management and Marketing Journal, 6*(1), 43–50.
Lassa, J. (2011). Institutional vulnerability and governance of disaster risk reduction: Macro, meso and micro scale assessment (With case studies from Indonesia). Ph.D. thesis, Bonn.
Mayntz, R. (2003). New challenges to governance theory. In H. Bang (Ed.), *Governance as social and political communication* (pp. 27–40). Manchester: Manchester University Press.
Miller, S. (2011). Social institutions. In E. N. Zalta (Ed.), *The Stanford encyclopedia of philosophy*. Last update February 8, 2011. http://plato.stanford.edu/archives/fall2012/entries/social-institutions/. Accessed 30 Oct 2012.
Sanyal, B. (Ed.). (2005). *Comparative planning cultures*. London: Routledge.
Sperling, F., & Szekely, F. (2005). *Disaster risk management in a changing climate*. Discussion paper prepared for the world conference on disaster reduction on behalf of the Vulnerability and Adaptation Resource Group (VARG). Reprint with Addendum on conference outcomes, Washington, DC.
United Nations Development Programme (UNDP). (2004). *Reducing disaster risk: A challenge for development a global report*. New York: Bureau for Crisis Prevention and Recovery, United Nations Development Program.
United Nations Office for Disaster Risk Reduction (UNISDR). (2011). *Global assessment report on disaster risk reduction: Revealing risk, redefining development*. Geneva: UNISDR.
Venton, P., & Hansford, B. (2006). *Reducing risk of disaster in our communities*. Teddington: Tearfund. http://tilz.tearfund.org/~/media/Files/TILZ/Publications/ROOTS/English/Disaster/ROOTS%209%20-%202006%20version/Disaster%20risk%20reduction%20-%20full.pdf.
Work, R. (2002). *The role of participation and partnership in decentralised governance: A brief synthesis of policy lessons and recommendations of nine country case studies on service delivery for the poor*. New York: UNDP. http://unpan1.un.org/intradoc/groups/public/documents/un/unpan006230.pdf. Accessed 25 Oct 2011.

Could the Search for Sustainability Reinforce Socio-ecological Conflict?: The Mining Industry in Chile and Its Impact at the Local and Regional Level

Fernando Campos-Medina

1 Introduction

The beginning of ecological institutional modernization in Chile started officially with the implementation of Law 19,300 "Environmental Bases",[1] which established the first Nation Environmental Commission (CONAMA) in 1994. The implementation of Law 20,417, under which the Ministry of Environment, the environmental superintendence,[2] the environmental courts, and the environmental assessment and biodiversity services were created in 2010, is considered the final stage of Chile's ecological institutional modernization. Chile then joined international efforts to organize a national response to its ecological crisis within an international context of increasing sustainable global governance (Speth and Haas 2006; Levy and Newell 2005; Akhtarkhavari 2010; Winter 2006; Jasanoff and Long Martello 2004; Park et al. 2008[3]).

Interviews The interviews quoted in this article, belong to the authors PhD thesis fieldwork, considering more than 25 interviews with academics from the so-called Chilean traditional universities, policymakers, representatives of NGOs, the local community and the mining sector. In terms of keeping the anonymity of the interviewees the reference are only general.

[1] In Latina America, the environment institution transformation in the 1990s was enforced by different laws commonly known as the "environmental bases" because a UNEP report with this title influenced them strongly.

[2] This is a specific word used in Chile.

[3] See Speth and Haas (2006), as well as Haas et al. (1995) for an integrated perspective of the principal events and concepts that have influenced the global environmental agenda over the past decades. To observe global environmental governance from an international law perspective, which is prevalent in the process of social learning, see Akhtarkhavari (2010). See Winter

F. Campos-Medina (✉)
Jena Graduate School "Human Behavior in Social and Economic Change" (GSBC), Friedrich Schiller University Jena, Jena, Germany
e-mail: fcamposmedina@gmail.com

B. Werlen (ed.), *Global Sustainability*, DOI 10.1007/978-3-319-16477-9_14

This article challenges the official representation of modernization as a simple institutional improvement. Further, as an alternative thesis, it suggests that modernization has contributed to the increasing depoliticization of the society-nature relationship over the last 20 years (Campos and Larenas 2012).[4] We support this interpretation by showing that basing the ecological debate on the notions of the sustainable extraction of natural resources and sustainable pollution management control is insufficient to deal with the negative and pervasive consequences of the extractive industries on which the Chilean economy is based.[5] The argument implies that it is possible to improve sustainable production and concurrently reinforce the socio-ecological conflict at the local level if industries are extended regionally.

Following Gligo (2006), the environmental-economic history of Latin America can be described as different waves of productive restructuring occurring at the regional level across the entire continent. As an example, Giglo mentions the transformation of the pre-Columbian cultures' agricultural orientation into the extraction of silver in the Colonial period in order to supply the Spanish metropolis. Thereafter, despite the independence movements, the situation did not change substantially; the nascent national states sought integration into the global markets through the consolidation of mono-productive regions. In the eighteenth and nineteenth centuries, Europe relied strongly on the import of products such as cocoa, coffee, rubber, and sugar cane to sustain everyday life, while imported minerals, such saltpeter, sustained the industrialization process. In both centuries, the governors and the political elites carried on as if the resources were unlimited and as if any previous forms of the society-nature relationship – from the pre-Columbian cultures – were inferior to those of Europe (Gligo 2006, 11).

The famous Potosi mine in Bolivia is a good example of the regional economic restructuring in Latin America and its devastating socio-ecological effects. The Potosi Mountains were rich in silver ore deposits; it has been claimed that, during the eighteenth century, more than the 50 % of the world's mineral production was extracted there. At the time, the city of Potosi had a population of 200,000, thus competing with big metropolises such as Paris and London. Moreover, it was larger

(2006) for an explanation of global environmental governance as a system of multilevel institutions and how they adopt the Earth system analysis derived from the natural sciences. For an account on the connection between the global and local forms of environmental governance, see Jasanoff and Long Martello (2004). See Park et al. (2008) for a critical perspective of global environmental governance in terms of the connection between sustainability and globalization and the effects of the environmental policies' marketization.

[4] Campos and Larenas (2012, 6) regard the ecological modernization in Chile as depoliticization because it implies a "constant decline in the public and social character of the ecological debate. The new environmental institution represents modern procedures' access to the ecological management, as well as the exclusion, of previous forms of socio-ecological conflict regulation." (Translation by the author.)

[5] Mining activity was responsible for 10 % of the national GDP in 1993; it achieved a peak of 20 % from 2006 to 2007, becoming stable at about 15 % in 2010 (based on the Chilean Central Bank's (n.d.) data, historical statistics, and the percentage of the GDP calculated at current prices).

than any city in the Spanish kingdom. Very few studies have investigated the scale of this transformation in production on the continent and the pressure this change exerted on the environmental components.[6]

This article calls for a redefinition of the notion of sustainability in the context of large industrial expansionism and regional productive restructuring. At the local level, we propose that sustainability should focus on everyday practices that ensure the local community can subsist despite the risk of water rights monopolization, the high demand for energy, and the concentration of land concessions for exploration and exploitation in the northern area of Chile. At the regional level, sustainability should not merely focus on clean production, but should also consider the negative effects of the mining cluster's expansionism. These negative effects not only affect local ecosystems, but also lead to rapid rural depopulation and accelerated urbanization, which in turn increases the number of informal settlements in the country's medium-sized cities.

2 Productive Regional Restructuring and Socio-ecological Conflicts

The abovementioned process of territorial restructuring, which will be referred to as *regional production specialization* from a theoretical standpoint, is the result of the coincidence of a territorial and an action orientation deployed at the local level.[7] The economic literature has equated this territorial tendency toward mono-production – in which single projects bring about an aggregated effect on a regional scale – with the notion of productive regional performance by calculating the coefficient of localization, or the index of production specialization.[8] From this

[6] For a more detailed view of the analysis of socio-ecological conflicts in Latin America, see Galeano 2004.

[7] Here, we analytically distinguish between "action orientation" and "territorial orientation" in order to describe the difference between activities that primarily mobilize actions pertaining to entrepreneurialism and market-oriented logic, and those activities geared toward rearticulating the territory for regional specialization. Max Weber coined the action orientation concept in his classical book Economy and Society (2002), in which he defines the types of action orientation in the concepts as: usage, custom, and self-interest. Giddens's (1986) structuration theory provided possible variants in terms of action orientation. Giddens's (1986) book was one of the first sociological attempts to incorporate the notion of time-space in a theoretical explanation, using the idea of regionalization. This distinction is totally analytical because, from a comprehensive perspective, every action spacializes and every meaningful space is constructed by territorially unfolding actions. For this integrative framework, see Werlen's (1993) "everyday regionalization" concept.

[8] In both cases, the analysis determines a territory under study and a territory of reference as a total universe (national state or supranational level). The relationship considers the percentage of employment in a particular economic activity in that region in relation to the total universe. Values around 0 show the non-presence of the activity in the territory, a value of 1 indicates equal

perspective, the degree of specialization is a valuable index for economic performance and a good proxy for the degrees of development. Territorial public policy tries to reinforce this economic orientation without much reflection on its consequences at the local level. Human geography, regional planning, agriculture, and ecological studies give us a different perspective, from which specialization is always contrasted with regions' possibility to sustain other relevant activities for the local community. This is especially significant for small-scale agriculture, which has always been represented as an integral part of rural communities' spirit.

The Chilean territorial specialization – represented by large extractive industries, especially mining and forestry, but also energy production, large fishing companies, and salmon aquaculture – has been extensively criticized in the academic and public sectors. The mining and forestry industries are criticized for engendering water shortages, thus contributing to losses in agricultural activity and the depopulation of rural areas. The following excerpts explain how academics and policy makers assess the socio-ecological problems due to the water scarcity in the Atacama Desert and due to the poor water quality in the southern rain forest. This socio-ecological conflict runs concomitantly with the struggle for land and the monopolizing of energy.

> The traditional activities, which justified the presence of rural human settlements and which generated communities' control of water resources, are disappearing [...] A community that does not have a water supply cannot exist as a community and is unable to work; therefore, it loses its social discipline, which implies protecting terraces, maintaining crops, cleaning irrigation channels, and distributing water in a socially equitable way. The community has no reasons for continuing to exist if this social discipline disappears; hence, most of its residents will move to the city (Academic State University).[9]
>
> In the southern lands [indigenous groups] cannot sell their land, but they are available for lease. So, what could I do? Forestry companies lease these lands to plant eucalyptus and pine forests, which implies degradation of the soil quality, as well as groundwater pollution. This problem occurs in communities such as Renaico and Collipulli where there are large forestry plantations; therefore, all the water is used, which makes impossible to cultivate anything in the fields close to these forests (Director of the National Agency)[10]

At this point, it is important to mention that the notion of territorial economies of scale, which underlies ideas such as economic efficiency and/or economic advantage, cannot solely explain the emergence of a regional specialization. This explanation, widely accepted in the economic literature, justifies the concentration of production clusters in specific areas. Furthermore, this explanation only highlights the benefits of economies of scale for the emergent companies without considering their negative effect on other activities in the territory.[11] On the contrary, and from

---

distribution of the activity in the regional and in the total universe, while values above 1 represent the specialization of certain activities in the region.

[9] Translation by the author.

[10] Translation by the author.

[11] This refers to the tension between the mining industries and agricultural activities; mining projects compete with preexisting agricultural activities because they are located in the same areas, or use the same water sources. The Latin American Observatory of Environmental Conflicts and

an environmental conservationist perspective, the emergence of regional production clusters[12] is due to the lack of mechanisms to measure the aggregated environmental impact of single megaprojects on a regional level. In this sense, it can be argued that the traditional environmental impact assessment mechanism is only applied to single projects. Consequently, it becomes a blind institutional tool for tackling productive restructuring at the higher territorial levels. We will argue that this form of environmental impact evaluation is incapable of explaining, measuring, and controlling the aggregated effects of the productive clusters emerging and expanding in different regions of the world.[13]

From this theoretical perspective, the regional productive restructuring is responsible for the reemergence and intensification of socio-ecological conflicts if its expansive nature is not counteracted. This restructuring – in the form of a mono-production specialization – only appears to be an active social reality under two circumstances in which capitalism unfolds: an inclination toward entrepreneurialism in the form of innovative enterprises and processes, and the intention to ensure and extend projects that have proven to be profitable. This discursive orientation becomes clear in terms of the mining industry and the consolidation of the regional mining cluster when companies are given national concessions to explore (innovation) and exploit (assurance) mineral deposits. It should be emphasized that the tendencies of extractive industries to explore/innovate and exploit/extend are an integral part, and not just a side effect, of the business model. The consolidation of big regional production clusters, even if they are based on green or sustainable measures, is an emergent social reality with an unknown magnitude and consequences.

When the capitalistic orientation of entrepreneurialism and assurance coincide regarding territory, strong pressure is exerted on the productive inputs: land, water, energy, and work. This intensifies tendencies to reorganize the environment and people's living conditions at the local level. In this context, we understand *innovation* as the development of new resource extraction areas. Complementarily, *assurance* is understood as the intensive exploitation of successful enterprises already anchored in the territory.

In respect of the mining industry, a methodology that combines a discourse analysis of national environmental laws and interviews with key actors will show that the advancement in sustainability is linked to the strong pressure on

the Latin American Observatory of Mining Conflicts (OLCA/OCMAL 2012) have identified 25 environmental conflicts that the mining industry has caused in Chile. All of these conflicts have occurred in the Atacama Desert region. In Latin America, 161 conflicts are recognized, affecting 212 communities; 5 of these conflicts occur on transnational borders. See OLCA.

[12] Here, the cluster notion is used in a very broad sense as the aggregated effects of different single mines located in the Atacama Desert region. For a critical perspective of this concept, see Martin and Sunley (2003).

[13] Among many others, the uranium exploitation in Argentina and that of the oil sands in Canada are examples of other socio-ecological conflicts emerging from different types of mining activities. These activities require large regional production restructuring and have important consequences for the local communities.

particularly scarce resources in the Atacama Desert. The sustainable extraction of copper and different improvements in mineral waste treatment cannot solve the mining industry's high demand for water in the world's most arid desert, or ensure that water will always be available for human consumption in the northern cities of Copiapo, Antofagasta, and Calama,[14] and especially for the Atacameños and Aymara[15] in rural areas. Similarly, the mining industry cluster's demand for energy requires a redefinition of the entire national energetic matrix through the construction of mega hydroelectric power stations in the south and new thermoelectric coal plants in the north of Chile. The following quote from a local radio interview suggests that the water conflicts are provoked by the incorporation of economic rationality to manage the distribution of water, which has been government policy since 1981.[16]

> They polluted our water in 1998. In 2001, the Chilean State declared the Loa River exhausted and since then nothing has been done to resolve the problem. Those responsible for this pollution are Codelco Chile and Soquimich. These companies still continue purchasing water rights, taking advantage of people's financial needs. Currently, we do not have water resources for agriculture. Quillagua received polluted water from Calama. I am angry with the Chilean State (President of an Aymara community).[17]

The following sections deal with two issues: The first discusses *ecological modernization* as an analysis perspective with a focus on the central characteristics that link environmental institutional modernization processes and the search for sustainable production. The second section presents an analysis of the mining industry cluster in the north of Chile in terms of: (i) ecological institutional modernization as a form of place achievement, (ii) the unequal distribution of environmental roles in national and international contexts, (iii) the construction of a spatio-temporality in environmental issues that reinforces economic entrepreneurship, and (iv) a goal-oriented understanding of sustainability linked to the assurance of highly profitable enterprises.

3 Ecological Modernization as an Analysis Perspective

The notion of ecological modernization has two opposing meanings. On the one hand, the modernization process is often understood in everyday language as an initiative to improve efficiency and efficacy in institutions. If so, this improvement

[14] For an in-depth discussion of the water conflict in Chile, see Larraín and Poo (2010).

[15] The Atacameños and Aymaras are two pre-Columbian native groups; the Atacameños inhabit the Atacama Desert in Chile and Argentina, while the Aymaras inhabit the Andes in North Chile, South Peru, and Bolivia. It is estimated that the Atacameños have been around since the fifth century while the Aymaras' origins can be traced back to around 2,000 years BC, even before the Inca Empire ruled the region.

[16] Chilean water code.

[17] Radio Station, University of Chile (2012). Translation by the author.

should be applied to environmental institutions by promoting the centralization of their control and supervision mechanisms. In Chile, the underlying hypothesis is that environmental responsibilities and norms are spread across a large number of state departments and legal bodies. Consequently, it has become essential to coordinate control and supervision in order to gain efficiency. This argument was used during the two major waves of environmental institutional modernization in Chile, and explicitly mentioned in the presidential speech introducing Law 19,300 of 1994 and Law 20,417 of 2010.

> However, this study also proved to be a large-scale diffusion, incoherent, and lacking management of the current sectorial legislation and its multiple amendments; therefore, it has caused great ignorance of its scope, uncertainty about the validity of the original texts, and a serious breach of law. Additionally, it has been found that the environmental protection public authorities are scattered across many different agencies, which operate in a non-organic way and whose responsibilities run in parallel and are ambiguous (Aylwin 1992, 10)[18]
>
> The re-design seeks to rationalize the competencies in order to have an authority which can issue policies and regulations for the protection of environmental resources; the sectorial competencies will therefore be organized and the coordination facilitated within the public system (Bachelet 2008, 11)[19]

In a similar vein, Mol and Jänicke (2009) argue that, originally, ecological modernization began as political projects in Germany and the Netherlands. There, the concept was approached as the "social scientific interpretation of environmental reform processes at multiple scales in the contemporary world" (Mol et al. 2009, 4), especially in projects in which caring for the environment do not restrict economic development but actually produce a benefit. On the other hand, certain groups of scholars use the notion of ecological modernization to criticize environmental public policy's new orientation, which was internationally implemented in the early 1980s and is still applied today. The main characteristic of this policy orientation is the extended belief that it is possible to maintain the economic system and simultaneously comply with environmental goals. In other words, satisfying the need for growth and protecting and improving environmental conditions[20] are feasible if new technologies and market mechanisms are incorporated into the sustainability development framework. This critical perspective was formed because no international environmental protection goals have been fully realized in the last two decades. For example, global warming has worsened (Intergovernmental Panel on Climate Change 2007, 2013),[21] the territorial roles and the inequalities between the northern and southern regions have been strengthened

[18] Translation by the author.

[19] Translation by the author.

[20] In some ways, all the critical literature on ecological modernization recognizes this characteristic.

[21] "Each of the last three decades has been successively warmer at the Earth's surface than any preceding decade since 1850 (see Figure SPM.1). In the Northern Hemisphere, 1983–2012 was *likely* the warmest 30-year period of the last 1,400 years (medium confidence)" (IPCC SPM: 3).

(Roberts and Parks 2007; Gonzáles et al. 2007), and socio-ecological conflicts seem to increase in many countries.[22]

Both the above interpretations of environmental institutional modernization are based on the idea that certain social institutions need to be changed in order to tackle the environmental crisis. However, the main difference lies in various social actors' reasons for the transformation. According to its advocates, modernization should improve every social life sphere in order to ensure a more rational use of resources. Modernization should help environmental institutions incorporate economic rationality in order to coordinate decisions; that is, they should become more efficient and effective, and the social costs they incur in meeting their objectives will be minimized.[23]

In contrast, for their detractors, modernization processes always have pros and cons (Van der Loo and van Reijen 1992; Rosa 2009); the ecological institution is no exception. The pursuit of the rational use of resources may lead to a liberal economic rationality's colonization of the ecological sphere that confuses the process's means and ends. In this sense, social, economic, political, and environmental institutions are changing in order to face the ecological crisis, but with the implicit intention to protect the economic imperative of growth, and not basically to defend nature, the local ecosystem, communities, and biodiversity. From this critical standpoint, the subordination of nature during the last three or four decades has been reinforced by the ecological modernization process in Latin America (Altieri and Rojas 1999; Guimarães 1999; Gudynas 1999).

Our argument is based on a critical perspective of modernization. The main question is whether, given the many examples of the devastating human consequences of environmental changes around the world, it is possible to merely conceptualize institutional modernization as social improvement. Megaprojects related, for example, to raw material extraction, energy production, and waste deposits – especially in developing countries – should be conceptualized as a specific form of global integration in these regions, which runs parallel to other waves of environmental issues' globalization regarding green production, decontamination processes, and the combatting of global warming.

In this article, ecological modernization is understood as the modernization process that all governments promote and which is applied to environmental institutions at the national state level. This conceptualization challenges the idea that ecological modernization is a process that describes the situation in developed countries with a market economy and a welfare state; that is, especially the primary OECD countries. Ecological modernization has been the socio-political answer to

[22] In respect of Chile, see Campos-Medina and Campos-Medina (2012).

[23] Ecological modernization gives rise to a particular form of rationality, which some authors have subsumed under the economic rationality concept, while others have considered it independent of other social spheres. Nevertheless, the particular characteristics of this "ecologic rationality" should be the object of empirical research. Describing the integrative character of all the material and symbolic interactions between society and nature, Leff (1986) argued that "[e]very society creates its own environmental rationality."

the national-level environmental crisis since the 1970s, regardless of the country's political, economic, or social situations. We can thus assume that, even if the industrialized countries were the forerunners, every country in the world has currently undergone a transformation of its environmental legal frameworks and institutions in order to address its ecological challenges. In other words, it is highly improbable that there is a country that has not mobilized resources of every kind to address these problems.

Since the ecological modernization process is limited to certain successful institutional transformations, it assumes that peripheral countries' modern environmental progress follows the same predefined patterns as those of developed countries. In this sense Mol asks: "[W]hat are the opportunities for non-triad countries in particular, to pursue development paths that differ from the dominant Western models of environmental reform, which seem so strongly connected to the ideas of ecological modernization?" (Mol 2003, 63). We conversely propose that every country represents a different form of modernization that responds to a different international integration imperative, whether ecological and/or productive.[24] The international level plays a fundamental role because multinational agencies influence the management of ecological policy, help shape the environmental crisis narrative, and facilitate the setting of priorities and the defining of the procedure.

In summary, ecological modernization is the transformation of the society-nature relationship at the national level, but in the context of a global transformation during which different roles are redefined. Lastly, we regress from the conceptualization of ecological modernization and sustainable development as the two main ways in which environmental institutional transformation could occur. From our perspective, even if the two orientations for action – ecological modernization and sustainable development – are not conflated (Langhelle 2000), it is theoretically and empirically impossible to separate them as two distinct tracks. The notion of sustainability is highly embedded in the modernization rhetoric, and the movement toward sustainable forms of production is therefore a central part of the modernization process in all analyzed international cases.

---

[24] The arguments presented here are based on sociological critique of the theories of modernity, which regard the modernization in Central Europe and North America as an unavoidable track that every other society has to follow. The modernization process can differ from society to society because the core of this sociological explanation is not the result, or the last stage, of modernization tendencies, but, conversely, the requirements that every society confronts in a growing and unavoidable modern context (see Touraine and Bixio 2000). In another variant, this argument is based on the idea of universalistic integration as presented by Chernilo and Mascareño (2005). Here, environmental international integration is compelled by universalism and particularism in a normative, as well as functional, context.

4 Sustainability as a Normative Referent and Factual Phenomenon

From a normative perspective, ecological modernization strives for an ideal state of sustainability by harmonizing the society-nature relationship. Even though the idea of sustainability as a normative referent dates back to the middle-ages (Bosselmann 2008), it is still considered a non-achieved stage, a goal to be socially pursued, which evolves in the same way as ecological consciousness. As a normative referent in the context of ecological modernization, we understand sustainability as a social ideal or, in other words, as an evolutionary orientation toward present and future actions. From a different perspective, sustainability has historically materialized in various ways. It may therefore represent different factual strategies and measures within the ecological modernization process, which covers different waves of sustainability deployment across time. Consequently, we consider sustainability a factual phenomenon.

A normative referent and factual phenomenon are closely interwoven; their differences can only be reflected in theoretical terms. None of their actions lack further orientation, in the same way that no meaning is socially relevant if it has not materialized, even if this materialization is at the level of language and narratives. Nevertheless, confusing the two dimensions, or, even worse, rejecting one dimension by favoring the other, complicates analyzing the ecological modernization process from a perspective in which the factual unfolding of sustainable strategies has positive and negative effects. Given the distinction between sustainability as a normative referent and a factual phenomenon, we consequently propose analyzing the negative effects of the ecological modernization in Chile. This ecological modernization is primarily oriented toward supporting the extractive industries and supplying materials to global markets, and only secondarily toward the protection of ecosystem and local communities.

The next section shows how ecological modernization oriented towards achieving sustainability has, in the Chilean mining industry, increased socio-ecological conflicts with the local communities. The analysis is based on four dimensions: (i) the expansionist character of copper exploitation in the industry and the tendency to monopolize all other production inputs, (ii) the consolidation of the Atacama Desert area's regional production role as a place of copper exploitation to the dramatic detriment of other traditional activities, (iii) the spatio-temporality of the mining activity, which reinforces the entrepreneurialism orientation, and (iv) the goal-oriented sustainability regarding pollution management and the rational use of resources while excluding all references to understanding and incorporating the negative consequences for the local ecosystem, the biodiversity, and the living conditions.[25]

[25] *Traditional living conditions* do not refer to a conservative perspective. On the contrary, it means that, in order to support extensive industrial activity, ecological modernization dramatically and daily reorganizes social practices. Examples are the transformation of the labor market, the

Our thesis proposes that, together, these four orientations form a comprehensive narrative about sustainability that has become key in Chilean ecological modernization. We argue that these orientations are the drivers behind the extractive industries' expansion over the last 20 years. Sustainability has become the centerpiece of the consolidation of certain economies and regions as suppliers of raw materials to international markets. International pressure to adopt green forms of production is without a doubt a positive transformation. Nevertheless, the incorporation of new technologies will increase the extraction of natural resources and further increase pressure on the so-called environmental services. In this sense, we have ascertained that the search for sustainability in the extractive industry – the main target of the Chilean ecological modernization process – will lead to the intensification of socio-ecological conflicts with local communities in the country's vast rural areas.

5 The Mining Industry Analyzed from a Territorial and Action Perspective

5.1 Place Achievement as Territorial Expansion: A Key Part of Ecological Modernization in the Context of Sustainability

Many attempts have been made to understand how economic development takes place. These actions are not casual; economic accumulation is based on very specific patterns of spatial restructuring, in which the territorial expansion process is simultaneously a condition for, and a consequence of, further capitalistic developments. The traditional Marxist definition of capitalism is rooted in what has been labeled land acquisitions or, more accurately, the German concept of *Landnahme*. Following Klaus Dörre's (2011) theory, *Landnahme* describes a process of constant territorial expansion, in which capital reproduction is not only produced by the internal movement of goods and money, but fundamentally also by acquiring new places for capital development "at the cost of none-capitalistic ones" (Dörre 2011, 91). As soon as we accept that capitalistic expansion does not fully comprise an internal improvement of capital efficacy, we are drawn to search for other strategies in which capitalism acquires strength and in which spatio-temporal restructuring can deliver useful insights. In this regard, it is possible to connect the improvement in sustainability with negative territorial expansionism.

Based on Dörre's conceptualization (2011), we propose environmental reforms that explain acquiring place by means of two parallel and interrelated processes. These processes have been labeled *Landnahme as territorial acquisition* and

intensification of immigration flow patterns, increasing housing prices and informal urbanization, as well as the trade and supply of products, energy, and services.

Landnahme as domain acquisition. The former is the obvious and empirically demonstrable expansion of the territorial dimension on which productive clusters operate, while the latter involves the incorporation of new activities, which capitalism's socio-economic principles determine.

These two phenomena were clearly observable in the copper industry in Chile between 1989 and 2012. In terms of territorial acquisition, the national copper production could be regarded as a proxy for territorial expansionism. Figure 1 shows that Chile's participation in the international copper market grew from 17.8 % in 1980 to 32.4 % in 2010.[26] This increment in the production was mainly due to the incorporation of private investments in order to explore and exploit new reservoirs in the form of megaprojects. These megaprojects involved the five big mines that were operative in 1993 but grew to include to more than 25 in 2012. In the mining industry, it has become standard practice to take on big private projects like the Doña Inés de Collahuasi (Developed by Anglo American Company 1999) and Los Pelambres (Developed by Antofagasta Minerals Company 1999), which were undertaken by some of the biggest mines in the world.

Figure 2 shows how the increment in the production of minerals was distributed over the territory between 1993 and 2012. In a temporal comparison, the production in the traditional mining area in the Atacama Desert grew by almost 200 %, while it increased almost five times in the fifth and metropolitan region. The only region that showed a shortage was the sixth, where the biggest underground copper mines in the world are located, which should provide a constant production during the next 50 years. The third region, which traditionally comprises small or middle mines, almost tripled its production. Importantly, in this period, the first and fourth regions began exploiting their biggest investment. In the southern regions of the mining cluster, this investment is not fully dedicated to copper extraction, but primarily to other minerals, such as silver and gold at the Pascua Lama (Barrick Gold 2011) mine – a big mine operating on the national border between the Argentinian San Juan province and Chile's third region, Atacama.[27]

From a domain acquisition perspective, *Landnahme* should be understood as the ability of capitalism to always refer to outside areas in the form of "non-capitalist [social] formations or some sector within capitalism, such as education, that has not been proletarianized" (Dörre 2011; quoting Harvey 2003). The interconnection between the political and economic systems that promote the extractive industries inside the Chilean environmental legal framework is a good example of this domain acquisition. The first central environmental law in Chile (Law 19,300 of 1994) could be considered a landmark in the territorial expansionism of mining activity, as well as in the industry's domain acquisition. We aim to prove that, since 1994,

[26] In order to better understand the relevance of Chile in the copper international market, it can be compared to the world's petroleum production, with the two main producers, Russia and Saudi Arabia, only contributing around 12 %.

[27] Currently, the operations in the mine are closed and Pascua Lama could be considered an emblematic case of socio-ecological conflict in Argentina and Chile.

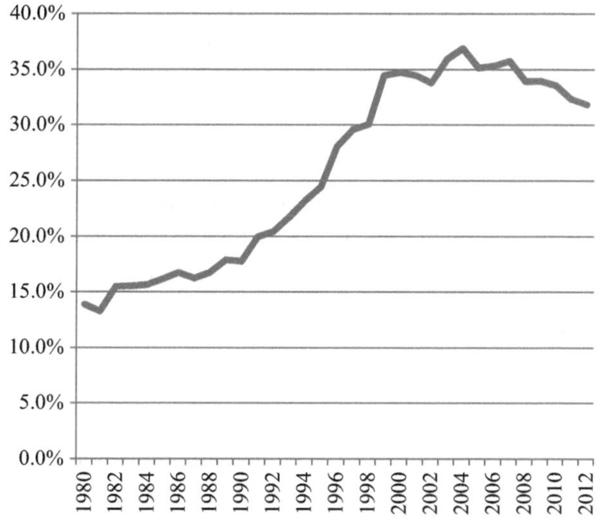

Fig. 1 Chilean participation in the world copper production. Self-elaboration using data from COCHILCO (Chilean Copper Corporation (n.d), Statistics)

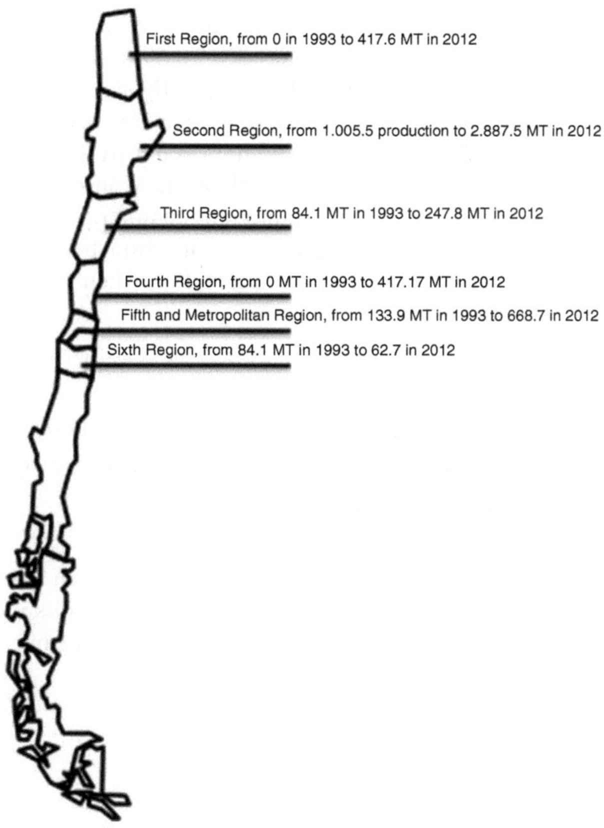

Fig. 2 Evolution of copper production by Chilean regions. Self-elaboration with data from COCHILCO (Chilean Cooper Corporation, Statistics)

market orientation has determined the environmental policy and that two political events were particularly relevant in this period: (i) the modification of Law 18,097[28] of 1982, which opened mining activity to private investors in the form of exploration and exploitation since 1990; and (ii) the presidential approval of a new environmental legal framework, which became explicit in his executive speech to introduce Law 19,300[29] in 1992. As the excerpt from the speech below suggests, the new framework's incorporation of sustainable ecological measures would be gradual and would not hinder economic development.

> Simultaneously, environmental protection cannot be considered in a restrictive sense. Our country should satisfy the increasing need for housing, health, education, electric power, etc. This means using the resources that it owns [. . .] this project has gradualism as its inspiring principle. It does not immediately require the most demanding environmental standards, or try to subjugate all the activities in the country without considering their importance for the evaluation process of the environmental impact. (Aylwin 1992, 16).[30]

At the beginning of the democratic transition in Chile,[31] the political system's explicit support of the extractive industries was the result of the expansionist character of the mining activities over the previous decades. Figure 3 shows the accelerated and sustained growth of Chile's GDP between 1986 and 2010. This exponential growth is largely explained by the development of mining. From 2000 onward, mining activities exceeded 10 % of the GDP, achieving a peak of 20 % between 2006 and 2008, and subsequently stabilizing at 15 % of the GDP.

In summary, if the perspective used here is accurate, the forms of place achievement through territorial and domain acquisition are strongly interrelated because territorial expansionism is based on the consolidation of a strong institutional framework that creates certainty for long-term investments. In empirical terms, this means highlighting the interconnection between social action, institutional transformation, and territorial orientation when explaining highly effective social phenomena.

5.2 Territorial Environmental Role Assignment: A Form of Unequal Distribution at the Core of Ecological Modernization

The unequal environmental role distribution is an interesting notion that complements the theory of ecological modernization. Unlike most theoretical approaches, it places the unfair allocation of specific functions in a territory at the center of the

[28] Constitutional Organic Law on Mining Concessions.

[29] The environmental bases law.

[30] Translation by the author.

[31] For a complete analysis of the Chilean democratic transition in the context of the socialist government, the coup d'état, and the dictatorship regime, see Boeninger (1997).

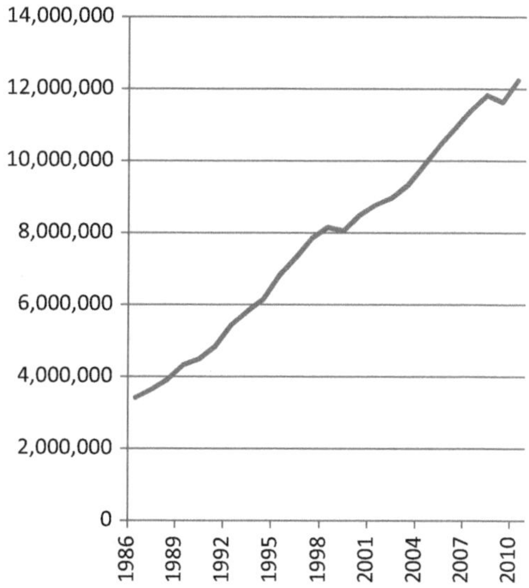

Fig. 3 Evolution of Chilean GDP. Self-elaboration using data from the Chilean Central Bank (n.d.), Statistic. Money valued at 1986

modernization process. This analysis reestablishes modernization as a powerful driver of spatial restructuring and criticizes the general assumption that sustainability is always a positive form of entrepreneurship.

Building on Smith's theory of "uneven development" (1984), we argue that ecological modernization (like capitalism) has a structural orientation toward the consolidation of uneven development patterns during the capital accumulation process, whether or not it is based on green technologies. Particular attention should be paid to the dynamics of the unequal territorial distribution of environmental roles, as well as to the distribution of the burdens and benefits. Here, the most relevant categories are found in a threefold production-consumption-deposit structure, which forms the basis of territorial differentiation. For example, places are defined in order to concentrate consumption and avoid dealing with the dynamics of production and deposits; this can be considered the key distinction between northern global cities and the periphery. Furthermore, using the same example, when the principal activity in a big metropolis is the consumption of products, merchandise, services, energy, and water, the common discourse on ecological modernization pays no attention to the territorial relations that support this privilege (consumption). Similarly, certain areas in Africa are internationally solely presented as non-treated waste deposit sites, while certain places in Latin America are frequently presented as raw material extraction areas. China, India, and other countries in Southeast Asia have acquired a specialized role as industrial producers, while Western Europe and North America have become known as knowledge creators.

The principle of territorial rearticulation, which Smith (1984, 4) considers "structural rather than statistical," is an important characteristic of uneven development. This means that the consolidation of uneven territorial patterns is not a side effect of development, but a very basic way in which ecologic modernization rearticulates territories. This phenomenon is also observable at the national level, for example, when developing countries consolidate an economic development pattern, in which large, uninhabited, and remote regions are responsible for producing and supplying energy and environmental services for megacities that, in many cases, house more than a third of the national population.

While there is no doubt that ecological modernization is an important driver of this development pattern, it cannot be seen as the only cause of this territorial rearticulation. The government's new environmental institutions have reinforced two notions: territorial speculation and territorial concentration. The first notion is to profit as much as possible by investing capital in a territory on the basis of future transaction projections. The second notion addresses the inclination to selectively concentrate either positive investments, or undesirable activities, in certain areas. When a speculative orientation is applied to a territory, there are very limited incentives to invest in deprived areas because the desire to maximize utilities leads to large investments in more profitable places. Similarly, when a territory is associated with a non-desirable activity, there is a growing impulse to allocate more such activity to the same place, because the area will increasingly lose value. If there is no strong political control of the territorial articulation of the environment, it is possible that large mono-oriented regions will emerge at the end of this cycle.

In the case of the mining industry, three situations are generally relevant: (i) consolidation of the mining cluster in the Atacama Desert region, for which important growth is foreseen in 2020; (ii) strong pressure on productive inputs, especially energy and water, but also land and the workforce, which, together, weaken traditional activities such as subsistence agriculture and livestock production on the Andes Plateau; and (iii) the unequal patterns of territorial allocation, especially concerning energy production and residual deposits, which damage the vulnerable population by favoring the mining industry.

The mining cluster was responsible for more than 15 % of the national GDP in the first decade of the twenty-first century. Figure 4 shows the stability of the mining sector and the extent (in percentage) to which it has contributed to the national economy. In the most important extractive areas, the mining industry achieved a record of 60 % participation in the regional GDP. With this data in mind, and given that mining activity not only involves extractive practices, but also the industrial process needed to transform minerals and the construction of large-scale infrastructure, it is possible that the highly specialized mining sector will exert pressure for the establishment of mono-productive structures at the national and regional levels. The extension of the mining cluster and the high level of profitability do not compensate for the declining relevance of various other activities in the region. One consequence of this poor quality of life in the mining region is the large number of mine employees who commute to the principal mining city of Calama, principally from La Serena and Santiago (respectively 1,097 km and 1,573 km away).

In Fig. 5, it is apparent that there is a link between the Chilean GDP and electric power production. Given the mining activity's large contribution to the national GDP and its energy consumption, it can be assumed that the increasing production

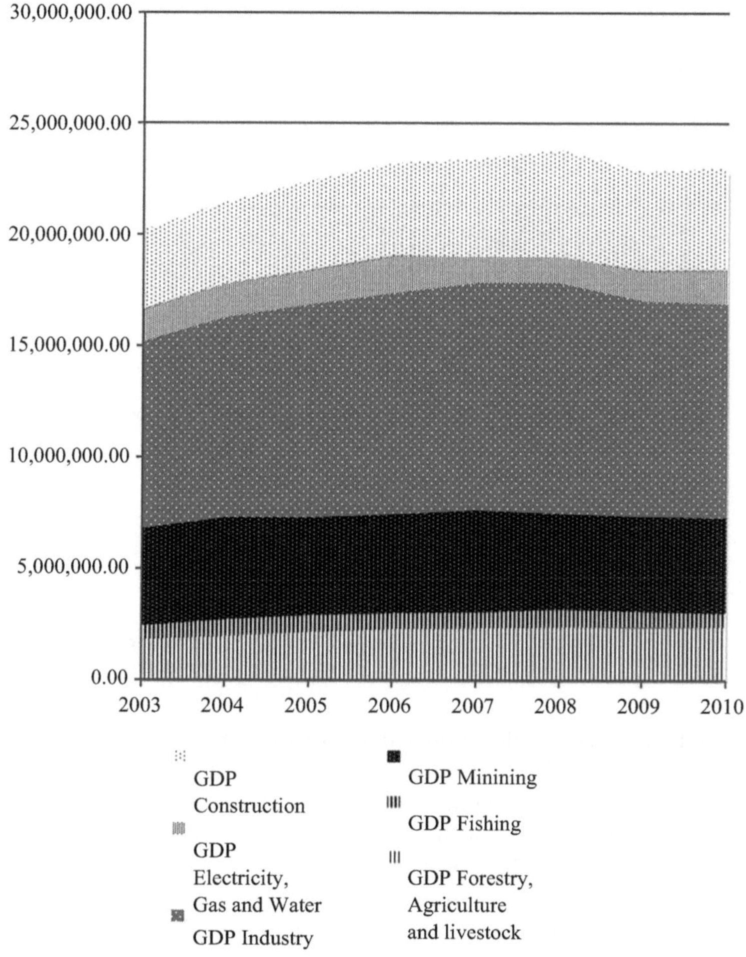

Fig. 4 Chilean GDP of relevant environmental activities. Self-elaboration with data from the Chilean Central Bank (n.d.), Statistic. Money valued at 2003

of copper between 1990 and 2011, which was the country's main source of revenue, was due to an excessive use of electric power. This tendency to require higher amounts of energy in order to ensure stable levels of production is another way in which the mining cluster pressurizes the territory, consolidating a mono-productive region in which no other forms of local activities are possible. Figure 5 suggests that the Chilean economy's politically reinforced tendency to promote extractive industries is dangerously dependent on energy, which, in the case of the mining industry, is generated by thermoelectric stations that rely on the burning of coal.

Unlike hydroelectric power plants that require large water sources, the location of thermoelectric power plants is independent of territorial conditions. Thus, thermoelectric plants are built in areas with low land prices, leading to a

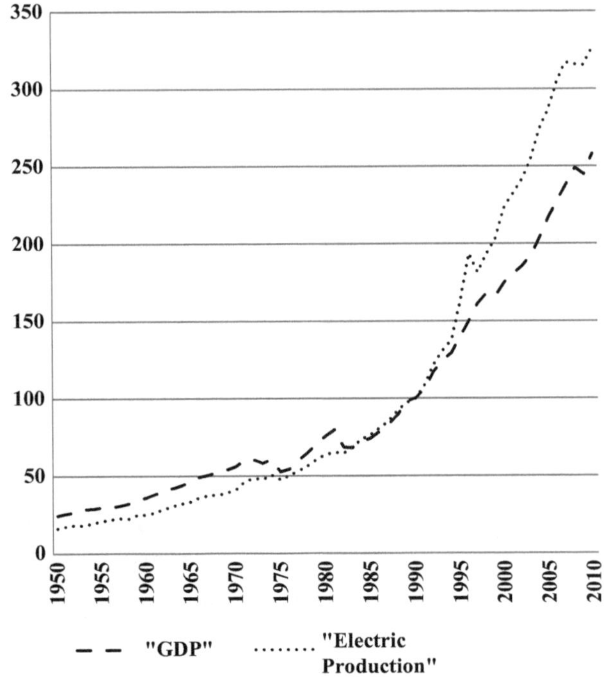

Fig. 5 Relationship between GDP and electricity production. Self-elaboration with data from the Chilean Central Bank (n.d.), Statistic

concentration of projects in the poorest municipalities of the northern region, where the most socially vulnerable populations live.[32] The territorially uneven distribution of roles associated with the mining industry affect local communities in two ways. On the one hand, the mining industry affects the communities on the Andes Plateau detrimentally by depriving them of the water resource necessary for agriculture. On the other hand, by allocating thermoelectric stations in poor municipalities, these communities are becoming increasingly socially excluded, thus aggravating other social problems.

5.3 *Constructions of a Spatio-Temporality Oriented Toward Economic Entrepreneurship at the Core of Ecological Modernization*

Introducing his understanding of time-space and his central concept of "regionalization," Anthony Giddens stated that "[m]ost social analysts treat time and space as mere environments of action and accept unthinkingly the conception of time, as measurable clock time, characteristic of modern Western culture" (Giddens 1986,

[32] For example, the coastal city of Tocopilla where two big hydroelectric plants are located.

110). Later, Giddens (1986, 110) took a more radical stance, arguing that, with the exception of some geographers, "social scientists have failed to construct their thinking around the mode in which social systems are constituted across time-space". However, what would such failure comprise? In his initial attempts, and following Hägerstrand's time geography concept, Giddens stresses that the "routinized character of the daily life" presents time and space as constraints to social action. From this point of departure, the notion of regionalization refers to zones "of time-space in relation to routinized practices" (1986, 119). However, the intention of this section is not to discuss this approach in depth, but to point out that the development of action theory was the first attempt to conceptualize space and time's capability to constrain action, and, in doing so, it reduces the possibility to account for actions' ability to structure time and space. In other words, actions are regionalized to such an extent that they lose their power to regionalize.

In his research project to develop action theory, Benno Werlen's (2005, 47) notion of everyday life regionalization helped fill the gap between the role of time and space in social action. He states: "...thus the *bordering of spaces*, in the broad sense, which should be seen as a means of everyday activities, never as their aim." (Werlen 2005, 47) and, in a more explicit formulation: "*space* is an element of *action* and not *action* an element of *space*, as a majority of geographical approaches – implicit or even explicitly – still claim" (Werlen 2005, 48).

The notion of *everyday regionalization* is central to analyzing the spatio-temporal restructuring that every new wave of ecological modernization requires. The project to expand the extractive industries in Chile to a major level in the international market, and maintaining them in this position for decades, requires a particular type of action, which we argue should have the ability to organize time and space and should not only be constrained by them. We propose that the intensification of the extractive orientation in the Chilean economy is rooted in a form of entrepreneurialism that overcomes the restrictions of time and space and instead orients actions toward stabilizing a particular representation of the two categories. Similar to large and long-run extractive enterprises, this representation is central in the production and reproduction of legal environmental procedures.

It is possible to highlight three discursive movements in the new Chilean legal environmental framework that promote a particular understanding of time and space in accordance with economic entrepreneurialism: (i) the intention to gradually incorporate ecological measures into productive processes; (ii) the spatio-temporality of nature's regenerative capability as the last frontier of economic speculation; and (iii) the ability of the economic and political systems to define a pollution-free environment as already contaminated.

When having to comply with new environmental standards, the idea of gradualism appears reiteratively in political discourses and industrial declarations. The official narrative highlights the gradual formation of the environmental crisis, or, in other words, its slow-motion consolidation, to propose a measured incorporation of higher environmental requirements in the productive processes. This discursive orientation forms the basis of the expansive wave seen in Chile's extractive industries after the country's environmental modernization. As a technical tool, the environmental impact assessment legitimates the sustainability of the project under evaluation. Nevertheless, due to the gradualness of the actualization and the application of the normative, an intermission was generated,[33] during which entrepreneurialism blossomed in the form of environmental speculation. The following translated excerpt from the Presidential Speech of 1992 defines gradualism.

> In this regard, there are two features of the environmental policy that have to be emphasized. These principles allow the basic goals conceived for this environmental bases project to be understood. In the first instance, one of its characteristics is gradualism. The country's environmental problems are the result of decades of applying policies, in which the environment was not a relevant aspect. Thus, reversing the causes of this degradation and searching for methods to combine a great development, progress, and environment protection, require structural modification that transcends measures in the short term (Aylwin 1992, 11).[34]

From the 1990s onward, the extractive industries have functioned under a narrative that strengthens the spatio-temporality of nature's regeneration as a last step in the productive process. The level of mining extraction activity cannot rely on preserving deposits, but on the ability to find new mines to replace exhausted ones. Water and energy requirements have been managed to extract all existing mineral deposits in such mines, because when the concentration of minerals decreases, the demand for productive inputs increases. From a different perspective, the mineral extraction narrative rejects the possibility of moderate or gradual exploitation. On the contrary, in the official discourse, extractive industries are expected to work fast while the market conditions are favorable, avoiding unfavorable substitution, as happened with Chilean saltpeter in the first half of the twentieth century during what was referred to as the saltpeter crisis. In this period, the development of the synthetic version in Germany made further saltpeter exploitation unprofitable, resulting in large unexploited deposits and completely abandoned towns. In sum, it can be argued that, as an action orientation in the mining activity, entrepreneurialism overcomes the limits of a sustainable production at the local, as well as the regional level. Therefore, this action can only be accepted if it is grounded in a redefinition of the spatio-temporality, according to which the environmental institution defines the society-nature relationship and accelerates extractive activity, without any intention of slowing it down.

Finally, and closely connected to the predominant spatio-temporality of nature's regeneration, the official discourse regards a pollution-free environment as already contaminated. According to Chile's legal definition of the concept, pollution-free ecosystems present contaminants under the legally defined levels (Aylwin 1992), or in levels that do not endanger human health. This situation, which has not gained

---

[33] The Chilean central government only passed the regulation to operationalize the new environmental framework in 1997; surprisingly, three years after Law 19,300 "Environmental Bases" was approved. It is important to take this into account if one considers the start of a new environmental institution a non-regulated intermission, during which the extractive orientation of the economy is exacerbated.

[34] Translation by the author.

much attention, plays a fundamental role in a national economy based on extractive activities. The mining industry produces significant levels of contamination during the mineral extraction process, for example, by separating minerals from the ground, and the later smelting process. Recognizing that no ecosystem is free of pollution implies that the mining industry cannot be held accountable for environmental damage, or the negative effects it has on other activities.

5.4 A Goal-Oriented Understanding of Sustainability Regarding the Assurance of Highly Profitable Enterprises

In growth-oriented capitalistic societies, the conceptualization of an ecological crisis should be translated into economic language to make it effective (Hajer 1995, 31).[35] In this sense, the terms entrepreneurialism, innovation, and industrial consolidation have become part of a narrative from which profitable enterprises are created to face environmental problems. We use the ecological modernization notion as a procedure through which political and scientific systems converge to create a social space that nurtures extractive enterprises.

This section describes the rational use of sustainable principles and technologies to increase the extractive industries' productivity in the 2000s. In Chile, one of the most dynamic economies in the region, the consolidated extractive orientation toward mega-regional clusters becomes apparent. This territorial formation has become part of, and is compatible with, the current wave of global productive restructuring.

The increasing level of production in the mining industry is based on the depletion of fossil water deposits and the growing demand for energy. There is increased competition for water between industries and local communities in the world's most arid desert, while the energy requirement has led to dramatic changes in the national energetic matrix.[36] This process of industrial expansion in terms of regional mono-productive clusters is a result of the first wave of ecological modernization outside the developed economies. In the peripheral regions, and particularly in the case of Latin America as a supplier of raw materials, environmental institutional modernization has had an exploitation-exportation function. To achieve economic growth, the democratic government (following the military

[35] "Ecological modernization uses the language of business and conceptualizes environmental pollution as a matter of inefficiency while operating within the boundaries of cost-effectiveness and administrative efficiency" (Hajer 1995, 31).

[36] In Chile, The energetic matrix is based on thermoelectric plants, which supply energy to the Big North Interconnected System (SING), and hydroelectric plants that do so to the Central Interconnected System (SIC).

dictatorships) reinforced the existing extractive enterprises and linked them to a national success rhetoric.

The biggest open mine in the world and the first international producer of copper[37] is part of Chile's national pride. Regardless of their political orientation, different regimes have coupled this narrative to the idea of economic development. Consequently, there is a lack of critical reflection of the changing scales in the productive processes. The competition for land and water has led to (i) a loss of agricultural activity in the rural areas where the productive industries are located, (ii) the depopulation of small inland cities and towns, (iii) rapid urban growth in mid-sized cities, mainly in informal settlements, and (iv) unstable national and international migration flows.

It is impossible for regional industrial clusters to emerge without integrating the "sustainable mechanisms" in the extractive activities. Moreover, it cannot be argued that Chilean mining companies' extraction and processing methods have become worse in terms of the amount of pollution caused, or that they have been exploiting resources irrationally over the last 25 years. The future development of the industry, which intends to double its production in the next decade, will lead to the increased incorporation of green technologies. International markets are increasingly applying pressure on industries to extract resources and produce products in a socially and environmentally friendly way. Similarly, international agencies constantly advocate incorporating market mechanisms to regulate contamination and ensure the rational use of resources.

> The mining industry is a large consumer of energy and in the next years their generation of new non conventional renewavel energies [NCRE] will not meet the demand for new projects to be developed between now and 2020 [. . .] therefore, the industry will at present continue to rely on an energy matrix in the SING [Big North Interconnected System] that is primarily based on the use of coal. This [is applicable] regardless of several mining companies' efforts to use NCRE supply in specific sectors of their operations. (El Mercurio 2012)

Nevertheless, the future growth of the mining industry cannot merely depend on the incorporation of green technologies. As the abovementioned quotation suggests, NCRE would not be the main sources of the energy supply. Scholars have argued that, in Chile, renewable energy is incorporated solely to support the thermoelectric plants in the desert region. Thermoelectric plants will be responsible for the additional energy required to desalinize water, on which the mining industry will rely when the area's fresh water has been exhausted.

> The solar projects to be installed on the Salar de Pintao, surrounding the Pica Oasis, will cover 5000 hectares [. . .] These issues are not solved; again, each of these projects is evaluated regarding its individual effect, but not for their total agglomerative effects [. . .] these 5000 acres of solar panels are designed to generate energy to operate thermoelectric plants, whose main objective is to desalinate seawater for the mining activities and cities, and which are due to large mining investments [. . .] all plants or mining facilities are planning to desalinate seawater in case of water resource depletion, but this requires large amounts of energy, which we do not have in the north. Installing these panels, which simultaenously support the series

[37] Chuquicamata is the world's biggest open mine and belongs to the National Copper Corporation of Chile (CODELCO).

of thermoelectric plants on the coast is not irrelevant; millions of cubic meters of sea water will be desalinated to support this activity. (Academic State University)

Different organizations have predicted that there will be significant growth in mining activities by 2020. Therefore, they suggest that in just 11 years – from 2009 to 2020 – the investment will deliver greater returns than in the 32 years from 1977 to 2008; increasing from USD 38.665 million to USD 42.448 million. Copper production will increase by 36.8 % in 2020, with private investment representing 53.5 %, and the participation of the National Copper Corporation of Chile representing 45 % of the total (Minería de Chile 2012).

It can thus be assumed that since the political system had in the 1990s ensured that the consolidation and maintenance of the extractive industries would remain the central drivers of the national development, the national environmental institution's role is to create prime conditions for its expansion at the regional level during the twenty-first century. In the 1990s, the main mechanism for ensuring the industry's sustainability was defining the levels of pollution and resource exploitation that the ecosystems could possibly endure. It is highly unlikely that this social goal was accomplished holistically. Moreover, these industries' degree of compliance was based on a restrictive definition of the environmental impact, which only applied to individual industries and never to an aggregated level. In the first decade of the twenty-first century, the main strategy to promote industrial expansion was the incorporation of market mechanisms to regulate environmental problems. It was believed that the low degree of success in terms of auditing and controlling the productive industries would be solved through compliance incentives based on market logic. Furthermore, the idea was that the pollution levels, as well as the extraction quotas, would be naturally achieved by consolidating the markets to deal with these issues, much like the proposed international CO_2 emission.

From this perspective, the national environmental institution's role was to secure investments that would foster a responsible attitude toward the ecosystem. The constant undertaking of sustainability measures in terms of pollution control management and rational resources exploitation has promoted only a very restricted understanding of environmental measures. To reverse the negative impact that regional extractive clusters have had on the environment and people, effective limits should be applied to industrial expansionism. As neither of these actions is currently implemented, industrial expansionism on a regional scale is responsible for the reemergence of ecological conflicts.

6 Conclusion: Overwhelming Confidence in Economic Sustainability

There is no all-inclusive definition of sustainability. The term has various meanings, which are often even contradictory. At least two interpretations of sustainability are relevant for this research and were therefore applied in this study:

economic and environmental sustainability in the process of ecological modernization. The Brundtland Report (1987)[38] coined the economic sustainability concept, which is focused on preventing situations of scarcity for the following generations. In this context, one might say that development, on its own, only restricts economic growth if it puts the future availability of resources at risk. However, this conceptualization does not consider the ability of specific ecosystems – with their natural and human dimensions – to support a constant form of "resources exploitation," or what could be referred to, in economic terms, as an overdemand of environmental services. Conversely, Gligo (2006) coined the concept "ecological sustainability," which is based on the preservation of ecosystems. In this context, and with references to the contribution of the Economic Commission for Latin America and the Caribbean (ECLAC) in the late 1970s, Gligo proposed that sustainability should be understood as the maintenance of constant levels of "matter and energy" in any ecological systems.[39]

An important difference between these two definitions is the implicit "limitations" and "orientation" of their social actions. In economic sustainability, social action is primarily oriented toward economic growth, which is the central goal of social actors. In the same context, environmental protection becomes a passive restriction of the socio-economic decision. In environmental sustainability, environmental protection in the form of ecosystem maintenance is the major socially defined goal, and, consequently, concrete actions are required and deployed to reintroduce the matter and energy that were extracted from all the ecosystems. The manner in which sustainability is socially defined has important repercussions for the "orientation and limitations" of social actions and territorial orientations. This argument formed the basis of our argumentation and allowed us to critically review the way in which sustainability is materialized through different social practices in a specific national context. We subsequently highlighted that many enterprises oriented toward economic sustainability – that is, they do not make compromises about their future needs in terms of the rational usage of natural resources or pollution control – have exerted strong pressure on ecosystems in terms of the natural and the human dimension. This has drastically changed these ecosystems' properties, often leading to their destruction, and, subsequently, reinforcing historical and/or latent socio-ecological conflicts.

This article tested this argument by examining Chile's mining industry empirically. We found that pressure on the environment and the reemergence of

[38] "Sustainable development is development that meets the needs of the present without compromising the ability of future generations to meet their own needs" World Commission on Environment and Development (1987, 43).

[39] "Ecological sustainability is achieved [. . .] where there is recomposed and/or introduced information, matter, and energy while the volumes (biomatter), the rate of change, and the circulation rhythms are kept constant, all of which characterize a constant system" (Gligo 2006, 18); translation by the author.

socio-ecological conflicts are the most common result of the environmental institutional reform in Latin America. Thereafter, to better account for the conclusion derived from the case study analyzed in this article, we incorporated a notion that we call *sustainability over-confidence*. Sustainability over-confidence is the social belief that it is possible to overcome the ecological crisis by merely restricting economic development environmentally in terms of (i) the rational exploitation of natural resources and (ii) the rational management of pollution by incorporating new green technologies. Here, two forms of criticism are raised to challenge rational management as an instrument with which to overcome the ecological crisis. First, as many authors have proposed, it is no longer possible to overcome the ecological crisis by refining the institutions that caused the problem in the first place. In other words, a new economic orientation is required that rejects the congruence between the devastating forms of production and consumption in Western societies. Second, socio-ecological conflicts are not merely a question of clean production; in many situations, such conflicts are the result of opposing visions of particular ecosystems' role with the orientation to convert them into a source of profit having to be weighed against preserving their natural state.[40]

The socio-ecological problems are broader and deeper than contamination control, clean production, green economy, and global warming issues, to mention just a few. Socio-ecological conflicts appear when two opposing orientations for action meet at the same places in a scenario, in which the element under pressure will (re)define the future living conditions in those regions. We argue that conflicts "meet at the same places" to emphasize a spatio-temporal coincidence, which is impossible to disentangle, because these conflicts are explicitly about ways of doing in a particular historical context/moments and territory. Simply put, in terms of sustainability, it is possible to summarize that over-confidence occurs when a society has confused the questions of *what* it wants to do in a territory with *how* it wants to achieve this goal. Economic sustainability only responds to the second question, suggesting that regardless of how the territory will be used, the resources should be used correctly, avoiding material and energy losses, and not compromising future reserves. Nevertheless, economic sustainability is incapable of building social consensus on the society-nature relationship with ecosystems and territories. Put differently, economic sustainability cannot collectively define the forms in

[40] As many authors suggest, this conflict demonstrates the struggle between an understanding of nature's "value of use" and "value of change," to use classical Marxist terms. Here the criticisms of environmental modernization rely on the pressure required to stabilize an understanding of nature in terms of the value of change, thus rejecting any reference to substantive and transcendent values. Many of this research project's interviewees stated that the Chilean ecological modernization stabilizes a form of abstract monetary compensation that is incapable of internalizing or reflecting all natural values – value of use – for the local and native populations, while, it is simultaneously also incapable of considering the total amount of environmental services that nature delivers.

which the society-nature relationship will be consolidated at specific historical moments. Weber's explanation of the modern confusion regarding the "means and ends" and the neutrality of science to state value judgments (Weber and Finch 2010), will be used to expound the main finding from an analysis of the environmental reform in Chile.

This article has shown how, in the mining industry, the question of sustainability has acted as an effective tool to maintain the level of copper exploitation and production in Chile, but that – although significant improvements have been made – it has been less successful at internalizing the effects of environmental pollution. However, when we observe the complete chain of productive inputs, the sector's state indicates a completely different performance. The monopolization of water resources and the redefinition of the energetic matrix are examples of the expansive orientation, which is a key characteristic of business orientation and not a mistake or side effect thereof as some analyses have suggested. This expansionist character continuously affects the living conditions in the Atacama Desert in different ways. For example, (i) it is impossible for rural communities to develop forms of agriculture due to the lack of an adequate supply of water, and (ii) the region's productive roles have been totally redefined due to the high wages paid by the mining industry, especially for permanent employees.

Taking a critical perspective of the sustainability orientation, this study discussed four interrelated levels of ecological modernization in Chile: (i) sustainability is viewed as a driver of place achievement, i.e. how economical sustainability constantly increases its influence in a territory and how this expansionist process aggravates socio-ecological conflicts; (ii) sustainability is considered a dispositive to define environmental territorial roles, i.e. how the pursuit of economical sustainability determines the uneven distribution of territorial benefits and burdens, which specifically prevents self-sustaining territorial dynamics and localizes undesirable activities in fragile ecosystems; (iii) sustainability is believed to motivate economic entrepreneurship by redefining environmental issues according to their spatio-temporal aspects by acting as a dispositive to legitimize and reinforce the increasing exploitation of mineral deposits; and (iv) a goal-oriented understanding of sustainability as fostering entrepreneurialism by restricting the definitions of ecological crisis and problems, and incorporates new and clean technologies as the most efficient solution. However, these new technologies are incorporated in order to support natural exploitation and, in turn, reinforce the socio-ecologic conflicts.

In this context, to ensure a development based on territorial sensibility, it has become necessary to redefine sustainability in social, political, and economic terms. This is the only way to control industrial expansionism and regional productive restructuring, which accompany other forms of global environmental concerns, such as global warming, technological transfer, and pollution control, in the global context. The four mechanisms discussed in this article have an inherent tendency to monopolize social resources. This exploitation can be limited if the everyday practices anchored at the local level regain relevance in the environmental discourse. To illustrate this, the article describes the great economic relevance of the mining industry for Chile's GDP and for the industrial employees; nevertheless, the rest of the workforce, who are not connected to this sector, experience accelerated instability. At the regional level, the industrial cluster's expansion cannot be sustained without a massive monopolization of all productive inputs, especially water, energy, land, and workers. This regional productive restructuring is occurring across all the extractive industries – and some plantation industries – in Latin America. This exponential growth would not be possible without the paradoxical incorporation of sustainable measures into the processes. In order to gain deeper insights into this phenomenon, we propose studying how sustainability has been coopted by a speculative rationality, which constantly manages social actions in order to accelerate the extraction of ores.

This brief account of the Chilean mining industry has shown how particular territorial and action orientations have interacted in order to mobilize the national environmental institution as a central driver of the consolidation of perhaps the biggest and most profitable extractive cluster in Latin America. While this article is a first step toward restoring the role of social science in the analysis of one of the most successful and liberal forms of entrepreneurship on the continent, this can only be fully achieved with a real account of the socio-ecological results of two decades' incorporation of sustainability measures into the policy agenda. Focusing on sustainability's economic-natural or ecological-anthropogenic orientations should allow a more factual analysis of ecological modernization. This is the only way in which we can prove that socio-ecological conflicts should be considered the main concern of environmental institutional modernization in a context in which contradictory forms of global integration interact.

References

Akhtarkhavari, A. (2010). *Global governance of the environment: Environmental principles and change in international law and politics*. Cheltenham: Edward Elgar.

Altieri, M., & Rojas, A. (1999). La tragedia ecológica del "milagro" neoliberal chileno. *Persona y Sociedad, XIII*(1), 127–141.

Aylwin, P. (1992). Mensaje Presidencial Ley 19,300 Bases Medioambientales. Historia de la Ley 19,300/1994 Biblioteca del Congreso de Chile.

Bachelet, M. (2008). Mensaje Presidencial Ley 20,417 Creación Ministerio del Medio Ambiente. Historia de la Ley 20,417/2010 Biblioteca del Congreso de Chile.

Barrick Gold. (2011). *Barrick annual report 2011*. http://www.barrick.com/files/doc_financials/2011AnnualReports/Barrick-Annual-Report-2011.pdf

Boeninger, E. (1997). *Democracia en Chile: lecciones para la gobernabilidad*. Santiago de Chile: Editorial Andrés Bello.

Bosselmann, K. (2008). *The principle of sustainability: Transforming law and governance*. Burlington: Ashgate.

Campos-Medina, F., & Campos-Medina, L. (2012, October 5–6). *The environmental institution in Chile, a political representation of the ecological crisis*. Paper presented in the Berlin conference on the human dimensions of global environmental change, Berlin. http://www.berlinconference.org/2012/wp-content/uploads/2013/01/Campos_Medina-The_Environmental_Institution_in_Chile_A_Political_Representation-330.pdf

Campos-Medina, F., & Larenas, J. (2012). La despolitización del conflicto socio-ecológico en la legislación medioambiental Chilena a partir de los años 90s. *Bitacora Urbano Territorial, 21*(2), 45–56.
Chernilo, D., & Mascareño, A. (2005). Universalismo, particularismo y sociedad mundial: obstáculos y perspectivas de la sociología en américa latina. *Persona y Sociedad, XIX*(3), 17–45.
Chilean Central Bank, Historical Statistics. (n.d.). http://www.bcentral.cl/estadisticas-economicas/series-indicadores/index_aeg.htm.
Chilean Cooper Corporation. (n.d.). www.cochilco.cl.
Dörre, K. (2011). Capitalism, Landnahme and social time régimes: An outline. *Time & Society, 20* (1), 69–93. doi:10.1177/0961463X10394965.
El Mercurio, Ediciones Especiales. (2012). http://www.edicionesespeciales.elmercurio.com/_portada/index.asp. Accessed 25 Jan 2012.
Galeano, E. H. (2004). *Las venas abiertas de América Latina*. México: Siglo XXI.
Giddens, A. (1986). *The constitution of society: Outline of the theory of structuration*. Berkeley: University of California Press.
Gligo, N. (2006). *Estilos de Desarrollo y Medio Ambiente en América Latina, Un Cuarto de Siglo Después*. Santiago de Chile: Publicaciones CEPLA-United Nations.
Gonzáles, J. A., Montes, C., & Santos, I. (2007). *Capital Natural y Desarrollo: Por una base ecológica en el análisis de las relaciones norte sur*. Papeles N° 100 2007/08.
Gudynas, E. (1999). Concepciones de la naturaleza y desarrollo en América Latina [versión electrónica]. *Persona y Sociedad, XIII*(1), 101–125.
Guimaräes, P. (1999). Aspectos políticos y éticos de la sustentabilidad y su significado para la formulación de políticas de desarrollo [versión electrónica]. *Persona y Sociedad, XIII*(1), 157–183.
Haas, P. M., Keohane, R. O., & Levy, M. A. (1995). *Institutions for the earth: Sources of effective international environmental protection*. Cambridge, MA: MIT Press.
Hajer, M. (1995). *The politics of environmental discourse: Ecological modernization and the policy process*. Oxford: Oxford University Press.
Harvey, D. (2003). *The new imperialism*. Oxford: Oxford University Press.
Historia de la Ley N° 18,097. (1982). Ley Orgánica Constitucional sobre Concesiones Mineras Biblioteca del Congreso de Chile.
Historia de la N° Ley 20,417. (2010). *Creación Ministerio del Medio Ambiente*. 20,417/2010 Biblioteca del Congreso de Chile.
Historia Ley N° 19,300. (1994). *Bases Medioambientales*. 19,300/1994 Biblioteca del Congreso de Chile.
Intergovernmental Panel on Climate Change (IPCC). (2007). *Climate change synthesis report*. http://www.ipcc.ch/publications_and_data/ar4/syr/en/main.html. Accessed 1 Oct 2012.
Intergovernmental Panel on Climate Change (IPCC) (2013). *Climate change the physical sciences basis; Summary for policy makers*. http://www.climatechange2013.org/images/uploads/WGIAR5-SPM_Approved27Sep2013.pdf. Accessed 15 Oct 2013.
Jasanoff, S., & Long Martello, M. (2004). *Earthly politics local and global in environmental governance*. Cambridge, MA: MIT Press.
Langhelle, O. (2000). Why ecological modernization and sustainable development should not be conflated. *Journal of Environmental Policy and Planning, 2*(4), 303–322.
Larraín, S., & Poo, P. (2010). *Conflíctos por el Agua en Chile. Entre los Derechos Humanos y las Reglas del Mercado*. Santiago: Ediciones Programa Chile Sustentable.
Leff, E. (1986). *Ecología y Capital. Racionalidad Ambiental, Democracia Participativa y Desarrollo Sustentable*. México D.F.: UNAM.
Levy, D. L., & Newell, P. (2005). *The business of global environmental governance*. Cambridge, MA: MIT Press.
Martin, R., & Sunley, P. (2003). Deconstructing clusters: Chaotic concept or policy panacea. *Journal of Economic Geography, 3*(1), 5–35.

Minería de Chile. (2012). http://mineriachile.com/2012/01/proyeccion-de-inversiones/. Accessed 1 Oct 2012.

Mol, A. P. J. (2003). *Globalization and environmental reform: The ecological modernization of the global economy*. Cambridge, MA: MIT Press.

Mol, A. P. J., & Jänicke, M. (2009). The origins and foundation of ecological modernisation theory. In A. P. J. Mol & D. A. Sonnenfeld (Eds.), *The ecological modernisation reader: Environmental reform in theory and practice* (pp. 17–27). London: Routledge.

Mol, A. P. J., Spaargaren, G., & Sonnenfeld, D. A. (2009). Ecological modernization: Three decades of policy, practices and theoretical reflection. In A. P. J. Mol & D. A. Sonnenfeld (Eds.), *The ecological modernisation reader: Environmental reform in theory and practice* (pp. 3–14). London: Routledge.

OLCA/OCMAL Latin American Observatory of Environmental Conflicts and the Latin American Observatory of Mining Conflicts. (2012). http://transparency.globalvoicesonline.org/listing/observatory-mining-conflicts-latin-america. Accessed 1 Oct 2012.

Park, J., Conca, K., & Finger, M. (2008). *The crisis of global environmental governance towards a new political economy of sustainability*. London: Routledge.

Roberts, J. T., & Parks, B. C. (2007). *A climate of injustice: Global inequality, north-south politics, and climate policy*. Cambridge, MA: MIT.

Rosa, H. (2009). Social acceleration: Ethical and political consequences of a desynchronized high-speed society. In H. Rosa & W. E. Scheuerman (Eds.), *High-speed society: Social acceleration, power, and modernity* (pp. 77–112). University Park: Pennsylvania State University Press.

Smith, N. (1984). *Uneven development: Nature, capital, and the production of space*. New York: Blackwell.

Speth, J. G., & Haas, P. M. (2006). *Global environmental governance*. Washington: Island Press.

Touraine, A., & Bixio, A. L. (2000). *Crítica de la modernidad*. México: Fondo de Cultura Económica.

University of Chile Radio Station. (2012). http://radio.uchile.cl/afondo/89184/. Accessed 1 Oct 2012.

Van der Loo, H., & van Reijen (1992). *Modernisierung: Projekt und Paradox*. München: Dt. Taschenbuch-Verlag.

Weber, M. (2002). *Economy and society 1*. Berkeley: University of California Press.

Weber, M., & Finch, H. A. (2010). *The methodology of the social sciences*. New Brunswick: Transaction Publ.

Werlen, B. (1993). *Society, action and space: An alternative human geography*. London: Routledge.

Werlen, B. (2005). Regions and everyday regionalizations: From a space-centred towards an action-centred human geography. In H. Houtum, O. T. Kramsch, & W. Zierhofer (Eds.), *Bordering space* (pp. 47–60). Burlington: Ashgate.

Winter, G. (2006). *Multilevel governance of global environmental change: Perspectives from science, sociology and the law*. Cambridge: Cambridge University Press.

World Commission on Environment and Development. (1987). *Our common future*. New York: United Nations (Also known as Brundtland Report).

Political Decision-Making and Scientific Insights: A Comment from the Political Arena

Matthias Machnig

Professor Werlen, Professor Dicke, Professor Abler, ladies and gentlemen. Thank you for the invitation. It's a great honor and pleasure to be here. My people have prepared a long speech. I put it aside to make a couple of remarks based not only on principles, but also on my experience.

For 4 years I was Germany's key negotiator at international climate conferences. And these experiences were very interesting. I want to bring them to the table. But let me start with the following remark. If we look at sustainability and the debate we have on it, we have to admit the following: at the start of the twenty-first century, the question of sustainability, the question of climate, was somehow at the center of the international debate. There were many international conferences, and everyone looked to Copenhagen, because Copenhagen should be the conference where we get a post-Kyoto agreement, where we have an international regulatory framework for how to deal with climate change.

We know what came out of Copenhagen. To be honest: nothing. And I say this because Copenhagen was in 2009 and if we look at the debates that started in 2008, in 2012 we must admit the following: The focus of the international debate has totally changed. We are now talking about economic crisis, we are now talking about financial crisis, we are now talking about employment crisis, we are now talking about growth crisis, but we are no longer talking about sustainability. So my first remark is, and, of course, as economic minister I must deal with the problem of growth and labor development, but what we must do is we have to put the question of sustainable development back at the center of the national and international debate.

The question of sustainability is no longer a question of environmental policies. It's an economic question. And we must give economic answers to this question. What does it mean? Today we are about six billion people in the world, in a number

M. Machnig (✉)
Bundesministerium für Wirtschaft und Energie, 11019 Berlin, Germany
e-mail: matthias.machnig@bmwi.bund.de

B. Werlen (ed.), *Global Sustainability*, DOI 10.1007/978-3-319-16477-9_15

of years we will be nine billion, maybe ten. It's true that today we have about two to two and a half billion people living in industrialized countries. This is going to double in the next three to four decades. It's true we have about two billion people who have no kind of access to energy, etc. And these different countries and different regions – such as China and India – they are going on a clear growth path, because they say – and this is understandable – we need economic growth to solve our social problems. But this growth is based on what I would call the old industrial notion and the old notion of industrialized development.

Industrialized countries tend to do the following: They have high energy needs, they have high resource needs, they have high land needs. So if we take the old industrial development path, we need at least two earths to make this possible. To be honest, we need something I call a third industrial revolution.

What is a third industrial revolution? We have to re-invent industrial development based on efficiency. Efficiency is the key to industrial development. This means more energy efficiency, this means more resource efficiency, this means more efficiency in land use, water use, and so on. We all agree in science and politics that we have to reduce CO_2 by 80–95 % by 2050. Now I will tell you what this means for a country like Germany. If we want to reduce CO_2 emissions by 80–95 % by 2050, this means we have to have an energy sector with zero emissions, we have to have a housing sector with zero emissions, we have to have a transport system with zero emissions, and then we have about 10 % of CO_2 emissions that are allowed for industrial processes. From today's state-of-the-art, we don't know how to drive these without exceeding CO_2 emissions. This presents a dramatic change. And this makes clear what the challenges are.

Look at the negotiations we had at Copenhagen: Five elements were important. First, do we have something like a binding mid-term and long-term target for industrialized countries? The EU is able and willing to go for a 20–30 % reduction target by 2020 and to go for up to 80–95 % by 2050. But some countries, such as the U.S. and some others – Japan, Russia – they did not want to go along with this process. Second, what are the contributions of developing countries? Because if you look at China, China is today the region or country with the highest CO_2 emissions in the world, and they are building day by day about two to three coal plants and we all know what kind of impact coal plants have on the climate. So the question was, is a country like China willing to contribute on CO_2 reduction, and the answer was no.

Third, how do we organize technology transfer? The problem developing countries have is that they don't have the technologies, the sustainable technologies, to solve their energy problem. And technology transfer also means, fourth, what kind of financial funding do we have? Photovoltaics is a sustainable technology, but compared to other energy production systems such as coal, it's quite expensive. And the question is, how can we help developing countries to build up an energy system with modern technology, but also based on economically efficient structures, and with prices that people are able to pay. Therefore, funding is a huge problem. Of course, the funding must come from the developed countries. What kind of funding mechanism do we have? So the Europeans made a proposal for an emissions trading system. By auctioning the permissions for CO_2 emissions, the money earned can go into a funding system. But it doesn't make sense if only the European Union has an emissions trading system, while other countries are not going to go there.

Fifth, how can we help developing countries to build up institutional capacities to deal with the problem? It is hard to find agreement, because there are totally different economic interests. Therefore, we have to look for the following, and I hope this conference will help with this question: How can we build up an economic system that makes growth possible but also enables us to make these structural reforms? And how to build up new structures to make an industrialized country more sustainable and also help developing countries to solve their social problems, and their poverty, and their infrastructural and other problems?

These are key questions. I want to give you some thoughts on what this means. My first thesis is: This process will not be market-driven. The market will play a role, but if we only look for markets we will not succeed. We need regulations to make the markets work on sustainability. I give you one simple example. With an emissions trading system – so you give CO_2 a price – you have a clear signal. The problem in Europe at the moment is the following: The price of CO_2 is too low. We have a CO_2 price of about 7 euro. We should have a CO_2 price of, let's say, 20 plus euro. This would be an economic incentive to invest in a more sustainable energy infrastructure. Therefore, we need economic instruments that provide incentives to go into more sustainable development. This means you have to talk about taxation and you have to talk about systems such as emissions trading systems.

Second, we have to build up standards for industrial production or industrial products. To give you a very simple example, I said the transportation system must take the direction of zero emissions. In Germany, as one of the top producers of cars in the world and Europe, we had a long debate on how to do this. There was much discussion with the car industry, which talked about self-commitment: we are fulfilling targets. The truth is, this has never been done. So what did we do? We made it a law, which was established in 2012. A car producer's fleet is allowed to have CO_2 emissions of 120 g of CO_2 per kilometer on average, and we are going to go down to 95 g by 2020. But reducing to 95 g of CO_2 per kilometer is no longer possible with the standards we have today. So what do we need? We need hybrids, we need e-mobility, we need fuel cells, etc. So you make innovations possible; you bring R&D into the market.

Third, we also need other standards. If you look at certain products we are using today, we all know the following: There are so-called rebound effects. Some of the products we bring onto the market are more efficient, but the total energy consumption is higher. So what we need is an industrial standard that, in 5 years, a new fridge coming onto the market is only allowed if it has the efficiency of the most efficient fridge we have on the market today. This is also an industrial standard that gives markets and companies a clear signal where to go.

Fourth, we have to look for totally different finance systems to build up the needed investments. We need a different funding system, because all we are doing in this direction needs much investment. To give you an example: Germany is phasing out nuclear. In 2022, the last nuclear plant is going off the grid. Now we have started what we call *Energiewende* or energy turnaround. We will have to change our energy system, and this is the biggest industrial project in Germany since the war. This is the biggest industrial project we have ever had, because if you are talking about new capacities for energy production, if you are talking about new grids that must be built, if you are talking about storage capacities that are important, you must also talk about more.

Solely for the investments to modernize and to make our grids able to solve the problem or to solve this restructuring that is in progress, in the next years we will need about 60 billion euro in investment. And this, of course, must take place without a development where we get energy prices that are a social or an economic problem.

Therefore, we need an economic debate on how to lead industrial countries into more sustainable development. We need new economic thinking, we need new economic instruments, we need new legislation and regulation to make this possible. This will not come just from free markets. We need markets, but we need regulated markets, in terms of giving them the right incentives. Ecology has to be the economy of the twenty-first century; otherwise we will not make this possible. It is possible because one thing has changed.

The question of sustainability is no longer a question that belongs to people from the environmental movement. This is an economic business now. So what has changed is that people are discussing sustainability, and the economic sector is going to understand that it is not only something about costs. This is a huge business opportunity. In 2008, Roland Berger, one of the best-known strategy and economic consultants in Germany, estimated a volume for green tech in the world of about 1,400 billion euro. In 2020, they said this will be 3,400 billion euro. So this is more than a doubling of a market that I would call one of the future markets.

I am going to end with a short sentence of a CEO from General Electric, Jeffrey Immelt. He said ‘green is green’. Green is green is very simple. If you are going to invest in green technologies, you are going to earn a lot of green dollars. So if we want to make this possible, we need new economic thinking, we need smart regulation, we need smart investment, and we must understand that this is not only a challenge, but also a huge opportunity, and also for industrialized countries. And if this conference can help us in this debate, then you have done a wonderful job. Thank you.

FSC
www.fsc.org
MIX
Papier aus verantwortungsvollen Quellen
Paper from responsible sources
FSC® C105338

If you have any concerns about our products,
you can contact us on
ProductSafety@springernature.com

In case Publisher is established outside the EU,
the EU authorized representative is:
Springer Nature Customer Service Center GmbH
Europaplatz 3, 69115 Heidelberg, Germany

Printed by Libri Plureos GmbH
in Hamburg, Germany